Reflections on Net Assessment | *Interviews with Andrew W. Marshall*

Edited by Jeffrey S. McKitrick and Robert G. Angevine

Andrew W. Marshall Foundation, Jaffrey, NH
and
Institute for Defense Analyses, Alexandria, VA

Institute for Defense Analyses
730 East Glebe Road
Alexandria, VA 22305
www.ida.org

The Andrew W. Marshall Foundation
1 Woodbury Hill Road
Jaffrey, NH 03452
www.andrewwmarshallfoundation.org

ISBN 978-0-578-38422-1

Library of Congress Control Number: 2022935866

Printed in the United States of America

Cover and interior design by Carolyn A. Leonard, Institute for Defense Analyses

Jacket photograph by Joyce Ravid

Contents

Acknowledgements vii
Editors' Comments ix
1 Themes 1
Outline for 24 September 1993 Talk 2
Themes 1 4
Outline for 6 October 1995 Talk 20
Themes 2 22
2 The RAND Years: 1950–1969 45
Early 1950s 48
Outline for 29 October 1993 Talk 55
1950–1969 58
3 The Nixon-Ford Administrations: 1969–1976 101
Outline for 14 December 1993 Talk 104
1969–1975 108
Outline for 8 April 1994 Talk 140
1973–1976 142
1969–1973 152
1969–1973 166
4 The Carter Administration: 1977–1980 187
Outline for 8 April 1994 Talk 190
1977–1980 191
5 The First Reagan Administration: 1981–1984 213
Outline for 26 July 1994 Talk 215
1981–1984 216
6 The Second Reagan Administration: 1985–1988 247
Outline for 16 December 1994 Talk 248
1985–1988 249
Outline for 25 January 1995 Talk 277
1985–1988
7 The Bush and Clinton Administrations: 1989–1993 283
Outline for 25 January 1995 Talk 285
1989–1993 286
Index 313
Acronyms 321

Acknowledgements

The editors would like to acknowledge the contribution of several organizations and individuals to the production of this book.

The Smith Richardson Foundation provided the initial funding that supported Kurt Guthe's interviews with Andrew Marshall. Guthe also transcribed the interviews, and as an aid to future researchers and analysts, he also conducted extensive additional research to identify referenced reports and documents.

The Andrew W. Marshall Foundation and the Institute for Defense Analyses (IDA) funded the final editing and publication of the Marshall interviews. IDA also employed the book's editors, as well as research analyst Patrick Savage, who converted the Marshall interview transcripts into documents suitable for editing and who proof-read each document. Barry Watts prepared the initial version of the index. IDA technical editor Carolyn Leonard conducted the final copy edit, designed and laid out the text, and prepared it for publication.

And of course, we would be remiss if we did not acknowledge Andrew W. Marshall's willingness to undergo many hours of questioning during more than a dozen interviews spanning six years. His patience, persistence, and strategic thinking will be sorely missed. This book is an effort to capture some of that thinking for posterity.

The Editors

Editors' Comments

This book is based on the transcriptions of a series of 12 interviews with Andrew W. Marshall conducted by Kurt Guthe between September 1993 and July 1999. Guthe prepared the transcriptions and edited them to improve their accuracy and readability, not their literary quality. Where comments could not be heard clearly on the recording, Guthe inserted the notation "[indistinct]." Where the need to change the tape caused a portion of the conversation to go unrecorded, Guthe inserted a notation indicating a gap in time. Guthe also used brackets to insert comments or indicate edits to the interview content; the editors have left those unchanged, and have done the same using the notation **[Eds:]**.

The editors of this collection of interviews have lightly edited the transcripts for readability and relevance. In addition to Guthe and Marshall, a number of former staff members in the Office of Net Assessment participated in the interviews. The words of all participants other than Marshall, including Guthe, are prefaced by **Q:**, followed by a number to differentiate the participants. The editors have also arranged the transcripts as a chronological account of Marshall's professional career up to 1993, divided largely by presidential administration. In some cases, more than one interview is relevant to a particular period or one interview will cover several periods; in those cases, the editors have included the relevant portions of each interview in the chapter focused on the appropriate historical period. The editors have also written an introduction to each chapter to contextualize the discussion that follows.

Guthe created the outlines beginning some interview excerpts from the handwritten outlines Marshall prepared for some talks. The endnotes for the interviews were prepared by Guthe based on sources and information he uncovered in his research. The endnotes for the chapter introductions were prepared by the editors, and are prefaced with **Eds:**.

The views expressed in this publication are those of the author and do not necessarily reflect the official policy or position of the Department of Defense or the U.S. government.

1 Themes

Introduction

At the beginning of the interview process in September 1993, Andrew Marshall met with Kurt Guthe to discuss the book project and the themes Marshall intended would guide the subsequent interviews. Marshall and Guthe met again in October 1995 when the interviews were complete to discuss and refine the themes they had intended to explore.

The editors chose to include both of these sessions as their own chapter: while the topics in these thematic interviews necessarily duplicate some of the ones discussed in the chronological chapters, often Marshall describes them in different terms or he adds different nuances to the discussion. Consequently, the editors concluded that these sessions further illuminate Marshall's thinking on a wide variety of issues. Any attempt to incorporate the two sessions into the book's chronological chapters ran the risk of having to eliminate portions of the interviews to improve "readability" and continuity. Such an elimination, the editors felt, would do violence to the overriding principle to record and report Marshall's words as delivered during the book project and to provide a complete primary source document for future generations.

Outline for 24 September 1993 Talk

[Note: Italicized items were not discussed during the interview.]

Themes 1

1. Importance of resource constraints/macroeconomic perspective
   — Criteria problem (early '50s)—dollars reflect resources/engineers' mistakes
   — Use in forecasting Soviet programs
   — Military burden—importance of for long-term competition
   - Leads to interest in Soviet underinvestment in health care and pollution—> future claims on resources, permanent handicap because of deep human problems, e.g., birth defects/life expectancy, etc.
2. Opponents not fixed—can't proceed by making forecast and planning against. Even using several alternatives not full satisfactory. But cannot assume worst (minimax/failure of game theory—hope of early RAND days)
   — Must try to understand character of others
   - Social-cultural
   - History—>organizations, bureaucratic organizational behavior

   — His/their weaknesses and your strengths—special competences; future of investment or social-cultural (Crozier) very important. Ignored in standard analysis (military) which focuses on his strengths (especially new ones), our weaknesses
   - All of these hard to change, which means can exploit his weaknesses. First net assessment story—why Soviet problems not worth reporting; reaction to weaknesses. Also means that suggestions for our improvement must take account of implementation problem (again most analysis does not—puts forward abstractly best solutions—weakness of most systems analysis)
3. Best to think of situations as involving:
   — Actors, each with strengths, weaknesses, resources, patterns of behavior, traits, etc., some of which may have trends of change over time, e.g., differential rates of GNP growth, etc.
   - Each has goals, strategies (often have to be constructed), memberships in coalitions, etc.
   - Context within which actors seek goals, compete
     – Important factors not under their control, or largely so, e.g., technology, demographics, climate (including Fletcher's 140-year cycles), new plagues, etc.
4. Importance in many cases of the long-term competition framework of analysis
   — Change of perspective, focuses attention on strategy for competing, forces attention to complex set of goals (often ignored in standard analysis). Focuses on specific decisions as moves in a long-term competition
   — Highlights issues of the complex consequences of specific choices and decisions (Watts, Roche)
5. Importance of nonrational aspects of man
   — Organizational behavior (limited rationality)
   — Ethology—new model of man—hardwired responses (e.g., eye vs. reason; Kahneman, Tversky, McGuire)
   — Analysis of perceptions, political use/importance of military power, display, etc. What do Russians focus on, give us good marks in, impresses them
   — Schlesinger/Ardrey—late '60s/RAND early '70s effort
   — Critique of standard deterrence theory
   - Rational actor
   - Organizational level of analysis—what assessments/advice from military staffs

(assessments). Soviet assessments of balance—role in net assessment—practices, character of analysis—their scenarios, assumptions, differences in calculations—*cratering, human vulnerability, for example (Hines's finding, their inferences from Minuteman silo spacing and their tests (vulnerability assumptions))*
    - Individual behavioral/gut responses (ethology). Small group behavior
  - Herb Goldhamer—change of values—early heroic warfare/glory vs. later
  - Unusual performance
    - Herb Goldhamer—ramming/criteria aspects of analysis
    - Early carrier battles vs. 5% rule (source of D-Day planning)
6. How to look at the future
  - Can't predict/false elimination of uncertainty—uncertainty has to be faced, managed not eliminated (can't be—much analysis assumes can be)
  - Some aspects of future surer than others—especially long-term economic trends (macro) (e.g., differential rates of GNP growth—but need right experts —Wolf/Bruce Scott), demography, maybe some aspects of technology
  - Value of scenarios—not as predictions but as way of enhancing our imagining of alternative futures (must be tailored to purpose of analysis). Peter Schwartz, Clem Sunter, Bruce Scott, Pierre Wack
  - Also use model above, e.g., actors (goals, strategies) interaction with context—path gaming (invention of)
7. Value of gaming
  - Early RAND impetus—use of human skills at game playing to discover strategies (good ones) for games resisting mathematical/closed solutions
  - Herb Goldhamer's political/crisis gaming (mid/late '50s)—> Helmer/Dalkey games—> SAFE (planning of forces game)
  - NSDG games/SDI—>change in way wars fought. Problem of standard analysis, usual criteria/*Nitze*
8. War, conflict normal
  - Robin Fox article (get). Profound belief in future of war, AWM
9. Only so much stupidity one man can prevent—AWM
  - Reason
  - Lots of stupidity
10. Strategy for influence
  - Don't take credit
  - Work on long-term problems, try for persuasion, change minds—shape way to look at problem (lots of competition on immediate issues—nature of U.S. government leads to waste of effort, decisions undone later)
  - Nature of U.S. government—no one can decide (Schlesinger/Somalia airfield case)—nature of process—extended social-political process/influence by inputs
  - Delusions of intelligence people
11. Addenda
  - Deterrence is an empirical problem
  - Centrality of the political/psychological role of military power
    - Herb Goldhamer
    - Schlesinger
    - Tiger/new model of man
    - Display, how to
  - Good questions, importance of (aphorism)
    - Enders Wimbush case/get right answer—DIA response/CIA indifference

Themes 1[1]

Marshall: I think that of the next couple of times that we meet, probably one session ought to be devoted to getting ready for the meeting that's at the end of October [29 Oct 1993], and the other one could be almost an interview—I could make some notes, you could create some questions on my early background or antecedents, and then maybe you could undertake to take whatever is said there and suggest how a chapter might be organized and then you could try to put something together. But for today, we can just talk about these intellectual themes. So maybe I should just start.

OK. I just have numbered them, and the first one—they occurred somewhat chronologically—is this importance of resource constraints and the macroeconomic perspective on many of these national security issues. It seemed to me to be a central theme of things I've been involved in. Very early on at RAND—once I'd gotten out to Santa Monica—I was involved in this so-called "criteria problem."[2] One of the things that economists were put to work on and one of the reasons that economists were brought to RAND was because it was thought that the RAND Corporation ought to have some economists to address this so-called "criteria," "military worth" issue. Clearly, resources were part of that. And one of the things that came up very early was that the engineers, who until that time had dominated all the studies, often made kind of intellectual mistakes in organizing or specifying the criteria. They also had some odd beliefs that the budgets or the dollars were just some arbitrary, bureaucratically imposed obstacle that they [the budgets] didn't have anything intrinsic to do with the problem. So one of the things early on was to try to persuade the engineers that budgets and dollars reflected resources, and they reflected choices that they were making, and that the costs in the macroeconomic sense were real.

Another way that this resource constraint issue matters or got involved was in the thinking about the ways to improve our forecasting of Soviet programs.[3] One of the problems had been that people made forecasts of how many missiles they could build, how many bombers, how many this or whatever it was, and there was no sense that if you added all the programs together maybe they [Eds: program costs] just outran all resources that they [Eds: Soviets] could possibly put into things. So one of the proposals that we made—and something I think some other people at the Agency were interested in toward the late '50s—was to begin costing out programs as best you could and then looking at what everything added up to.[4] That was, in some sense, the initial motive of the development of the costing of the military program that was undertaken by the Central Intelligence Agency. But anyway, that notion that overall resources are limited and need to be thought about arose in this context of trying to do a better job of forecasting and looking at the trade-offs the Soviets themselves would face.

Related to that is this whole notion of the military burden, and especially the importance of looking at that not only for our own side—which we didn't pay a lot of attention to, that is, people like Joseph Loftus and I—but from the Soviet side, because they were so obviously outspending us as a percentage of gross national product, and this meant in the long term that this would cause problems, difficulties for them, that they could do so only with penalties in terms of either current consumption or of investment, and that either one would, in the end, cause some problems to them, as turned out to be the case.

Later, my own interest in this issue was very much the source of my interest in the health care and pollution side of things in the Soviet Union. That is one of the reasons I started Murray Feshbach off...Well, the sequence was Feshbach's demography work showed this change in the life expectancy of people. First, I guess, it was just the increase in, I suppose, the late '60s, early '70s, the death rates from certain diseases, increased infant mortality, a variety of things, coupled with the changing ethnic balance in the Soviet Union, interested me in the problem of the future demographics.

But the thing that I thought would have the biggest impact in the end was this underinvestment in health care and the health of the population, because ultimately the population, its vigor... In the first place, just its vigor and vitality might be undermined and that would weigh on the long-term competitiveness of the Soviets. But in addition, it meant that when they tried to correct these problems, a new claimant of overall resources would surface and compete in some way or another with the defense expenditures. And also, this was something that couldn't be ignored in perpetuity, because the people themselves, I think, would eventually see that their health status relative to others would decline. In fact, at one time I tried to get the Voice of America to start broadcasting stuff like, "Did you know that your life expectancies have gone down?" and so on. But I couldn't persuade them to do it. I think it was stupid on their part.

Q1: Why [Eds: wouldn't VOA agree to broadcast]?

Marshall: I don't know, they just seemed indifferent to this.

The pollution stuff, I think, again, is another big thing where they have underinvested, in the control of pollution, and are going to pay a price for the next 100 years that will be higher in terms of higher expenses, people [indistinct]. And it's those kinds of failures of investment that create semi-permanent handicaps.

Alright, so I had the sense that these resource constraints are very important and had to be thought about in these matters. Just one anecdote: in the competitive strategies stuff, one of the ideas was this notion of cost-imposing strategies. It was interesting that one of the really most intelligent military officers, a man who was then the Chief of Staff of the Air Force [Lew Allen, Chief of Staff, 1978–82], wasn't sure about this, that it was a viable idea. First, because his view was that if we do something and we increase the amount of money they have to put into the appropriate counters in the defense area, they would simply always find the resources. In other words, that there were, in a real-life sense, endless resources available to them. The long-term consequences were not, I think, part of his thinking. And the second thing, which was a good reason, was that we might start something that we wouldn't stick with and so we would stimulate activities on their side which we weren't prepared ourselves to counter and to deal with. And that's a very legitimate concern.

OK. The second intellectual theme, I would say, is that the opponent is not a fixed thing, that is, that you can't proceed by making a forecast of what the opponent is likely to do over any extended period of time and then plan against that. Because his responses… He's not fixed in any particular shape. He has some constants limiting what he does, but not towards the… And even using several alternatives is not fully satisfactory.

On the other hand, I think one of the early mistakes at RAND was the tendency to assume the worst about the opponent. The thing that made game theory and the minimax kind of philosophy fail in the end is that it's wrong to, in some sense, assume the worst—that he will be the most rational, flexible opponent. And this led to a great interest on my part in trying to understand the character of other states, the needs of opponents, the social-cultural factors that influence their behavior, their choices, their [indistinct]. Also, I came to understand that history affects their viewpoint, but more than that, leads to their organizations having certain characters, certain structures, which themselves perpetuate certain kinds of behavior, certain kinds of processes, and so on.

A related theme is that you need in your analysis and your thinking to take account of his weaknesses and your strengths, which tends not to be the way at least the bulk of American analysis has tended to proceed. It focuses a lot more on our weaknesses and new or possible strengths that he might have. Clearly, you've got to take account of that, but the actual functioning of analysis tends to have a blind spot with respect to the opponent's weaknesses and how to exploit them.

Some of our strengths, or his weaknesses—our strengths in particular—come about because of special competences that are created by past investments, by past history and experiences. For example, all of the skills involved in aircraft carrier operations would probably take anybody starting from scratch 15 or 20 years to fully reproduce. So many of these strengths are the results of investment or they're social-cultural.

I was very struck at one point by Michel Crozier's book—I think called *The Bureaucratic Phenomenon*—where he showed how different cultures lead to different kinds of organizations and different kinds of relations within organizations of superiors and subordinates. He showed how these things are very related to deeply rooted cultural things that relate, for example, specifically to this whole matter of how—right from childhood—superiors treat subordinates and so on. The whole set of relationships that are inherent in the particular culture get reproduced in many ways in the organizations.[5] This means, among other things, that Soviet command-and control systems are Russian; the relationship of superiors and subordinates is very, very different than in the United States and most other Western countries.

Again, all of this is ignored in the standard analysis—well, that's repeating—I mean it tends to emphasize the opponent's strengths. Especially the intelligence community tends to focus on new strengths that he might create by buying X, Y, and Z.

The thing to note, I think, in a way, is that this means that these strengths [and] weaknesses are very hard to change on both sides, which led me, on the one hand, to feel that this usual contrast between capabilities and intentions was not adequate. Because the tendency is to see capabilities as relatively fixed and, therefore, intelligence people should focus on them, whereas intentions or the behavioral sorts of things could rapidly change. And there's some truth in that, but the fact is that behavior in many ways is also very hard to change, or the way you would fight, or the performance of your organizations, is very, very hard to change. It's not a matter of what kind of weapons you own, but the skills, the capacities of these organizations. And, yes, with a new leadership, they could change their views and the decision to go to war or not go to war might change, but how they would fight, how they would behave, in many other ways is far more fixed than this intentions-versus-capabilities simple paradigm suggests.

As I said, the intelligence community does not focus very much on weaknesses. The first net assessment that was done was started when I was still at the National Security Council staff, then I changed and moved to the Pentagon in the middle of it, and when it came to the end of the study effort, the intelligence people had been asked to talk about the strengths and weaknesses of the Soviet ground forces. And the initial [Defense Intelligence Agency] position was that there were no weaknesses. Then you asked them, "How could that be?" Their argument was, "Well, if they have a problem, they will notice it and fix it. So it's not worthwhile taking account of it in any real planning." I guess it was at that point that I said, "Would that we were like that." But Lt. Gen. Danny Graham, who was the head of DIA at that point, when I explained the thing to him said, "Nonsense," and went on and developed a list of weaknesses. This episode is symptomatic of the problem.

Another implication of all of this is that recommendations about new, good things that we should do to improve our posture, our performance, must take into account the problems of implementation by the organizations that are being asked to change. And again, most analysis does not deal with this at all, and tends to put forth abstract best solutions. This is another weakness of most systems analysis. When I was working in the mid '60s on this whole problem of organizational behavior and drew attention to this point, the notion was coined of "the missing chapter" [in RAND reports], that never really dealt with this very serious implementation problem.[6]

Alright, a third theme is a kind of model of how you ought to think about these situations that are typical in military and national security analysis. That is, that the whole structure of the situation should be seen as follows: there are actors—at one level they could be states, at another level they could be separate military organizations within states. The actors each have strengths, weaknesses, a stream of resources that supports them, patterns of behavior that come from their history, the culture that they are in, and so on. Some of these things may change gradually over time and be affected, for example, by differential rates of GNP growth in particular countries. Also, each of these actors has its goals and its strategies for attaining those goals. And in this sense, all of the actors need to be treated entirely symmetrically.

Then these actors are set within a context within which they seek to attain their goals and to compete among one another, or to form coalitions, whatever. This context has to be described in terms of factors that are important to the analysis, but not really under the control of any of the actors, or largely are so. These factors may include most aspects of technology. The demographics of even their own societies are seldom under the control of the leaders of countries. Climate is another factor. For example, Joe Fletcher makes a good case that there are something like 140-year cycles in the climate right now. Also there could emerge new viruses, other organisms, new plagues.[7] These factors represent the changing character of the context.

It's that way of thinking about things, the structure of the problems, that is very, very important. In some ways, this is related to the earlier theme of, essentially, the opponent isn't fixed. You can't just project his behavior and plan against it.

And then, the issue is on the strategies and other things that would determine the interaction between the opponents and the evolution of the state description of other [paths?]

A fourth theme in my life is the importance in many, if not most cases of looking at particular issues, within this long-term competition framework. While there was some of this kind of a framework that characterized some of the very early thinking that I was exposed to at RAND,[8] and certainly related to the sort of thinking that stemmed from George Kennan's analysis of Russia, the thing that I noticed when I resurrected it and formulated it in a way more adapted to thinking about the analysis of national security issues in the late '60s was the change of perspective that it gave you.[9] That is, this focus on the need to think through a strategy for competing over an extended period of time forced attention to, first, a complex set of goals that you really have in such a competition, which, again, often is ignored in the standard analysis, which tends to try to search for quantification, to narrow down the criteria to things that can be focused on in analysis. It then relegates what most people focus attention on, that is, some specific decision or program, to being really a move in the implementation of a strategy in some extended competition.

That together, at least for me, just transformed the way you ought to look at many issues, and I think, in some ways, has been really very central to a lot of my thinking ever since. And this thing itself stemmed from the work on trying to look at the Russians and understand them, which led to this focus on not only this resource issue, the constraints, but on the fact…[gap due to tape change] … focuses attention on the multiplicity of goals that you have to think about, trade off, to some extent, in making your decisions.

That's maybe the wrong time. I was talking about the organizational thing, which then led me not only to the Carnegie-Mellon University focus on the limited rationality of decision-making of organizations, but also to the Harvard Business School people, who were trying to characterize and understand organizations as having themselves strategies for competing and succeeding in the environment that they were in.[10] And indeed, that one of the critical factors in the management of a large organization was the correct specification of a strategy, which could then guide other kinds of decisions and so on.

So it meant that when one looked at a particular area, like the strategic nuclear systems and strategic warfare, intellectually you ought to think about what our strategy for competing in that area was, and then, from that, you would get a clearer picture of how to think about specific choices, like this bomber versus that bomber, or how many, the whole complex set of issues. It's also kind of related to some of the things that Jim Roche and Barry Watts raise in their criticism of measures of effectiveness ["Choosing Analytic Measures"].[11] That, again, if you look at real life, you see that particular moves or actions that one takes have a very complex set of consequences far beyond the kind of things that are reflected in these measures of effectiveness that narrow things down, are very reductionist, and tailored to try to allow quantitative analysis to proceed.

A next theme for me, I think, has been the importance of the nonrational aspects of man. Now, some of this is reflected in this interest in organizational behavior and limited rationality, the limited rationality not only of organizations, but of human beings themselves, who can't be fully rational in that there's too much to do. Then there's the aspect that comes from ethology, where you see that man is simply another animal, and is not fully rational, his action, his behavior is not dominated by the rational part of him. Indeed, Lionel Tiger's made the point that man is an "action animal," he is not a "thinking animal." Man was designed to survive in the jungle, where quick and reasonably astute decisions are what are important.

In some sense, we're living in a period where a new model of man is emerging. Most intellectuals hate it because it does not accord as high a place to rationality, and it raises what seems to be the issue or the insight that in some ways a lot or some significant part of the responses of man to situations are "hardwired" and have been placed there through the course of his evolution. I was talking once with Daniel Kahneman, the very interesting psychologist, who made the point that if you look at the eye, sight is, through evolution, one of the most elaborated and perfected of our senses. You can see when you look at the eye and understand it in more detail how it is kind of layer after layer of adjustments and adaptation to reach this level of accuracy and precision that we have. His notion was that the rational, the thinking part of man was still evolving and [had] not fully reached the state which sometime it will. You just have to look at the work of Kahneman and Amos Tversky to see how fallible man's thinking is, the kind of mistakes, from a purely rational perspective, that the mind is inclined to make.[12]

And then you look at the work of Michael McGuire and these people on one of man's nearer relatives—where the most detailed work has been done—the vervet monkey.[13] You see how the whole blood chemistry changes by the social status of the animal. I mean dominant animals have much higher serotonin levels and so on. There even is some reason to believe their mental functioning is different. And we know that in other animals we have similar kinds of things. Perhaps the most dramatic is a story about a lizard, which we can get into.[14] And there have been some tests on humans. Although, again, humans are frightened by and don't want to know about some of these things, but it's clear that the blood chemistry of humans is, to some extent, related to their social status. The importance of understanding, taking account of what are the nonrational aspects of man and his behavior then leads to this interest in the analysis of what in the '70s began to be called "perceptions," that is, perceptions versus reality. This is central to understanding, say, the political use of force and its effectiveness, the political use of military displays, the impact of the possession of military power on the part of one actor on the behavior of the other.

I got into this work on perceptions, particularly this ethology aspect of it, because when Jim Schlesinger and I were at RAND—he had arrived in the early '60s and I had returned after living in Paris for a couple of years [1963–65]—one of the topics that came up between us was the work of Robert Ardrey.[15] Ardrey developed this notion of the "territorial imperative," this very visceral response to certain kinds of encroachments that people have. And this gets later into this whole issue of how do you think about deterrence and so on. When I worked for Henry Kissinger on the NSC, one of the things I tried to get the intelligence community to do—which they did assign one guy to do some work on—was to try to understand what it was about us and our behavior that the Russians paid attention to and focused on, presumably because they thought it was important, what gave us good scores in their books, and what things didn't. The idea behind this was that one of the things that we could do to enhance deterrence and to make the influence of our military capabilities the highest was to understand this. And a little progress was made, but it was seen as novel to even raise some of those kind of questions.

The fact that you want to raise them and move beyond these abstract discussions that have characterized so much of the analysis at that point was driven by this sense that these other people, in the first place, may be different than us.

Secondly, what they pay attention to, react to, is not simply a matter of calculation, as much of the analysis up to that time had tended to suggest. If you look at the standard deterrence theory, it has this tendency to think of states as not only unified, but rational actors. Again, the supposed unity and rationality of states had come up very early in this work on organizational behavior and was a bad assumption, because the behavior you were thinking about was often the behavior of an organization, not of a single actor. A state might be made up of several organizations, and there's this concept of the limited rationality of those organizations.

And you might think of it in terms of, say, like the Russians, wanting to know what advice, what assessments, what evaluations the General Staff gives to the leadership. And even the leadership is not necessarily a single individual. This led to a great deal of emphasis in the work in this office on the Soviet assessment, the assessment of the Soviet General Staff, of the state of the balance, as being a very central issue which is going to be heavily influenced by the specific character of the analysis that they undertake, the choices of scenarios that they [make], various other kinds of assumptions, which can lead to differences in specific calculations that are part of any evaluation. And we have some evidence that there are some very significant differences there. Some later work that John Hines has done suggests that back in the '60s, their thinking about us was dominated by some inferences they made from particular things we did and nothing about what we said. I mean, particular choices they made.

The other thing—coming back, moving beyond this organizational level of things—is, just at the individual level, there's a lot of evidence that man is not a fully rational animal. People have always understood that men have emotions and that emotions can get in the way of their thinking, but that model tends to assume that at the center of everything there is this rational calculation, and it's just that these emotions cloud things or distort the functioning of this rational machine. But what this ethological and evolutionary view suggests is that there is no perfect calculator at the center of this thing, that man's ability to be rational is itself probably, as Kahneman would suggest, still in evolution and has built-in short cuts, quick algorithms and so on because of the bias towards action and behavior. Again, the work of Kahneman and Tversky shows how easily, by framing issues and questions different ways, you get different answers to what are logically the same question.[16] Then apart from the fallibility of the central rational mechanisms, there are simply these physiological responses, the changes in blood chemistry, which may change the behavior and the thinking of people, that all need to be taken into account.

Because of Schlesinger's interest in all of these kinds of things when he became Secretary of Defense and I moved over to the Pentagon, one of the things he asked me to work on, in particular, was this whole matter of essentially this political-psychological impact of military capabilities, military actions, displays, posturings. So I undertook to do this, got some money from the Defense Advanced Research Projects Agency, and much of the program was done out at RAND under Herb Goldhamer's direction. There was an early meeting that people came to, including Lionel Tiger, whose work I had read.[17] He and Robin Fox had just put out this book, *The Imperial Animal,* which I found very good and I think still is an excellent piece.[18] So we started down this road and had some general thinking.

We had specific studies using *The Economist* as a surrogate for elite views in England and *Le Monde* as a surrogate for French opinion to try to look historically and see what events, what kinds of things seemed to change and shift the assessments of the military balance between ourselves and the Soviets. Goldhamer did that work.[19]

Herb began to write a book which, unfortunately, he never finished, but the first part of it has come out as this piece, *Reality and Belief in Military Affairs.*[20] After he died, we collected together and put out in a publishable form all of his notes for the rest of the book.[21] He was a very meticulous gatherer of material and so on. A lot of interesting insights come out of that. One of the more interesting, still further things that came up in some of Herb's work is, if you look at the changing cultural environment, you see a change in values. You see that early warfare was an opportunity for individual men to perform heroically and to attain glory. Later, war became far more a matter of having other goals. It wasn't an occasion for the display of individual courage and so on, but was related to some goal of a king, a prince, a state. And one of the most striking illustrations he gives—it was taken from a poem, I think—was a case of two opponents separated by a stream, and one of them says, in effect, "We'll let you cross the stream so we can have a really good fight."[22] Now, maybe it's too romanticized, but I think he's right, that at an earlier time that probably was true. So it changes your sense of what war and warfare are all about.

Another thing—which, again, came up in my memory of Herb Goldhamer—is the point that we discussed, going back to these very early days at RAND, when we were looking at this whole "criteria problem" and related things, thinking about the kind of studies RAND was doing. Herb made this very interesting, important point that in all of these analyses, certain behavior is assumed on the part of the men involved and, therefore, what the performance of the systems would be and so on.[23] The point he made very early was that in real life, there would be some cases of unusual self-sacrifice that would alter all of these calculations. He gave as his example, if all else failed, pilots ramming enemy bombers carrying nuclear weapons and approaching the cities they were defending. And I think that's very real and almost all the analysis just does not capture that. It's hard to know how important that really is, but it needs to be thought about.

Previously, I mentioned one of the early kind of anomalies in the analyses' criteria was that in the early RAND bomber studies, they assumed that you couldn't go above five percent attrition per sortie. I found that arbitrary and tried to understand the source of it, which really was World War II and the D-Day planning constraint. And then to try to persuade people that this was not the way you ought to look at it, I tried, from whatever memoirs I could find, to calculate the attrition rates in the early aircraft carrier battles, and I believe that they're close to 50 percent, at least at the strike sortie... It seemed to me that in the kind of situation which was then being envisaged, where both sides had some nuclear weapons, that if it came to that situation, the casualty rates of the bombers could be set very high.

Still another theme—I guess number six on my list—is my particular notion of how you try to look at the future. I'd say, first, that you really can't predict the future farther out. If you try to give a single forecast it's kind of a fake, it tends to eliminate uncertainties that can't be eliminated. And in general uncertainty has to be faced, managed, dealt with in some effective way and not simply made to go away in some make-believe way. Unfortunately, much analysis proceeds on the opposite principle.

But there are some aspects of the future it's possible to say something about. That's, I think, particularly true of gross, long-term economic trends, things like rates of growth of GNP, especially differential rates of growth of GNP. But again, it's a matter of picking the right experts. In that respect, Charlie Wolf has done some really good things, and Bruce Scott of the Harvard Business School has been absolutely miraculous with his kind of insights into how things were going to tum out for Europe, say, in things he wrote as early as the early '70s.[24] Another area where there's some chance of prediction that makes a difference is demography. And there may well be also some aspects of technology, not all of it.

In general, there is a value in sketching out, thinking through some alternative scenarios of the future, not as predictions, but as a way of enhancing our appreciation of the alternative directions in which things might go. But this creation of scenarios has to be tailored to the purpose that the analysis is going to be used for; there are no general-purpose scenarios for the future. And there are some people around who understand that and do good work of this sort: Peter Schwartz is an interesting person; a man named Clem Sunter, a South African; Bruce Scott; this man Pierre Wack.[25] A number of them were associated with the Royal Dutch Shell Group strategic planning activities in London for a while.

The other thing I would say about looking at the future is to recommend thinking of it in terms of the model that I discussed earlier of actors with goals and strategies, interacting with some kind of context. For that purpose, some form of path gaming might be useful. I've been instrumental in trying to get that kind of gaming created and done, beginning in, I guess, the middle '80s or so.

Alright, a seventh theme, I would say, is the value of gaming as a technique.[26] There was a really early RAND impetus towards gaming. It grew in a funny way out of the focus on game theory. People soon found that there were games they wanted to look at or think about that were just too hard mathematically to look for solutions. So an idea arose, and I think it was suggested by a man named Merrill Flood, who was at RAND at the time, and subsequently at the University of Michigan. He just made the observation that human beings are very good at playing games, and can find effective strategies, reasonably effective strategies, and maybe the way to handle some of these games that we can't solve in closed solution mathematically would be to have them play and gain experience and devise strategies and so on, and that could give us an insight into what good strategies are.

So there were games that came from that, that were strictly kind of artificial, small games that couldn't be solved. But that soon led to other games—I'll mention several—some of which were then applied to much more complex military problems. Another thing that it led to was political gaming. Herb Goldhamer was an early innovator in political, or kind of crisis, games which I think the first one of that sort of thing was probably played at RAND somewhere in the mid-'50s [1955–56]. Herb ran that.[27] And that led to much more elaborate and ambitious games that Olaf Helmer and Norman Dalkey ran, big political kind of games, and also led to a SAFE [Strategy and Force Evaluation] game, which was focused around the allocation of budgets and kind of planning and programming for forces, which was around the late '50s [1961–62].[28]

Anyway, I've been very interested in games from the point of view of an initial way of getting some insight into the strategies that people might adopt in various situations. That led both to my interest in and use of some games in the middle '80s to explore such things as the impact of an SDI [Strategic Defense Initiative] system on the actual character of warfare, the results of which I think were quite interesting in that they showed the problems of much of the standard analysis, which then just missed, in my view, the really big impact of SDI. I mean the impact apart from the things that people now talk about—the resource and psychological impact it had on the Russians—which was that people changed significantly the way in which the whole war was fought. It was as though its [the SDI system] main effect was to change the character of the game that was played. So measuring just what it was going to do to the attrition rate didn't capture what it was really doing. It also turned out that, contrary to some of the arguments of the time, both sides, when they had some SDI-type systems, felt the situation was more stable, which wasn't what people tended to say.

I think the role of these games, as I say, is an initial, exploratory tool to try to understand how changed weapons, capabilities, or something will change the kind of operational concept, strategies that people will employ.

An eighth theme or thought is that war and conflict are normal and natural. Robin Fox had a very interesting, recent article in *The National Interest* ["Fatal Attraction: War and Human Nature"].[29]

At one point—I don't know how or why I said it—but Jim Roche then made a wall poster out of it, I said, "I have a profound belief in the future of war." Fox's article is particularly interesting in that he talks about how war ought to be seen like sex, as part of life. It's not something that's unnatural that we're going to get rid of any time soon.

A ninth theme or saying of "Chairman Marshall" is there's only so much stupidity one man can prevent. You live in big organizations in which a lot of not-so-great things are happening most of the time. And only so much of that can be prevented. So you have to be very modest about what you think you can do.

Q1: What rules of thumb have you used in picking and choosing which idiocies to combat?

Marshall: I think in the near term it's hard to change almost anything. I think the things that you can change are longer term... This is really item ten, the strategy for influence.

At the tactical level, a lot of the things are obvious, like, give other people credit. If they steal your stuff—terrific! The main thing is to change the way people talk about, think about the problems. So my notion is that you work on long-term, more or less permanent kinds of problems and issues, and try, through giving people better ways to think about those problems, to persuade them that this is the best way of looking at it, these are the kind of considerations that should be taken into account. And in changed minds and shaping the way in which they look at problems, you gain influence. And for a lot of that, you can tell whether you're being successful when people adopt your terminology.

One of the reasons that this works, I think, in these organizations is that there's an enormous amount of competition for controlling or influencing the immediate issues. It's the nature of this government. Now, I think in a different kind of government structure, the strategy I'm suggesting would be less fruitful. But in our current government, not simply the current administration, but the whole nature of the U.S. government is that it's designed so that nobody can decide anything, and a decision that seems to be won one day is reversed the next day or next week or next month. It's hard to make any specific decision stick.

And much of this competition to determine what's going to happen tomorrow is, in my view, a waste of effort. Nonetheless, that's what 90 percent of the people are involved in. And it's interesting that I think many of the people in the government don't understand that.

An example I would give is when Schlesinger was here. This was the time when some Russian ships were tied up and some barges were tied up near a Somali airfield. A runway, I think, had been lengthened, and there were some buildings there that people thought maybe meant they're going to start emplacing some missiles.[30] Schlesinger asked me to go out to the CIA and to ask them to provide their analysis of exactly what was happening. But in addition, he wanted them to go back and look at some other cases, because the hunch was that the whole nature of the way in which the Russians obtained access and made use of foreign facilities was rather different than our own, and that being able to explain that would allow one to make a better case about what we ought to do, think through what we ought to do, and so on.

Well, the people I went to talk with said, "Yes, yes, well, of course we'll do something, but what decision is this information that we will be providing an input to?" And I said, "Look, Schlesinger doesn't have in mind any particular decision. He thinks it just would be good to have this kind of analysis and insight into the way in which the Russians go about acquiring access, making use of some of these foreign facilities. And by the way, don't you understand that it isn't as though there's some decision that's going to be made next Thursday, because that's not what this is all about. This is an attempt to help educate the Congress and other people. And what you're involved in influencing is a long-term, gradual, social-cultural-political process. It isn't as though a decision is going to be made next Thursday.

But people don't understand that. They're very captured by this notion that there are decisions that are made. Now, there are situations where there are decisions that get made, as in actual operational things. I mean, yes, you do send that fighter over there, or you commit the attack on Baghdad, and so on. Right. But most decisions aren't like that. They're really part of a sequence of things [indistinct]. There's this long, complex process, and that's what you're participating in. So there are delusions on the part of people, like those in the intelligence community, as to what they're actually doing, what they're contributing to, and so on.

Going back to an earlier point on the deterrence thing: The important thing is to see the deterrence issue as an empirical problem. That is, here is this set of people, or this country that you're trying to persuade or dissuade from doing this or that or these kinds of acts under these circumstances. Then the problem is finding out, what is it that is distasteful, frightening, deterring to them? And that's an empirical issue, it's not something you sit down and think about. People just don't look at that.

Another point is this centrality of this political-psychological role of the military, because most of the time that's what it's doing. It isn't as though you buy the military only because a war is going to happen. We've just been through a long 40 years, when the whole function of the military was to deter the other side.

Now, it does it via its capabilities to fight, but the deterrence thing itself is the result of the calculations and the reactions of this other set of people. So understanding that is absolutely central. And there are people, as I say, like Herb Goldhamer, Jim Schlesinger, Lionel Tiger, and some of these other people who have this new model of man, that all have something to contribute...or the cultural anthropologists. And again, most analysis treats the issues of deterrence as entirely ones of, "If there is a war, how many targets are we going to take out?", which maybe isn't entirely irrelevant, but it doesn't get you focused on the aspects of the thing that are probably the most important.

Another addendum—it's really a separate kind of theme, something that I've certainly believed all my life, at least for the time that I can remember—is the notion that poor, mediocre answers to good questions are more important, more useful than splendid answers to poor questions. That means that getting the questions right is very, very important. Most analysis spends far too little time on what the real questions are. Also, of course, most analysis tends to gravitate, to focus on the questions the analysts can answer, not what the real questions are. There certainly are dramatic cases of that. I think much of the whole effort on the Soviet economy fails on that grounds, when you look at what people really focused on. Joe Loftus's and my experience at RAND was that the specialists on the Soviet economy failed to provide useful analysis. We were trying to make these improved projections of the Soviets' military programs. We would ask these guys—who were supposed to be the experts on the Soviet economy—the questions that we thought were relevant. They had no answers to them, did not ever promise to give any answers on those things, but went merrily along answering the questions that they were focused on.

Q1: What were the types of questions you and Loftus would ask them?

Marshall: We were interested in the effect of the military burden on the long-term prospects of the economy, and what the trade-offs were that they had to make to support this large effort, things of that sort. And we just never could get anything.

Now, the other thing I would say on the "good questions" thing is that what's also striking is that even if you have the answers to good questions, nobody may listen. An example, I think, of that is that because of, initially, Feshbach's work showing that one of the things that was going to be happening to the Soviet conscription cohort was that the ethnic composition was going to be shifting and there were going to be many, many more of the Asian ethnics, the questions arose in my mind, "Is that a problem? Are there ethnic tensions, difficulties? Do these people really speak Russian? What are the implications of this?"

So I had some people out at RAND, headed by Enders Wimbush, start interviewing emigres who had served in the military who could give their own personal testimony about the kind of tensions, and also how the conscription system really worked, in the sense of, after you were conscripted, how was it decided what your assignment was? A lot of interesting stuff came out of that, including the sense that, yes, they were going to have a lot of trouble, not only because the language ability of these people was not good and, if anything, was tending to decline, but, secondly, there were enormous tensions and brutality, and therefore, significant internal problems would result if they had this. The RAND group also found out, first, that these people had been almost entirely, at that point, sent to the construction battalions and forces, and to some of the MVD [Ministry of Internal Affairs] divisions, because they might be needed to shoot Russians and were much more likely to do that. Second, that when they were sent to regular combat divisions, they tended to be cooks, bottlewashers, and not in the actual combat teams, say, within the tanks and so on. This was the case at that time; later that changed. So if you had a lot more of them, they probably were going to have to move into some of these other jobs.[31]

The people in the intelligence community, principally DIA, just disputed the fact that they, the Soviets, were ever going to have any difficulties, that they were simply just going to manage and handle this and there was just going to be no problem.

Q1: With no further explanation? They just asserted this?

Marshall: No, just by assertion.

That's the end of my themes. Maybe you can think of others later.

Q1: Those are very good. I just made a checklist for myself and I think that everything that I got here has been covered. Oh, I know... In reading some of the materials yesterday, there was a series of correspondence that you had with former Secretary of Defense Donald Rumsfeld about "centers of excellence" and you kept returning to this need for these "centers of excellence" to bring up younger analysts, training them. And that's something that you addressed in *On Not Confusing Ourselves*.[32]

Marshall: Yes, right.

Q1: We've talked about what makes for good analysis, but how about what makes for good analysts?

Marshall: I think, and it's reflected in that chapter ["Strategy as a Profession for Future Generations"] that what makes for good analysts, apart from their being bright and so on, is probably early exposure to some broad set of issues, some broad questions in some kind of group activity, where real answers are required, where they can work as a kind of team member, without a lot of hierarchy, where it's not a matter of, you know, the senior professor that does all the talking and comments from the younger people are not appreciated and so on. So I think that. Now, I do worry that there are some professional backgrounds which may kind of inoculate you against ever having the right kind of broad look at things. So it seems to me that the way it can mainly happen is through some early work experiences. And I don't know that there are lots of good places to have that happen right now. I may be wrong.

Q1: You think there have been more opportunities for that in the past?

Marshall: Well, there have been some. The RAND environment, I think, was certainly like that. And I've tried to run this place in that way. And I suppose that Herman Kahn's place [Hudson Institute] was probably a good place for that. But I don't think there are lots of good places. I might be wrong about that.

Q1: One of the major themes you had in the original, extended outline was the evolution and outcome of the Cold War, how it turned out the way that it did. Obviously what you've been talking about today touches on all of that...

Marshall: No, what I was trying to do here was to discuss the kind of perspectives, issues, attitudes that I think come up repeatedly and sort of characterize things.

Q1: That's in a different category.

Marshall: I think so.

Q1: And then a lot of the examples that you gave bear on this issue of the interaction of analysis with the real world.

Marshall: Yes.

Q1: But in particular, the theme of your strategy for influence, I think, deals with that.

Marshall: Yes.

Q1: Would you mind if I made copies of your notes?

Marshall: No, if you want to. And then there's this thing [Arthur Schlesinger, Jr., "The Future Outwits Us Again," *Wall Street Journal,* 20 Sep 93]. I clipped this out because it seemed to me—I usually don't agree with him—but he had some things that go with this theme of what you can predict about the future or how to look at…

OK. Perhaps one final thing which I think about, and one of the most important things I think I learned when I was at RAND early on is that the objective in any analysis is to do the best that this country can do, not just the best that RAND or whatever organization you're talking about can do. One of the very good things about RAND in the early days was that there were these problems, but a lot of people from the outside were drawn in, both in the special summertime period when people from universities and others came out, but also there were a lot of visitors throughout the year. There was always a constant stream of people like John von Neumann, and this one and that one, or this and that mathematician or physicist or economist. Since we were focused, at least in the Economics Department, on how you would bomb the Soviet economy, what was the most effective use of the limited number of nuclear weapons, people who had been involved in that kind of activity in World War II, like Sidney Alexander and Carl Kaysen and others, were brought out.

So there was just this whole attitude that, alright, here is the problem, what is the best answer? What is the best solution? How should you think about it?

You see other organizations where they just don't do that. They'd die rather than bring in anybody else. If they're given a problem, it's, "What can we do?" So this attitude is very important—and little practiced, in my view. And some of the government organizations are very bad on this or have changed. For example, when I first knew the CIA, it was a very open place; it looked for outside help, ideas, and so on. Unfortunately, by the late, maybe it's the middle '60s, it had decided that it didn't need any help.

Q1: What precipitated that?

Marshall: I don't know. Some of it may have been precipitated by the fact that, in the beginning, of course, the organization itself had been brought in, newly formed, people still had lots of external contacts, and later it had become much more to be dominated by people whose whole careers had been in CIA. I don't know. Now, also, of course, that was a time of some estrangement from the universities because of the Vietnam War. But I think it was related to this kind of maturing of the organization.

Q1: Yeah, maybe you could talk a little about that, because that's something that comes up in your writings, it's part of a cycle.

Marshall: I think that that's true. I think it's happened to RAND, too. I don't have a lot to say about it, but I think that it's very real and deserves some serious study. And it's one of the reasons, for example, going back to this thing you were mentioning, my interest in forming new centers of excellence, that I would start them new. I would not say, "RAND, we want you to reform yourself. Why don't you do these things." Part of the argument I had with Rumsfeld and others was, no, you can't just… Look at some of these places. They ought to be small. They ought to be embedded in some friendly atmosphere and not have to provide all of the administrative infrastructure that goes with an organization, like health programs and retirement plans and all of that. And [they] maybe could be embedded in a place that has some technical expertise; it could be used for consulting or something. But I would start these organizations new, precisely because of this aging kind of effect.

How and why it takes place I think is something worthy of study, but I don't have anything profound to offer about that. It's very clear in the case of the CIA. It changed its character almost entirely, I think. And also, organizations suffer traumas and they go through periods of low morale. Certainly when Schlesinger went out there to become Director of Central Intelligence. We ought to talk about that later. I sent him a handwritten letter.

Q1: Oh, I think you mentioned that. You mentioned it to me the last time we talked, but you didn't go into any detail.

Marshall: Yeah, right.

Q1: OK.

Marshall: But it was, when he went out there, already a very low morale place. Again, it's sort of a similarity between RAND and CIA. In the beginning, people worked long hours, they were very immersed in their work. Later, it becomes much more routinized, and so it's an eight-to-five thing. And Schlesinger sarcastically—sometime after he'd been out there, I think reflecting on some of the things I'd told him in the letter—said, "I don't know why they put lights in this place. You could be trampled on the stairways at 4:30, because everybody's bailing out." Organizations go through that.

Q1: But you don't know of any work that's been done about this problem in organizations…

Marshall: No, I don't… No. Maybe we'd have to call up some really good people in the organizational theory, behavior area, somebody like Jim March or others, and say, "What do people know about this?" But I think it's very real.

Outline for 6 October 1995 Talk

[Note: Italicized items were not discussed during the interview.]

Themes 2

1. Importance of long-term assessments/long-term competition
    — Structural factors/usefulness of
        - Social/cultural competences
        - Organizational/historical competences
        - Economic growth rates
        - Demography
        - Lead to strengths, weaknesses
    — RAND, early '50s
    — RAND, 1969–72—>early focus of Net Assessment
    — Competitive strategies
    — Soviet Union endgame
2. Importance of resources
    — Economics matters, at least in long run
        - Feedback into long-term competition
        - Limits on actions
        - Long-term consequences—e.g., Soviet Union's health and environment
3. Knowing the opponent is important
    — RAND—AWM/limited usefulness of minimax. You always know something about the opponent
    — Don't assume rationality
        - Intellectuals escape from reality; focus on the actual facts about the world
        - Exploiting weaknesses, developing strengths central to doing well/winning
        - Long-term interest in sources of other models
            –History—behavior of states, people—role of honor, emotion, etc.
            –Bureaucracy, organizational behavior—late '50s search for insight into Soviet military behavior
            –Ethology, etc.
            –Robert Ardrey
            –Lionel Tiger and Robin Fox
            –Daniel Kahneman and Amos Tversky
            –Fits with long-term interest in evolution, man as product of evolution. Rene Dubos/C.H. Waddington
        - Belief structure of others may be very different from yours
            –Soviets: *Fuel-air explosives*, nuclear effects
            –Columbus
        - Deterrence as empirical problem
4. Importance of reality as reference
    — Work with Joseph Loftus in the late '50s: What are we trying to forecast—behavior of Soviet military/programs/forces? Product of complex processes within Soviet government/military organizations/industry—>What do we know about organizational behavior?
    NATO, mid-'50s
    [note in left margin of p. 2 of handwritten outline]
    — Long-term competition concept emerges from relook at long-term competition in strategic forces in 1969. Ask what are choices of new programs—look at as moves in continuing competition

- — Herman Kahn-AWM: Model-oriented/theory-oriented vs. reality-oriented
    - Physics/Edward Teller—experimental focus
    - *Economics—deli*

Beliefs

5. Some progress can be made on almost any question/issue
    - — RAND Soviet economists refuse to answer [AWM-Loftus questions]. First sign of fakeness of effort/mid-'50s
    - — Even not-so-good answers to good questions better than good answers to poor questions
        - Feasible
        - Obviously better if can do something
6. Character of man—not the liberal version
    - — Product of evolution, limit to rationality
    - — Role of honor, passion—Herbert Goldhamer on bravery/sacrifice/air defense
    - — Status/dominance
    - — Lionel Tiger
    - — Faith in the future of war/conflict
7. There are real experts
    - — Some people are smarter than others—some [___?] exceptional
        - Herman Kahn on Hans Bethe
        - Find them/get them if possible (RAND in '50s/*cost of wartime [protection?]*)

Themes 2[33]

Marshall: Well, the idea of this meeting is in some ways a recapitulation of something we did earlier, although while we covered some of it earlier I kind of did it independently and then looked back to see a little bit about how much overlap there was, and happily there was a fair amount, so there was some level of consistency. But the idea was to try to summarize what you might think of as "intellectual themes" over the years. So let me go through this. Now you're, of course, free to interrupt at any time, but then you will have maybe other things to add or say, "No, no, no. I see you as very different," or something like that.

So, I would say the following; it's my notion of things that have been major distinctive themes in the way I've thought about things. The first one of these is this importance of essentially looking longer term, anyway, taking a fairly long-term perspective, both in terms of various kinds of assessments, or this long-term competition notion. I guess certainly this notion of competitions is very central, and certainly at the beginning, early days at RAND there was this notion of looking at ourselves and the Soviets as being in a very extended competition. That wasn't very distinctive, because I think in the early '50s, Eisenhower and various people had much of this. So it wasn't as though there was something very distinctive and special about it. But, that if you're going to look at that kind of thing, it's very important to consider social-cultural aspects of the competitors; organizational arrangements and the degree things have derived from historical experiences, historical arrangements within that particular society; economic growth rates over the long-term are very important and are something to look at; demography.

Because all of those things lead to comparative strengths, weaknesses that determine one society versus another.

And, as I've already mentioned, there was this early RAND thinking that was along this line. It unfortunately vanished in the middle '50s, or the second half of the '50s. Then, for reasons I've discussed elsewhere, it recurred in my own thinking in the period '68, '69, and led to this formulation of this notion of the long-term competition and the need for a strategy to succeed in these extended competitions, which got transmuted later into competitive strategies, and certainly affected my notions of the Soviet Union endgame, as it evolved, particularly the impact of economic, demographic, and other factors. So, anyway, the first item on my listing of themes for myself was this importance of the long-term, the importance of looking at extended competitions as the context and all that went with it.

OK, the second thing on my…

Q2: Before you leave that, though, I think apropos of your remark that there was really nothing unique about that in the early '50s, because the Eisenhower Administration was looking…

Marshall: Well, or generally at RAND. There were a number of people…

Q2: Yeah, at RAND. That may be so, but I think one of your major contributions is that you kept with that.

Marshall: Yeah.

Q2: It was almost like a strategic fad for everybody else. They were like, "OK, what's the strategy du jour," or "Oh, my gosh, that doesn't work, let's do MAD [Eds: mutual assured destruction], or let's do massive retaliation," or, other constructs. But, you held, and have held, constant to that view for years. We've come back to that as a nation from time to time, like with competitive strategies, and you've been this kind of constant reminder that if you're going to really think seriously about strategic relations with other countries and things that you need to hold this kind of view.

Marshall: Right, although even in my own life there was certainly this emphasis on it through the '50s, maybe a little longer than other people, and then this recurrence in the late '60s, of going back to it as very important.

OK, a second thing is in some ways related. It's really just this importance of resources; that the economics of the situation really matters, especially in the longer run. I raise that because it's been my sense that there aren't very many people who actually seem to really believe that. I know in the discussions here I've mentioned before that in my conversations with military officers that came up, particularly in connection with the competitive strategies notions as they developed in the late '70s, early '80s, there was a tendency on the part of these officers to see the Russians as never having any resource constraints, that they could just add on indefinitely. So the notion of cost-imposing strategies as having much payoff was, I think, something very hard for other people to buy. And it has a lot of ramifications, because every society has cost or resource constraints, and the military establishments have resource constraints.

Then it also, of course, led to, again, as part of this long-term perspective on the Soviet Union, my interest in Murray Feshbach's work, first on the demography and the shifting ethnic balance of the conscript cohort of the Soviet Union, which meant there were going to be fewer and fewer Russians, and people speaking reasonably good Russian, and people with needed skill levels, and so on, and all the things went along with that.

And it also led me to get him interested in and push him into this look at the Soviet health care system, and pollution, and the environmental problems, because it seemed to me in the longer term that no society could really ignore the implications or the consequences of the neglect in those areas, because in the end it leads to a weaker, less-typical population. I even used the crack of, "A farmer may not care whether his cows are happy unless it's related to the output of the milk," and even a totally ruthless nomenklatura in the end has to worry about the productivity and capacity of the population below.

Then there's the fix-up costs that then become competitive to other allocations and resources to the military, and that that would be an important thing to develop, not so much from a humanitarian or "Isn't it awful?" point of view, but from a kind of strategic point of view and perspective. Indeed, that it would be good to try to publicize what was happening to the health and the life expectancy of the Soviet population, to heat up the demand for doing something about these things, as a drain from the resources available for the military. Now, nothing ever much happened about that. It proved impossible to get the U.S. propaganda people, even in the Reagan Administration, to take such things seriously.

Q2: Did you ever think that the declining health care and adverse demographics might not only affect the Soviet capabilities in terms of degrading them, essentially, but also might affect their interests in terms of creating internally generated strategic imperatives to take other courses of action? I mean, not just creating lebensraum, but actually, "OK, we've got a lousy health care system over here, let's go see if we can acquire one in a hostile takeover somewhere?"

Marshall: No, I didn't think it would likely lead to that. I did think that it might lead to dislike, distrust of the elite and that might have some longer term kind of consequences. And there were people, for example, particularly there was this whole school of writers—I think they were called the "Siberian school" of writers—who were Russian and, in some sense, very nationalistic at one level, but at another level were very concerned, increasingly concerned about the environment and what was happening in Siberia. And hence, this might eventually lead to some pressure to turn inwards and spend monies on these things.

OK, the second theme that struck me in thinking about it is this one of "knowing the opponent is important." That means not treating the opponent as some abstract, rational calculator that we don't have any data about, but which you speculate what some clever other fellow would do. I always say, "I don't know why I felt all this," and in some of this I don't know the answer to, "Why was I interested in these things?" I guess some of it's just reading history. I think you can't come away believing in rational opponents and actors if you read any of that.

And so, I remember early on being very dissatisfied with some of the game-theory enthusiasts at RAND and the limited, in my mind, usefulness and relevance of minimax solutions. It seemed to me that you always either knew something about the opponent and how he might behave, or you could find some information about it that would be very useful, and so settling for this absolute assured kind of outcome, independent of knowledge about the opponent and his preferences, was not really very interesting.

Certainly you don't want to assume rationality. You need to focus on the actual facts about the world and how people behave. Also, it seemed to me—again, in this game-theory context—that exploiting weaknesses and irrationalities were part of the game, and that game theory, as structured, tended to drive you away from that.

And then for some reason, I've always had this long-term interest in other kinds of models of behavior, other than this rational model. Where'd that come from? I don't know.

Well, one source was certainly an interest in history. The history wasn't that detailed. I mean, reading Arnold Toynbee, I think, had a big impact on me. But clearly just observing the world you see that the behavior of states and some individuals' emotions, honor, pride, are not captured very well in this sort of other thing.

Then, fairly early on I got very interested in bureaucracy, organizational behavior. This interest developed in the late '50s and was driven, as I've mentioned before, by dealing with the puzzles of the Soviet military, and the kind of behavior that you saw in terms of the kinds of systems and programs the Soviet military had. I remember working with Joseph Loftus and saying, "Well, what is it that we're trying to predict the behavior of? It's big organizations. And what do we know about that?" Then you go off and try to find what people know about that. And there already were some things that were being done. Fairly soon you get on the track of Jim March and other people at Carnegie-Mellon University and they have interesting ideas about organizations and how they behave and [their] limited rationality...

Then in the '60s, I discovered ethology, initially through reading Robert Ardrey. Jim Schlesinger was at RAND then and he had read Ardrey and we spent a lot of time talking about that. Then I discovered our friend Lionel Tiger, who had written a book on *Men in Groups* in the late '60s. And when Schlesinger became Secretary of Defense I even undertook to start corresponding with Ardrey and we had a meeting with Schlesinger and Ardrey and me on one of his visits here. He was living in Rome, I guess, and then he moved to South Africa.

And then, somewhere along the line, I discovered Daniel Kahneman and Amos Tversky, and how we're probably hardwired to make certain kinds of mistakes.

And then, for some reason, even much earlier I had been very interested in Darwinian notions of evolution and read books by this fellow Rene Dubos and another book by a man named C.H. Waddington about that.[34]

So this whole notion that Ardrey and...that goes back into the '40s... that Ardrey and Tiger and these people were putting forth...or for that matter, when you talk personally with Kahneman and Tversky and you ask them why these things are the way they are, they go back to the way the brain actually functions and so on. So it was just very natural for me to see human beings as the product of evolution and of having certain patterns of behavior.

I also had an awareness—I don't know quite where it came from—of the need to be willing to accept that the belief structures of others might be very different from your own. That's certainly true when you start looking at different cultures. I think I've mentioned to some of you—I've never checked its claims independently so I don't know—but one of the most striking recent articles was this article about Columbus and his supposed motivations. I mean this... Schama...

Q3: Oh, Simon Schama.

Marshall: Yeah. He wrote this very interesting article in *The New Republic* reviewing, back in '93 [1992], a whole bunch of books about Columbus and making the point that Columbus's view of things is so different from anything we could imagine.[35] I mean his view of the world, which Schama claimed—I have no idea whether you could prove that this was Columbus's view—meant that, in part, Columbus's exploration was not just a matter of gold and this and that, but it really was about saving souls, which most people these days could not really imagine, but then there was a different religious climate and a sense that the world would end sometime soon and there would be the Final Judgment, and it's imperative to get as many people into the Christian Church as possible.

Now that was a view I've never even heard before—I don't even know whether it's true—but it seems, given my general notion that there can be surprising differences in perspective, it's not implausible. I mean, Schama is a respectable, serious historian, and I think he's reporting on one of the books about what was Columbus's view of the world and of life at the time. A modern American person would find it very difficult to look at things that way and therefore, unless he knows that this person has different perspectives, he would not even imagine anything like that.

The last little note I have under this importance of knowing the opponent is that it really means that deterrence or influencing the behavior of others is a very empirical matter. Deterrence is not treated very well in the standard literature. We tried to apply this theme of knowing the opponent very much in the stuff we did on the Soviet stuff, in the targeting revision in the '70s [the Nuclear Targeting Policy Review]. And John Hines's stuff shows even at the level of the professional military in the Soviet Union these surprising differences of assumptions, going to such basic things as how big a crater does a nuclear weapon create and all the ramifications of that.

My fourth general theme was the importance of reality as the reference. Again, this somewhat duplicates the other things. As I said earlier, if you're trying to understand the Soviets and speculate about them and make projections you have to certainly try to bring to the analysis as much as you know about the Russians and so on, and about large organizations, Russian organizations—if you can know anything—if you're really going to do that. Sitting back and speculating without any data about it is not very helpful.

There was this very interesting RAND study, which I referred to in other talks, that in the middle '50s did just that and said, "Well, where are the Russian intercontinental bomber bases?" and put them all deep in the country along the Trans-Siberian railroad at the time when Loftus and I knew where they were and we were not allowed to tell the other people that. Instead, we would say to them, "No. Where are ours? Well, a lot of them are where they were back in the 1920s, when the range of airplanes was pretty short and where we put them near the coast, near our borders because we were supposed to be defending ourselves, or where the good weather was."

Even this emergence or reemergence of this long-term competition thing really comes from a relook backward at what had been happening up to 1969 and you see this kind of pattern of behavior on both sides in the strategic nuclear area, and you see that to some extent you're engaged in an extended kind of move-countermove sort of situation, and at looking at that and thinking what the strategy for that would be is an interesting thing to do.

Also, Herman Kahn and I felt that it was possible to classify people into people who were model-oriented and those who were reality-oriented and most of our colleagues at RAND were model-oriented. People get dragged into that and most of academia is designed to turn people into model-oriented individuals. This is certainly true in economics.

Q3: Models are simple.

Marshall: And…

Q4: Would you elaborate for a second or clarify? In this case, by model I could either mean simulations, etc., or I could mean simple-minded assumptions about how the world operates.

Marshall: No, it's more general. It tends to be that people start with some sort of notions about how the world works and then they get elaborated and people end up being focused on the nuances of the next little elaboration of the model, but the elaboration is not necessarily connected with reality. Say in economics, where people first start out assuming that all of the supply and demand functions are continuous mathematical functions, and then they start saying, well, can we… And then, therefore, you get to certain kinds of conclusions about how demand matches supply and so on. Then they will start working on, What if they're discontinuous in some minor way? And then, What if they're even more discontinuous? Can you still… [gap for tape change]

…And they become indifferent to data and what the real world is like. The rewards go to people who can prove theorems or maintain findings under still more and more relaxed mathematical conditions. I remember Herman saying that what was interesting to him in physics was that even people who were primarily theoreticians, like Edward Teller… And one of the good things about physics, was that there was another class of people who did experiments and the theorists paid a lot of attention to the results of the experimenters. And that wasn't true in a lot of other fields, where there may not be a separate class of experimenters, the equivalent of experimenters. Secondly, even if there were, they might be considered such low creatures that one didn't pay any attention to anything they did. Or they're off in some periphery of the science and so they don't count.

Anyway, I ended there in my notes on themes, but then I had one page of kind of "beliefs."

Q2: Before you start that, as somebody who has invested millions and millions of taxpayers' dollars in models…

Marshall: Yes.

Q2: …maybe we don't want to leave the impression that you're a complete skeptic on the utility of models, and I take it that the point is that, I mean, one, you should always keep coming back to reality to kind of validate the model…

Marshall: Yes, yes.

Q2: …or verify or adjust, etc.

Marshall: That's right.

Q2: But also not to get caught up in the model for the model's sake…

Marshall: Right.

Q2: …but run the model as an analytic tool to help get insights or guide analysis or something and not get caught in the beauty and the intricacy of the model for its own sake.

Marshall: Yes, I think that's…

Q2: I mean, clearly you don't have total disdain and distaste for models.

Marshall: No, no, no, obviously not, because, again, as in the physics case where it kind of goes along in a healthier way, there have been phenomenal successes.

Q5: Even in these much fuzzier areas you…when push comes to shove, you can't get along without models of some sort…

Marshall: That's right.

Q5: …you have assumptions, you have concepts, you have ideas about how things work…

Marshall: Right.

Q5: Those are. in effect, models. You may not have three million lines of computer code or something like that, but you're still operating on models. But I think the point you correctly brought out is you got to keep going back to the empirical realm and check it against reality.

Q4: I like particularly the part about people who were modelers had different brains from real people.

Q2: It's a left side, right side issue.

Q3: Two points. One, I think the distinction between modelers and non-modelers is very important. And I understand what you two are trying to do: rehabilitate modelers. Let me reinforce the pathological tendencies of the models. If you start with a model, you are consciously choosing to ignore a huge aspect [indistinct]. What's the model do? Focuses attention: I know what the key variables are, I'm going to look at those.

Marshall: Yes.

Q3: And I have to do that, the rationale is, because I can't look at everything. Right? What does Andy do? He looks at everything. The other way to approach it is: I don't know what's going on. I mean this should be the caricature of a study. You start out without any preconceptions—and by the way, that includes all them, because you always have preconceptions. It's better to have those preconceptions, model their explicit [indistinct] and focus on them and then you can come fixate on them. But I think there is this ideal of having the best of both worlds, having some rigor, in a sense [indistinct] what you're saying, instead of having this… But it does have this tendency to go off [indistinct] for reasons which have probably something to do with the nature of people's heads.

But if somebody met you, kind of at the age of 25 or 30, my guess is that they would bet you're a modeler.

Marshall: That's probably right.

Q3: Because you like math, you like chess, you like these abstract puzzles. And by personality you are somewhat reticent, not diving right into the maelstrom of human activity.

Marshall: Yes.

Q3: And you would think this is a classic mathematician, an abstract thinker.

Marshall: Yes, yes.

Q3: So what happened? I mean just because you read history books?

Marshall: Look, I mean, it's true. Early on, what did I read? Well, like every boy, I liked all kinds of military things. In addition, at that period, a lot of scientists were writing popular books. James Hopwood Jeans and Albert Einstein had books, and Ernest Rutherford and others had books, so I read all of those and I was very interested in them. And then I read about mathematics, and got all kinds of books and was very interested in mathematics. And then at some point I got also very interested, though, in literature. I read this book by this fellow Ford Madox Ford called *The March of Literature*. So I started to read Fydor Dostoyevsky and Leo Tolstoy and all of these things. And then I read Toynbee *A Study of History* when it first came out in the late '30s.

And I don't know, you know. You've asked me why; I don't know.

Q2: That's your answer. You just kept reading.

Marshall: Well, and I also read a lot of philosophy, like Alfred North Whitehead and George Santayana and F.H. Bradley and sort of general treatments of intellectual history and philosophy and so on. So I don't know exactly why.

Q3: I don't want to derail the conversation, but when there are people like you who are kind of good at thinking abstractly, it's so alluring, with a few simple assumptions and propositions and the strength of my own head, I can figure out the whole universe and I don't have to know anything.

Marshall: Yes, right.

Q5: That's the rationalistic, idealistic, fantasy nonsense.

Q3: No, no, but I see it all the time. It's what draws in some of the best students: "We'll have the most elegant model, the most parsimonious model, [indistinct] the one with the fewest facts in it."

Marshall: Yes, yes.

Q3: That's what they say.

Marshall: That explains the most.

Q5: Even physics, maybe in some ways especially physics, requires you, if you're doing real science as opposed to the silly stuff, you've got to go back to the data and you've got to take it seriously and you've got to keep [indistinct].

Marshall: But it's so alluring; I think the earlier proposition that the simplicity of models makes them attractive is absolutely right.

Q5: No, no, I understand the allure.

Marshall: That's why you get all these people to do that stuff.

Q5: I understand the allure. What I guess I'm pushing towards is a sense of what Newton and company got started really being about, as opposed to what it means. A really fascinating thing to spend the rest of your life fiddling around with, but isn't necessarily going to help you answer any of your questions?

Q3: No. I'm much more kind of [indistinct]. We will agree that, properly done, there's this good relationship between research and abstract modeling. We also agree that there are pathologies that people kind of get sucked off by the modeling approach into these kinds of austere and... at any rate, exaggerations from reality. But Andy didn't do it. For some reason it didn't happen to Andy, though if you looked at his professional background [indistinct] interest in mathematics, training in economics and just this kind of way he thinks about these things...

Marshall: Yeah.

Q3: You'd say, "This is going to be somebody who's going to be another Oskar Morgenstern or another Johnny von Neumann." And, you know, just kind of push game... He would be one of the premier game theoreticians. Like Tom Schelling, you know, used to go around and say, "You know, Andy Marshall was really one of the few people who really understood game theory." But it didn't work out that way.

Q2: No, it did work out that way. Just Schelling didn't understand how he [Marshall] understood game theory.

Marshall: Yeah.

Q3: [Quoting Schelling:] "I remember going out to RAND in 1959 and Andy Marshall was doing this kind of transcendentally ordered preference schedules as a basis for targeting the Soviet Union."

Marshall: Yes, yes.

Q5: Let me maybe go back to something much more familiar to some of the warm bodies here at the table, which is the issue of, you've now got the balance paper done, and it's 120 pages long, it has three thousand footnotes, and it's just wonderful, and Andy looks at it and says, "It may need a little restructuring." [Indistinct] And...

Q3: "Why don't you move the words around on the page a little?"

Q4: I never had that problem.

Q5: But it's the issue of openness to—There's a whole bunch of different ways all this stuff can be put together and lead to—And, you know, that's sort of—I mean that's the approach that you really ought to take, and that's the approach that the first-rate scientists, whether it's Darwin or whatever...

Marshall: Yeah. Well, there were two things. I don't know, going back, exactly why I had one of these characteristics. But there were two areas that had certainly reinforced this or maybe led to it. One was the work on organizations, I mean in the...

Q3: Yeah, that was what seemed to be coming to me through these—I have six themes—but whatever...

Marshall: ...particularly the stuff that people were doing at Carnegie-Mellon.

But then also the ethology thing, because there you saw in a very dramatic way that up to a certain point people had all of these beliefs about animals and how they behaved and so on, which turned out largely to be wrong. And how was it changed? Well, it was some people who just went and watched what really happened. And in fact, Konrad Lorenz edited a book that has a collected series of papers—it's not one of his major, mainline books—where he defends this notion that at a certain stage it's very important to just watch in great detail what happens, without preconceived notions, models and so on, just to see.[36] He's obviously responding to some criticism that people were maybe making at that time or had made a number of years prior. And what scientists discovered, of course, and are continuing to discover is both how rich, complex are the lives of even animals that don't seem to have any marvelous brains or even small animals or so on. I mean just how very rich, complex the whole thing is, particularly in the social relations among these animals.

So those two experiences of first getting interested in those for various reasons, but then you get reinforced this notion of what's really happening, how things really function.

Q2: Was this before you went to RAND?

Marshall: No, this was after.

Q2: So, in a sense, maybe these special intelligence tickets would have allowed you to view reality in the context of being around these guys creating models. You had a chance to do the sanity check on these things.

Marshall: Yeah, that may have reinforced it.

Q2: I mean I remember this one story of Alain Enthoven going up to you and finally going, "Ah! Now I understand what you were talking about," because you had access to these compartments that he didn't...

Marshall: Yeah, right.

Q2: ...so he was kind of wrong on a whole bunch of this stuff?

Marshall: Well, there certainly was that and then that particular aspect of my experience at RAND, I think. But somehow—I don't know, again, maybe it was living through World War II, I don't know—but I mean you just somehow get the sense that you got to look at things, and also that surprising things can happen that people didn't believe.

Q3: Right. One of the things... I mean early on in your talk you said something like, you said, "I've made this thing about knowing the other [indistinct]? It was always possible to find something out about these things." And most of us said, "Well, the Soviet Union is a closed society, it's got all these spies—so let's make it up." But this kind of assumption you could go out and do some work, you'll actually find out something about this, which is partial and incomplete and imperfect but better than...

Marshall: Yes, well that's at the top of my list on the "beliefs" kind of thing over here.

Q5: Well look, there's another dimension of this it seems to me, and this sort of occurred to me three or four pages back, at least in my notes. It's this issue of continuing to come back to very basic questions.

Marshall: Yes.

Q5: Is it an intellectual style, or is it just me making an observation? I mean I've had you drive me back to much more basic questions time and again, because I have a tendency to get on to the fancy, sexy nuance and that's the stuff that gets disconnected. I mean I have strong tendencies in that direction. But you have always and... Well, and the other dimension of it is just sticking with very basic questions for very long periods of time. And it just seems to me, observationally, those are fundamental intellectual characteristics, however they came about. I'm not sure it's knowable, although it's interesting to try to probe.

Q2: Yeah, I mean you have this sort of intellectual tenacity, you know, don't get distracted and sidetracked, etc. Just really stays... And that was my point earlier about you staying with the long-term competition when other people were sort of phasing that out.

Marshall: Yeah.

Q4: I think that at some point, Andy, when you go through editing, you probably need to say more about the economics issue and strike a balance between your comments early on that economics really matters, especially in the long run, and comments that you make later on about game theory, understanding the opponent, and your examples... some of your comments about how simple-minded, basically, economics can be.

Marshall: Yeah.

Q4: I drew a strong message over the years from you about the fact that economics really matter, but my sense was that economics as I was taught it, and raised, and in reading about it wasn't the kind of economics in 90 percent of the cases that you were really sort of thinking about.

Marshall: Right. It matters the most in the sort of broad and kind of rough way at this level of kind of macroeconomics. Of course, microeconomics is mainly nonsense. But the sense that resources are scarce and limited is important. That's part of microeconomics, too, but I mean that at this macro level the point is that there's only so much to go around and if it gets soaked up with one area it's not available for others and that that matters, and that there really are these serious limits. Some of them can be bypassed for a time, but then you have these longer-term consequences of having underinvested that got to get fixed later. So there's some very broad kind of accounting balance that you got to pay attention to.

Q4: Coming back to the point about always coming back to some basic... occasionally going back to some basic questions: Taking a look at economics the way you just described it is almost Samuelson Econ 101 thirty years ago, understanding those issues with more richness than the way they subsequently were developed.

Marshall: Well, right. But going back to the basic questions, I guess my belief about that is very much, I think, how useful that's been to really try to think a lot about the questions and what they really are. What the really basic questions are is very important.

Q4: Were there a set of economists that you found most interesting? As a class, for example, did you find economic historians interesting people, or interesting work?

Marshall: Well, again, I didn't really do that much in economic history, although I did some economic history. I thought when I was at the University of Chicago the most interesting man was Frank Knight. He had some classes that were… He was the person who was also very critical of much of economics, and so he would give examples about how economics was crazy. He said, "Every farmer knows that one boy is worth one boy and two boys are worth about half a boy and three boys are worth almost nothing," whereas an economist would say three boys are worth three boys. And that it was crazy because the family was really the most relevant unit, not the individual, and so on. So I thought he was terrific. Now I actually spent a lot of my time, however, in the kind of mathematical statistics and making models estimating… I wrote my master's thesis on a test for the accuracy of predictions of a model that had been created, and things like that.

Q4: At this point in time were you a devoted modeler?

Marshall: No, I was more interested in the statistics of the thing. And I was interested in kind of the logical positivists… and so on. I finally decided that it wasn't clear that these models were saying anything about the economy, really. What they were were predictions of the next number that would appear in the Commerce Department's summary of the thing. I mean that's as far as you could go.

Q3: Epiphenomena.

Marshall: Yes. Well, it's supposed to be related to the reality, but the Commerce Department number is what it was predicting, not necessarily anything about the economy.

Q5: Would you be inclined to say anything more about the issue of limits? I mean I have sort of, I guess, odd views about the 20th century, because a lot of people read the 20th century as there are absolutely no limits, there's nothing we can't do. There's an alternative reading in a lot of fields. There is, like mathematics that suggest that there are lots of limits out there, actually, as a matter of fact, if you pay attention. And some of them you can, say, you can wheel around to some extent. Some of them you can't do much with. I mean does that have any…

Marshall: No, I know about that, and certainly one of the most fruitful ways of thinking about things has been focused on these kind of limits, like conservation of energy and all these kinds of things, but then exclude lots of possibilities. So I've been aware of that, but I guess I wouldn't say that's played a big role in my intellectual life. I guess that what has interested me is the kind of narrowness of most of the [indistinct] in the political and economics fields and so on, the narrowness of most of the thinking about things, which tends to assume a whole bunch of things can't happen or are unlikely, which if you look at history, history says happen all the time. So there's that. So I've been mainly, I'd say, on the side that believes reality and the possibilities or "unlikely" events are far richer than what the theorizing and the standard thinking tends to suggest.

Let me go on with these beliefs. The first one, as we've already talked about, is that some progress can be made on almost any question, that even the hardest questions you can make some headway on or be somewhat better off than just sitting back and not doing anything.

Q2: That's not a prediction, however. That's an observation.

Marshall: Yes. Right.

Q2: We've shown that there are some areas that are remarkably resilient, like counter-command-and-control and those kinds of things in terms of analyzing…

Marshall: Well, I don't mean progress in the sense of fixing it, but intellectual progress.

Q2: I mean understanding.

Marshall: Understanding, yeah.

Q2: I mean we struggled for years and years to understand…

Marshall: Well, yes, there are some areas in which that's right. But on the other hand, I don't think really serious systematic efforts have been made.

Q3: You probably have some notion of just what a fearless statement that is.

Marshall: Yes.

Q3: Because I'm in the business about working with young researchers at the beginning and they would say, "No! What about the risk that it'll be a dead end? What about the risk that I'll have spent all these years and nothing will come out?" … and the lack of willingness to kind of take a chance and kind of… They're ruled by their fears.

Marshall: Yeah.

Q5: But you always learn lots from dead ends.

Q3: Absolutely true, but that's not the attitude that they take. They want safe topics which… there's an established method for finding out this stuff and you're guaranteed of some boring payoff, but you get some payoff.

Marshall: Yes.

Q4: Let me ask you this: Does that whole… Would you apply that also to Ph.D. students who live in fear that they will spend two years of their lives and find out their topic is dead?

Q3: I guess I'm contrasting the current crop of students that I work with, with some of the people that I knew, and Andy knew some of them… more interesting academic friends who said, "Yeah, I don't care. I mean it's some Interesting stuff. If I work hard I'll find out something."

Marshall: Yeah, that's right.

Q3: [Indistinct] "And I don't know exactly how I'm going to do it, but something will come out of it."

Marshall: But I'm also thinking of it a bit from kind of the practical side, I mean practical in the sense of putting you in a position to more intelligently speculate, a little bit, on one area. Again, one of the notes I have here says one of the things that first made me feel that much of the work on the Soviet economy was not what it ought to be was in the middle '50s, when Loftus and I had some questions about the economy and how it related to the defense effort. And we asked some of the RAND people that were at the center of this effort and they essentially refused to answer, to even try to answer these hard questions.

Q2: Was it too hard or...?

Marshall: Well, yes, too uncertain, too this, too that; "We can't make an estimate." And it was exasperating. My sense was that it was kind of unclear what this effort was all about if it wouldn't even try to answer these very fundamental questions. And, again, I talked about this before, that it seems to me that even not-so-good answers to good questions are better than good answers to poor questions. And that, along with this "it's possible to make progress on almost anything" belief is, I think, very important as a set of beliefs in the way of going about things.

Then I guess the second thing is—which overlaps a little bit with some of the other things I said before—about the character of man or mankind. I mean the liberal view of man is just unbelievable. I don't know where I came upon that, but fairly early I decided it was just not so. Man's character is the product of evolution. There are limits to rationality, and the role of honor and passions is important. In other cases I've mentioned this thing of Herbert Goldhamer's...

Q4: Is it, however, an area in which progress is still possible?

Marshall: About what? About understanding the character of man? I would say that progress in understanding the character of man is possible. I don't think it would end up going down the standard kind of view of things.

So, it's certainly related to my interest in the whole ethology area. But it goes beyond that, I think. Well, I suppose there's that, and I've mentioned I know before this observation of Goldhamer at one point early on in connection with these people who were looking at air defenses, doing all their calculations, and he pointed out that what was left out was that some pilots would ram enemy airplanes.

Well, at any rate, status, dominance—all very important. And so, I think all of this is very important in thinking about the general character of the future.

[gap for tape change, Marshall comments on his belief that war has a prominent future]

Marshall: And then the third one of the beliefs is, there are real experts. I mean that there are some people who really are exceptional. Some of it's just that they're smart—there are some people that are just dramatically smarter, more intelligent than others, or have some kind of training or feel for things that other people don't have. I think I've mentioned this one case of—again, it comes from Herman Kahn and came out of a discussion that he and I would get into. We'd concocted various estimates to prove the failure of the standard economics, which tended to say people are paid their marginal product. He was doing Monte Carlo calculations in connection with the bomb designs and noted that you could make a case here with Herman Kahn, four or five other people, lots of computing time—at a time when computing time was very expensive to do these things—and Hans Bethe, on the other hand, who could do these things with paper and pencil.

And it's almost as though there were some of these very best physicists that had a pipeline to God about how the world really was put together. Because all of these calculations that these people have to make are approximations—you have to choose the right approximation or the right…Many of them were, say, integrating under some complex surface and you could get the answer approximately by… because they have a lot of it concentrated in some particular spot. If you know where that spot is, then you just calculate in the area around it, but you got to know where that spot is in this complex surface, and there are these people who time after time know. But there are all different kinds and so it's very important to find these people and pay attention to them. And it happens throughout one's experience, I think. In the end, it seems to me that one of the things in one's life is to calibrate people and then pay attention to those people who have some significant amount of this superior insight into how things really are. The thing that was really wonderful about RAND in the '50s was there were quite an unusual number of those people that somehow or another had been brought there in the Social Sciences Division. You had people like Goldhamer or Nathan Leites, in his particular areas.

Q2: Well, they're all in particular areas.

Marshall: Yeah, right, they're not universal things.

Q2: So Goldhamer and Leites and who else would you put in that category?

Marshall: Well, I don't know who they are in some of the other areas. I mean like physics, you'd have to have somebody like Herman Kahn, who might have been able to identify them in that area. But the point is that there are people who really are just exceptional.

Q3: You said you'd calibrate them, by which I assume you meant you kept track of whether or not they were right more often than wrong.

Marshall: Yes.

Q3: Could you talk a little bit more about that? Because very often the people that you have found to be real experts were not the people who were nominally the experts…

Marshall: Yes, right.

Q3: …so obviously the conventional scoring and reward system doesn't…

Marshall: Yes, that's true.

Q3: I think Lionel Tiger's status among anthropologists is… he's not … he's too popular, he's too funny or something like that.

Marshall: Yeah.

Q3: So how do you calibrate?

Marshall: I don't know, exactly. Well, in some cases you would determine whether they're a steady source of insight in some things. But another thing is with Goldhamer and Leites, people like that, they would have written papers and so on.

Q3: I remember you once saying about Leites… You sort of had been keeping track of his views on the Soviet Union for 30 years…

Marshall: Yes.

Q3: …reasonably credible [indistinct]. So that was a more straightforward case.

Marshall: Yeah, right.

Q3: In some of these cases it's a little more difficult to score.

Marshall: Yeah, that's true.

Q2: I recall when I was here one of the things that I was always taken by was you would always find these guys that were, like, the Central Intelligence Agency analysts in the bowels of the Agency that, you know, they'd been working on this paper for three years and it never got published, because nobody could ever get consensus on it, but it was very, very important and seemed to be very, very interesting. Like Joe Backhofen and the work he was doing on Soviet fuel-air explosives.

Marshall: Right.

Q2: He could never get the darn thing published, but he knew enough to find you or you found him.

Marshall: Well, I think a lot of that is if you are obviously open to that, or you're in a high enough place—or people have the illusion you're in a high enough place—they seek you out. But it has to be associated with a willingness to listen and give a hearing to people that have a nonstandard kind of view or some new thing.

Q2: So you don't find them, you create an opportunity for them to find you? Then you have to be smart enough to figure out that they've found you.

Marshall: Right. At any rate, that's what I came up with in my list of beliefs.

Q4: Let me raise one. It would seem to me that you… as long as I've known you, have been very interested in what I would just call "strategy."

Marshall: Yes, very.

Q4: Now it is clearly built into some of your comments, like long-term perspective, the notions of competition…

Marshall: Right.

Q4: But it has struck me that you had—at least for the last… since I've known you—sort of this enduring interest in strategy, how you think strategically…

Marshall: Yeah.

Q4: …whether we are thinking about the right kinds… thinking about strategy the correct way, and so on and so forth. And yet you didn't really put that down as one of…

Marshall: Well, maybe I should have. Certainly I'm very interested in that.

Q4: As a belief?

Marshall: Again, I guess in a more visible, concrete way, it came out of the early RAND experience. Because in the immediate postwar period there was this whole issue of how are we going to deal with Soviets, what our strategy was going to be. And then I first got into this business of organizational behavior, and got onto Jim March and company—it was the very late '50s. But then I said, "Well, who else is working on organizations?" and went to the Harvard Business School. There I first came across these ideas that were being developed on business policy, business strategy, and got very interested in picking up what they were saying. That all continued and it had, again, a big impact on the longer run.

Q2: Did you develop a strategy for how you were going operate in the U.S. government? I mean, did you look at the U.S. government as this large, complex organization and develop a plan about how you were going to proceed if you were going to be successful? And if you did, what was the strategy?

Marshall: Well, to some extent I did. When I first picked Chip Pickett to come to work for me it was because he had been up at the Harvard Business School and I talked to C. Roland Christensen, who was in this strategy business. And I picked Chip because at that point one of the main tasks I had from Henry Kissinger was following the implementation of the Nixon reorganization of the intelligence community. So I wanted someone with this business school strategy background, and someone who was interested in organizations, because what we were going to try to do was change the behavior of these organizations. And so the question was one of sitting up there and listening to their lying protestations of how they worked for the President and, of course, they would do whatever he wanted, and so on. You were going to need some kind of strategy for changing them and you had to be clever about that and so on. So I very much had that view, based on my earlier interest in organizations and all of that. So I was interested in it in that way, that was very conscious, to have that kind of approach.

Then as we shifted over to the Pentagon pretty exclusively it was the assessment thing… Initially, we couldn't get the assessment stuff going when I was at the National Security Council because the Department of Defense really didn't want to do it and there was an elaborate rigamarole to even get the whole thing started. When I came over here, I don't think I developed a very systematic, conscious strategy, but I certainly knew that there were things that you didn't want to do if you were going to survive. That certainly affected this notion of sticking to diagnosis; that was a very conscious part of the strategy. Based also on my experience from the RAND period I also thought that if you're really going to try to affect things—while it's slower in some sense—the thing to aim for is changing the way people think about certain problems, having them adopt your vocabulary, and so on, etc. It also was important to take a longer-term view of things, and, in fact, not to be very anxious or focused on achieving some great big, immediate changes.

Q2: I think we could all sit down and write the Andy Marshall strategy: keep a low profile; don't mess with programmatics and budgets; find smart people in important places, whether it's your nominal boss or not, to service with your product; have a little office on the A-ring of the Pentagon that's got lousy furniture and have an office half the size that you're supposed to have; have a very small staff; none of the trappings of power. I think based on working here, we could all kind of write out… Now, did you sit down and think about those kinds of things consciously and purposefully and implement them, or did you just kind of evolve into those, or did you understand them intuitively?

Marshall: I think it was more I understood that intuitively. People used to talk about getting me up in the E-ring, which was farthest from my desire, I mean…

Q4: I never presumed you were actually thinking that way about it. I think that the view early on among us was just interesting [indistinct].

Marshall: Well, that's true.

Q4: I can remember secretaries that screwed themselves into the ceiling over the [indistinct] Mr. Marshall [indistinct]…James Roche also doing the same thing…. which subtly, probably it made you even more useful because you couldn't be interpreted as a threat.

It might be really an interesting… It's a great question. It would be almost an unwritable chapter, but it would be really interesting to speculate on why Andy Marshall as an institution stayed in the Defense Department for as long as he did.

I have an…always believing in one of your earlier precepts about [indistinct] should be based on fact and reality, I have one good solid data point to indicate that he actually had a fairly strong sense of risk. And that is that my next-door neighbor was the president of Cort rental furniture, and some years ago, while we were playing bridge, he said, "You know, I run a rental operation here in town." He said, "You know, we got a guy who's been renting furniture from us for 15 years." He says, "You can't [indistinct]. He's never bought it." He says, "You can't believe this." I say, "What's his name?" He says, "Andy Marshall." I said, "Hey, I know that guy."

Marshall: Well, I always thought that I would leave in a few years. I had no plan that would keep me here very long.

Q2: In ways that contributed to your longevity.

Marshall: Yeah, that's true, I mean…

Q2: Because you always were planning to leave and you had a home.

Marshall: And a related thing was that it certainly, was probably very important that I didn't have any children and always had a reasonable amount of money; it didn't matter to me whether I had the job or not. It was really a matter of, you know, I would feel badly about if I no longer were Director of Net Assessment, but it was not anything that was going to ruin my life or anything. But also, from everything I saw at RAND, where I was sort of in the same business of trying to do studies and change the way people look at things, I knew that you had to take a very long-term view. This notion that you could write something as though it's kind of like a best-seller just doesn't pay off. And so the main thing you could do to really change things was to change the way people thought about them, and that doesn't happen quickly.

Q4: I think that's another one of your observations that you didn't make, because I have it down here… fame, which is not… I mean I realize it's in a way brought up by your comment… But it's this notion that seems to be embedded in your thinking over the years… some people would call it patience, but my view is… you just articulated it... it's this notion that if you really want to change the way the intellectual horizon works, you have to be prepared for a long-haul effort.

Marshall: Yeah. Also, I guess it's an advantage not to have ambitions to have some prominent or truly high-ranking job. I mean it doesn't mean anything to me. It's not anything I would care about.

Q4: Let me draw you back to the strategic perspective. My sense is that the real sort of growth in your interest in strategy comes [indistinct] comes somewhere in the... begins in the late '70s, mid-to-late '70s, when Jim Roche is here and you actually... you do some things for the Chief of Naval Operations...

Marshall: Yeah.

Q4: ...when I start thinking of the examples of you getting deeply involved in strategy, strategic notions, concepts and so on and so forth, I didn't see that much evidence of it in the '72 to '75 time frame. Starting around '75 is when I think things began to occur.

Marshall: Well, I think that's right, except that there was this whole antecedent period and the stuff I did at RAND and so on, which was very much of that same sort. And as I mentioned, I'd been very interested in the stuff at the Harvard Business School, developing certainly in the '60s. So I was very interested in that. But it was sort of, I think, Jim, in particular, and Secretary of Defense Donald Rumsfeld and some of the questions he posed about the Navy that led to further work on strategy. We wrote papers for Rumsfeld regarding the Navy and strategy. And then Jim and I, just because we wanted to do it, wrote this paper in '76 [*Strategy for Competing with the Soviets in the Military Sector of the Continuing Political-Military Competition*] on, OK, supposing the Pentagon wanted to have a strategy, what would it look like? We wanted to make the case that the Defense Department ought to really do things of that sort. And it sort of went on from there.

Q2: Well, in part, I think that's because you'd had this notion that net assessment was about diagnosis and not prescription. And so, you know, there was always this... at least you said to me you were always concerned about people doing net assessments also doing the strategy...

Marshall: Yeah [indistinct].

Q2: ...because you always cook the books on the net assessments to get to the preferred strategy, and so when it's been...

Marshall: Well, not just strategy, but that they would have certain systems or...

Q2: Yeah, they're pushing pet rocks...

Marshall: Yeah, right.

Q2: ...and things like that. So when from time to time it's been posed that you would also take over sort of Department strategy functions, you've been very reluctant to do that. In fact, I think you refused to do it. I mean you sort of backed into a couple... on helping on competitive strategies or things like that, but really not taking the lead.

Marshall: Well, we certainly pushed the idea that one ought to do this, and to do experiments trying to spread the notion.

Q2: But not do them out of this office, with the same people doing the assessments also doing the strategy.

Marshall: Yeah, right.

Q2: That's a very good distinction to draw.

Marshall: Well, maybe we ought to adjourn, sign off on this one.

Endnotes

1 Eds: Interview conducted 24 September 1993.

2 For background on the "criteria problem," see Charles J. Hitch and Roland N. McKean, *The Economics of Defense in the Nuclear Age* (Cambridge: Harvard University Press, 1960), 158–181; R.N. McKean, "Criteria," in Edward S. Quade, ed., *Analysis for Military Decisions* (Chicago: Rand McNally & Co., 1967), 81–91. For examples of Marshall's work in this area, see A.A. Alchian, G.D. Bodenhorn, S. Enke, C.J. Hitch, J. Hirshleifer, and A.W. Marshall, *What is the Best System?* D-860 (Santa Monica: RAND Corp., 4 Jan 1951); A. Alchian, J. Hirshleifer, and A. Marshall, *Sound-offs on Pay-offs,* D-892 (Santa Monica: RAND Corp., 20 Jul 1950); A. Alchian, J. Hirshleifer, and A. Marshall, *Report of Cost Team Plus Panel,* D-829 (Santa Monica: RAND Corp., 20 Sep 1950); J. Hirshleifer and A. Marshall, *Proposed Payoff Criterion for MSA* [Multiple Strike Analysis], D-817 (Santa Monica: RAND Corp., 2 Oct 1950); J. Hirshleifer and A. Marshall, *Revised "Best Guess" Salvage Value Assumption for MSA,* D-820 (Santa Monica: RAND Corp., 5 Oct 1950); J. Hirshleifer and A. Marshall, *A Proposed Salvage Value Procedure for* MSA, RM-476 (Santa Monica: RAND Corp., 20 Oct 1950); A.W. Marshall and Stephen Enke, *The Substitution Value of U-235 in Two Assumed Strategic Campaigns (U),* D(L)-923 (Santa Monica: RAND Corp., 20 Apr 1951), SECRET//RESTRICTED DATA. On the subject of this report, see also Stephen Enke, "Defense Economics: Applying Economic Criteria," *American Economic Review,* 55, no. 2 (May 1965): 418–420; Stephen Enke, "Some Economic Aspects of Fissionable Material," *Quarterly Journal of Economics,* 68, no. 2 (May 1954): 217–232; and S. Enke, E.W. Lee, and A.W. Marshall, *The Deployment of Local Defenses Suggested to Barlow by Economics,* D-935 (Santa Monica: RAND Corp., 11 May 1951).

3 Marshall's effort, with Joseph Loftus, to improve forecasts regarding Soviet military programs was called "Project Sovoy." See J.E. Loftus and A.W. Marshall, *Research on Soviet Military Programs (U),* D(L)-6500 (Santa Monica: RAND Corp., 26 Jun 1959), SECRET; J.E. Loftus and A.W. Marshall, *On Improving Midterm Estimates of Soviet Military Capabilities for Use in U.S. Military Planning (U),* RM-2892-PR (Santa Monica: RAND Corp., Aug 1962), SECRET; J.E. Loftus and A.W. Marshall, *Forecasting Soviet Force Structure: The Importance of Bureaucratic and Budgetary Constraints (U),* RM-3612-PR (Santa Monica: RAND Corp., Jul 1963), SECRET [Submitted for declassification review, Fall 1994]; Andrew Marshall, "Propositions from Case Study on Forecasting Soviet Force Structure," 20 Jul 1970, in Hans Heymann, Jr., rapporteur, *Bureaucracy and National Security Policy: A Conference Report* (Draft), R-644-PR/ARPA (Santa Monica: RAND Corp., Feb 1971), 46–47; Andrew W. Marshall, "Arms Competitions: The Status of Analysis," in Uwe Nerlich, ed., *The Western Panacea: Constraining Soviet Power through Negotiation, Volume II of Soviet Power and Western Negotiating Policies* (Cambridge: Ballinger Publishing Co., 1983), 6.

4 The following RAND papers *may* be related to this effort: Andrew W. Marshall and Joseph E. Loftus, *Soviet Military Expenditures, 1955-1963: An Official Summary View (U),* 0-6365 (Santa Monica: RAND Corp., 27 May 1959), SECRET; Andrew W. Marshall and Joseph E. Loftus, *CIA Preliminary Estimates of Soviet Military Expenditures, 1947-1959 (U),* D(L)-7085 (Santa Monica: RAND Corp., 8 Jan 1960), SECRET.

5 Michel Crozier, *The Bureaucratic Phenomenon* (University of Chicago Press, 1964).

6 On the "missing chapter," see C. Wolf, Jr., "Working Group I--Improving Policy Analysis: Introductory Paper," 17 Jul 1970, A.W. Marshall, "Notes on Improving Policy Analysis: Attachment to Introductory Paper (Working Group I)," 17 Jul 1970, and "Working Group I: Improving Policy Analysis: Summary of Findings," 23 Jul 1970, in *Bureaucracy and National Security Policy: A Conference Report,* 57–58, 61–63.

7 William H. McNeill, *Plagues and Peoples* (New York: Quality Paperback Book Club, 1993); originally published in 1976).

8 B. Brodie, C.J. Hitch, and A.W. Marshall, *The Next Ten Years,* D-2700 (Santa Monica: RAND Corp., 30 Dec 1954).

9 A.W. Marshall, *Long-Term Competition with the Soviets: A Framework for Strategic Analysis (U),*

R-862-PR (Santa Monica: RAND Corp., Apr 1972), SECRET. See also A.W. Marshall, *Long Term Competition with the Soviets: A Framework for Strategic Analysis,* IN-21542-PR (Santa Monica: RAND Corp., 15 Apr 1971) (CONFIDENTIAL; Declassified 7 May 1980). Although published in 1972, R-862-PR was written between summer 1969 through the end of 1970.

10 Herbert A. Simon, *Administrative Behavior: A Study of Decision-Making Processes in Administrative Organizations,* second edition (New York: Free Press, 1965; first published in 1945); James G. March and Herbert A. Simon, *Organizations* (New York: John Wiley & Sons, 1958); Richard M. Cyert and James G. March, *Behavioral Theory of the Firm* (Englewood Cliffs: Prentice Hall, 1963).

11 James G. Roche and Barry D. Watts, "Choosing Analytic Measures," *Journal of Strategic Studies,* 14, no. 2 (Jun 1991): 165–209.

12 Daniel Kahneman and Amos Tversky, *Intuitive Prediction: Biases and Corrective Procedures* (Eugene: Decision Research/A Branch of Perceptronics, Jun 1977); Amos Tversky and Daniel Kahneman, "'The Framing of Decisions and the Psychology of Choice," *Science,* 211, no. 4481 (30 Jan 1981), 453–458; Daniel Kahneman and Amos Tversky, "Judgment under Uncertainty: Heuristics and Biases," *Science,* 185, no. 4157 (27 Sep 1974): 1124–1131.

13 Michael T. McGuire, "Social Dominance Relationships in Male Vervet Monkeys: A Possible Model for the Study of Dominance Relationships in Human Political Systems," *International Political Science Review,* 3, no. 1 (1982): 11–32.

14 For a discussion of the dominance systems of lizards, in which a "tyrant" exerts hegemony over subordinate males, see Edward O. Wilson, *Sociobiology: The New Synthesis* (Cambridge: Belknap Press of Harvard University Press, 1975), 296–297, 444–445. The fighting routines (displays and postures of submission) lizards use in their struggles over territory are described in Irenaus Eibl-Eibesfeldt, "The Fighting Behavior of Animals," *Scientific American,* 205, no. 6 (Dec 1961): 114–116.

15 Robert Ardrey, *The Territorial Imperative: A Personal Inquiry into the Animal Origins of Property and Nations* (New York: Atheneum, 1966).

16 Amos Tversky and Daniel Kahneman, "Rational Choice and the Framing of Decisions," *The Journal of Business,* 59, no. 4, (Part 2, Oct 1986): S251–S278; Daniel Kahneman and Amos Tversky, "Choices, Values, and Frames," *American Psychologist,* 39, no. 4 (Apr 1984): 341–350; Kahneman and Tversky, "The Framing of Decisions and the Psychology of Choice."

17 Paul Y. Hammond, Charles Wolf, Jr., and Peter de Leon, *Perceptions of the Military Balance: Report of a Workshop, August 26-27, 1974,* 1974, WN-8876-ARPA (Santa Monica: RAND Corp., Oct 1974), FOUO.

18 Lionel Tiger and Robin Fox, *The Imperial Animal* (New York: Holt, Rinehart and Winston, 1971).

19 Herbert Goldhamer, *The Economist's Perception of the U.S.-Soviet Strategic Balance,* WN-8991-ARPA (Santa Monica: RAND Corp., Feb 1975), FOUO; Herbert Goldhamer, *Le Monde's Perception of the U.S.-Soviet Military Balance, 1948-1973,* WN-9318-ARPA (Santa Monica: RAND Corp., Nov 1975), FOUO; Herbert Goldhamer, "The US-Soviet Strategic Balance as seen from London and Paris," *Survival,* 19, no. 5 (Sep/Oct 1977): 202–207 (summarizes WN-8991-ARPA and WN-9318-ARPA).

20 Herbert Goldhamer, Joan Goldhamer, ed., *Reality and Belief in Military Affairs: A First Draft (June 1977),* R-2448-NA (Santa Monica: RAND Corp., Feb 1979).

21 *Reality and Belief in Military Affairs: An Inventory of Additional Material in the File of Herbert Goldhamer—Volumes I-III,* assembled by Joan Goldhamer, N-1726/1-3-NA (Santa Monica: RAND Corp., Aug 1981).

22 "After the battle of Maldon, in A.D. 991, the Vikings requested permission to cross a ford unmolested in order then to attack the English. The request was granted by the English commander.

> The ground is cleared for you: come quickly to us, gather to battle…
> Then neared the fight, the *glory* trial…
> So they stood fast, those stout-hearted warriors at the *war play…*"

Reality and Belief in Military Affairs: A First Draft (June 1977), 13 (quotation from the poem "The Battle of Maldon", emphasis added by Herbert Goldhamer).

23 Herbert Goldhamer, *Human Factors in Systems Analysis,* RM-388 (Santa Monica: RAND Corp., 15 Apr 1950).

24 *The Future Security Environment,* Report of the Future Security Environment Working Group, submitted to the Commission on Integrated Long-Term Strategy (Andrew W. Marshall and Charles Wolf, Working Group Chairmen) (Washington: Department of Defense, Oct 1988), 3–10, 72–96.
25 Peter Schwartz, *The Art of the Long View* (New York: Doubleday, 1991). Clem Sunter, *The World and South Africa in the 1990s* (Cape Town: ltuman & Rousseau Tafelberg, 1987). Pierre Wack, "Scenarios: Uncharted Waters Ahead," *Harvard Business Review,* 63, no. 5 (Sep/Oct 1985): 72–89; Pierre Wack, "Scenarios: Shooting the Rapids," *Harvard Business Review,* 63, no. 6 (Nov/Dec 1985): 139–150.
26 See remarks by Marshall in SRI International, *Theater-Level Gaming and Analysis Workshop for Force Planning,* Vol. I, Proceedings, September 22-27, 1977, held at Xerox International Center for Training and Management Development, Leesburg, VA (Menlo Park: SRI Int'l., 1977) [Available from DTIC, ADA 068177]; A.W. Marshall, "A Program to Improve Analytic Methods related to Strategic Forces," *Policy Sciences,* 15, no. 1 (Nov 1982): 47–50.
27 Herbert Goldhamer and Hans Speier, "Some Observations on Political Gaming," *World Politics,* 12, no. 1 (Oct 1959): 71–83; H. Goldhamer, *Toward a Cold War Game,* D(L)-2607 (Santa Monica: RAND Corp., 22 Oct 1954); H. Goldhamer, *Summary of Cold War Game Activities in the Social Science Division* [First Round], D-2850 (Santa Monica: RAND Corp., 12 Apr 1955); P. Kecskemeti, *War Games and Political Games,* D-2849 (Santa Monica: RAND Corp., 15 Apr 1955); P. Kecskemeti, *Summary of Cold War Game Activities in the Social Science Division: May Experiment* [Second Round] D-2975-RC (Santa Monica: RAND Corp., 20 Jun 1955); H. Goldhamer, *The Political Exercise: A summary of the Social Science Division's Work on Political Gaming, with Special Reference to the Third Exercise,* D-3164-RC (Santa Monica: RAND Corp., 1 Sep 1955); E. Schnitzer, *Third Political Exercise: Summary and Documents,* D-3639 (Santa Monica: RAND Corp., 1 Sep 1955); E. Schnitzer, *Fourth RAND Political Exercise: Summary and Documents,* D-3163 (Santa Monica: RAND Corp., 25 May 1956); J. Goldsen, *The Political Exercise: An Assessment of the Fourth Round,* D-3640-RC (Santa Monica: RAND Corp., 30 May 1956).
28 This may be the SWAP (Strategic Air War Planning) series of games. See Olaf Helmer and Lloyd Shapely, *A Strategic Air War Planning Game,* RM-2105 (Santa Monica: RAND Corp., 27 Jan 1958) FOUO. T.A. Brown and E.W. Paxson, *A Retrospective Look at Some Strategy and Force Evaluation Games,* R-1619-PR (Santa Monica: RAND Corp., Sep 1975).
29 Robin Fox, "Fatal Attraction: War and Human Nature," *The National Interest,* no. 30 (Winter 1992/93): 11–20.
30 Schlesinger gave a briefing on these developments, using U-2 photographs, to the Senate Armed Services Committee on 10 Jun 1975. See Senate Armed Services Committee, *Disapprove Construction Projects on the Island of Diego Garcia,* 94th Cong., 1st sess. (Washington: GPO, 1975), 12–21. Schlesinger recounts the episode in Ben Rich and Leo Janos, *Skunk Works* (Boston: Little Brown and Co., 1994), 188–190.
31 Alexander R. Alexiev and S. Enders Wimbush, "The Ethnic Factor in the Soviet Armed Forces," in Alexiev and Wimbush, eds., *Ethnic Minorities in the Red Army: Asset or Liability?* (Boulder: Westview Press, 1988), 121–195.
32 Andrew W. Marshall, "Strategy as a Profession for Future Generations," in Marshall, J.J. Martin, and Henry S. Rowen, eds., *On Not Confusing Ourselves: Essays on National Security Strategy in Honor of Albert and Robert Wohlstetter* (Boulder: Westview Press, 1991), 302–311.
33 Eds: Interview conducted 6 October 1995.
34 The Dubos books might be *The Torch of Life: Continuity in Living Experience* (New York: Simon and Schuster, 1962) and *Beast or Angel? Choices that Make Us Human* (New York. Charles Scribner's Sons, 1974). The Waddington book might be *The Strategy of the Genes: A Discussion of Some Aspects of Theoretical Biology* (London: Allen & Unwin, 1957) or *The Ethical Animal* (New York: Atheneum, 1961).
35 Simon Schama, "They All Laughed at Christopher Columbus," *The New Republic,* 205, nos. 28 & 29 (6 & 13 Jan 1992): 30–40.
36 This paper by Lorenz might be "The Fashionable Fallacy of Dispensing with Description," in Richard I. Evans, *Konrad Lorenz: The Man and His Ideas* (New York: Harcourt Brace Jovanovich, 1975), 152–180.

2 The RAND Years 1950–1969[1]

Introduction

The end of World War II saw just two great powers remaining: the United States and the Soviet Union. Although the countries were wary, sometimes contentious allies during the war, that alliance was a temporary convenience. French political philosopher Alexis de Tocqueville had predicted in the 1830s that the United States and Russia would eventually become rivals.[2] Competition emerged in the 1890s when the two powers sought influence in Manchuria, and the 1917 Bolshevik Revolution in Russia added an ideological element to the rivalry. The United States did not formally recognize the Soviet Union until 1933. Following the defeat of Nazi Germany and Imperial Japan, the enmity between the two nations reemerged in the form of friction over the fate of central and eastern Europe. The Soviets, who had suffered 20 million dead during the war, sought to protect their western borders and to eliminate Germany's capacity for aggression. Soviet troops occupied eastern Germany as well as several states previously under German control, including Poland, Czechoslovakia, Hungary, Romania, and Bulgaria. The United States sought to rebuild a free and capitalist Europe with Germany at its center. From 1945 to 1948, Soviet leader Josef Stalin increased Soviet control over the occupied countries, effectively turning them into satellites. Soviet actions in Europe and around the world convinced U.S. leadership that the U.S.S.R. was intent on expanding communism and was unwilling to compromise with the West. The Truman Administration resolved to oppose any further extension of Soviet influence around the world.[3] Containing communism and countering Soviet influence became the focus of U.S. national security strategy and policy for the next 45 years.

To bolster Western European security and limit Soviet influence, the United States began to consider a European-American military alliance in 1948. After extensive negotiations, the North Atlantic Treaty was signed in 1949. In the agreement, the United States, Canada, Belgium, Denmark, France, Iceland, Italy, Luxemburg, the Netherlands, Norway, Portugal, and the United Kingdom agreed to consult each other about threats and defense matters, and, more importantly, to consider an attack against one signatory in Europe or North America an attack against all. The North Atlantic Treaty Organization (NATO) soon developed a military command structure and a permanent civilian secretariat. Throughout the Cold War, NATO played a key role in American strategy as the United States sought to counter the Soviet Union in Europe.

Another key area of competition between the United States and the Soviet Union during the Cold War was nuclear weaponry. When World War II ended, the United

States enjoyed a nuclear monopoly. Although the U.S. atomic stockpile was small, roughly a dozen bombs in 1947, and the bombers capable of delivering the weapons numbered less than 30, U.S. strategists were confident in the weapon's effectiveness and assigned it an important role in their plans. In 1949, however, the American nuclear monopoly ended when the Soviet Union tested its first atomic fission bomb. Soviet acquisition of nuclear weapons prompted the United States to initiate a crash effort to develop the more powerful hydrogen fusion (thermonuclear) bomb.[4] The United States first successfully tested a thermonuclear device on 31 October 1952.[5] The Soviet Union responded with its own thermonuclear test on 12 August 1953.[6]

In 1954, the Dwight D. Eisenhower Administration adopted a nuclear doctrine known as "massive retaliation." The new policy emphasized nuclear weapons even more in U.S. strategy. It sought to meet "the Soviet threat" without "seriously weakening the U.S. economy" by threatening massive retaliatory damage in response to Soviet bloc aggression. The intent was not to rely entirely on large-scale strategic bombing to deter and counter aggression, but to ensure that the costs always outweighed the gains.[7] The nuclear arsenals of the United States and the Soviet Union expanded rapidly. By 1960, the United States owned more than 18,000 warheads while the Soviet Union possessed more than 1,600.[8]

Meanwhile, the Soviet Union and the People's Republic of China, which had enjoyed amicable relations and extensive economic and military cooperation after the Chinese civil war in 1949, grew increasingly antagonistic. When Soviet leader Nikita Khrushchev denounced Stalin's "crimes" and initiated a process of de-Stalinization, Chinese leader Mao Zedong criticized the Soviet Union for straying from its Marxist-Leninist roots. When the Soviets ended nuclear assistance to China in 1959, relations deteriorated further. In 1960, the Soviet Union withdrew all of its advisory personnel from China.[9]

By 1960, critics were also attacking "massive retaliation" as no longer credible.[10] As a presidential candidate, John F. Kennedy argued that the doctrine left the United States with only one response to any Soviet aggression, a massive nuclear attack. After his election, Kennedy adopted a new doctrine known as "flexible response." The plan inherited from the Eisenhower Administration called for launching all strategic forces against cities in the Soviet Union, China, and Soviet satellite nations. Kennedy and his secretary of defense Robert McNamara, wanted multiple options to respond to any Soviet challenge. In particular, they wanted to focus more on destroying Soviet military forces and less on cities. They also wanted to consider the growing Sino-Soviet split and avoid destroying Chinese targets unnecessarily. Simultaneously, the Kennedy Administration sought to increase spending on non-nuclear forces in order to bolster U.S. capability to conduct limited wars without the need to employ nuclear weapons.[11]

The competition between the United States and the Soviet Union extended well beyond the Eurasian landmass. In an effort to deter U.S. interference in the affairs of a regional ally and to increase Soviet nuclear strike capability, Khrushchev agreed to deploy nuclear missiles in Cuba in 1962. U.S. surveillance flights soon discovered the construction of the missile sites, precipitating the Cuban Missile

Crisis. After discussing several options, ranging from a stern warning to airstrikes followed by an invasion, Kennedy ordered a naval quarantine of Cuba to prevent the delivery of further missiles, and demanded the sites under construction be dismantled. Tensions ran high for several days but negotiations eventually brought an agreement: the Soviets would withdraw their offensive weapons from Cuba and the United States would publicly agree not to invade Cuba. The crisis was the closest the United States and the Soviet Union had ever come to war.

The contest between the superpowers also spread to Asia. The victory of the communist-led Vietminh army over French forces at Dien Bien Phu in 1954 led to the creation of a Communist Vietnam north of the 17th parallel and a non-Communist state south of the line. The United States sought to stem the tide of communism and soon became the primary supporter of the South Vietnamese regime. By 1963, more than 16,000 U.S. military advisors were in Vietnam. In 1964, Congress passed the Gulf of Tonkin Resolution authorizing President Lyndon B. Johnson to use armed force to protect U.S. forces and to deter aggression from North Vietnam. U.S. combat forces subsequently flowed into the theater in large numbers: by 1968, nearly 550,000 U.S. troops had deployed to Vietnam. The significant commitment of troops and resources to Vietnam left fewer resources available for other efforts to compete with the Soviet Union.

Early 1950s[12]

Marshall: I came to RAND in January of '49, straight from graduate school at the University of Chicago. All through '49 what I worked on was a project with Herb Goldhamer.[13] The idea behind the project was that in both world wars, a substantial number of people were rejected for mental, psychological reasons. Seemingly, the raw data showed that there was a significant increase in the percentage that were rejected in World War II as compared with World War I. And the issue was—all being dressed in the notion that the next war, the most likely version of it, was very much another war like World War II with a major mobilization and so on—Is this a trend that needs to be taken account of when we look at manpower resources? So that was the focus.

Q1: Who asked for the study?

Marshall: It was generated within RAND, and was something that Herb wanted to work on. I think, realistically, it probably was also that RAND very much wanted Herb to join them, and this was what he wanted to work on, so have him work on that for a while. He had wanted some statistical help in doing that, and asked Allen Wallis, who then referred him to me. He interviewed me, and I came to RAND, both because I didn't know very much about RAND—or hardly anything; it was just being formed, in some respects—but also it had one important, interesting feature: the salary offer was about 50 percent above the other offer I had.

And Herb was obviously a very intellectually interesting guy. He was just a wonderful man and we became very close friends.[14] So I went to RAND, but it was kind of happenstance

It was in '49, while I was working on this project that, because of a man named Russ Nichols, who was an economist, the representative in the RAND Washington office of the Economics Department that was also just forming up, I learned more about what the rest of RAND was really doing. Then I began in early '50 to transition to some other things, and gradually moved into the mainstream of what RAND was doing.

We still had, at least during the first nine months of '49, a sense of a monopoly on nuclear weapons, a vision of what the war would be like if it were to occur. But this, of course, all rapidly changed during the '50s as the Russians got nuclear weapons; not only got a few, but both sides began to accumulate very substantial numbers.

Thermonuclear weapons were invented, which upped the potential scale of the destruction.

The progression of things that I got involved in, and the kind of intellectual problems that were encountered, goes from the general "criteria problem" for some of the RAND studies, and then a lot into targeting, and then into intelligence matters, strategic warning issues.[15] Then…well, then there was an interlude when I got married and went off to the University of Chicago to teach for a year.

Q1: What year was that?

Marshall: It was the academic year '53–'54.

Then when I came back, I got involved in this effort which led to *The Next Ten Years* paper.[16] I also worked with another man on tactical nukes and Europe and target—[17]

Q1: What was his name?

Marshall: Harvey DeWeerd.

And then I ran a study on reconnaissance, using Monte Carlo methods to look at the effect of reconnaissance on the efficiency of bombing.[18] Then Joe Loftus and I started this Project Sovoy, which has a long history, but was probably the first really big intellectually interesting thing I did.[19]

Q1: What does "Sovoy" stand for?

Marshall: "Soviet military" [Sovetskiye Voyska]. It was an effort to supply, I think as we discussed last time, better inputs for the RAND studies. And it led us into this question, "How do you make better projections of future Soviet programs?" Which, in turn, led you both to this further interest in the budgeting constraints, resource constraints and how to take account of those, as well as to the issue of organizational behavior, because what you were really looking at was the behavior of some large, complex organizations that created and operated these things.

And then I got involved in the Gaither Committee activities of '57, and wrote, when I got back, this Goldhamer-Marshall-Nathan Leites piece [*The Deterrence and Strategy of Total War, 1959–1961*], and then went to the first part of the Surprise Attack Conference in Geneva in '58.[20]

At the end of the period, I was involved in the so-called "SOFS" [Strategic Offensive Forces Study] study in '59, '60.[21]

Q1: Did you have much interaction at that time with other people at RAND?

Marshall: Not a lot. Herb and I made a trip out to RAND, probably from the middle of September through the middle of October. And then I got to know a lot of people, although the biggest mechanism for that was *Kriegspiel*. I think I'd mentioned earlier that Herb had been the junior chess champion of Canada. He and I used to play occasionally, but not very much. We'd have lunch or something like that and play some rapid transit games. But when we got out there, Herb had already met John Williams, who was really one of the leading, most interesting people at RAND. He was at the head of the Mathematics Department, although his own professional background had been in astronomy. He was the center of a very lively *Kriegspiel* activity, which went on not only during lunchtime, but at his home in the evenings. In fact, we spent maybe three or four nights a week at his house in Pacific Palisades playing. And I got to meet a few people, but the main purpose of the trip was related to the mental health project. You see, we had collected a bunch of data, and various statistical analyses and data processing was going on, but in Santa Monica. A lot of my work was writing out the instructions, from Washington, and sending the instructions to them about how the data should be broken down and analyzed. A great deal was done on that. Now, of course, there were always lunchtime discussions with people and then that was a lot more talk about military things.

Q1: Any recollections of what some of the subjects of discussion were?

Marshall: The thing I most remember was that the people at RAND had a rather different view of what the war plan, at least for nuclear weapons, ought to be like. The military establishment focused on a kind of the—Another fellow, an Air Force officer who said he wrote the first war plan, mentioned to me once that it was like the last "max effort" raid of World War II. The whole idea was to deliver as quickly as possible all of the weapons against—at that point, prior to their getting nuclear weapons—basically the Soviet economy. Whereas at least as early as '50, late '49, certainly no later than '50, there were people around RAND writing about another way of using these limited number of weapons, which, for example, would have been to have made a few attacks, but then distribute leaflets saying, "Evacuate the cities," or also getting the government to try to surrender.

Q1: This was the "Warbo" ["Warning and Bombing"] study?[22]

Marshall: Yeah. And, say, somebody like Herb, even though he wasn't directly involved in the Warbo study, had interesting things to say and talked with people about that.

Then, of course, you had—I don't know the exact timing of this—Bernard Brodie coming on board.[23] Now, he had, because of his earlier work, some access to the actual war plan, and had, I think, written, talked with [Gen. Lauris] Norstad [Air Force Deputy Chief of Staff] about it, and [Gen. Hoyt] Vandenberg [Air Force Chief of Staff], and had also been critical.[24] And clearly, later....One of the things you'll most want to read is, for example, his, I think it's about a '52 or '53 [Sep 1951] paper called "Must We Shoot from the Hip?"[25]

Another part of the discussion at RAND during these early years was the subject of preventive attack. There were a set of people, including Williams, who were for kind of preemptive war, just as there were some Air Force officers, and, in fact, for that matter, John von Neumann.[26]

Q1: What was the Stalking the Tiger ["Hunting the Tiger (and Other Aspects of the Active Life)"] paper?[27]

Marshall: And that's another one you want to read. It's by Williams.

Another interesting thing that was done during this time was that soon after the thermonuclear weapon, which proved successful, von Neumann was very interested in using it to affect the climate, because if you look at the world, if you can shift the temperature of the earth a couple of degrees, you, in effect, drive the Soviets out of agriculture.[28] So there was an interest in looking at Krakatoa and other large volcanic things. After these eruptions, you see the effects through a couple of years, and it does change things, and you get these things of the...I forget what...the "summer without sun" and various things.

Q1: These studies were done at RAND?

Marshall: Von Neumann was quite interested in this, and then some—I don't whether they were real studies, but I remember, though, people getting ahold of these old weather reports and looking at them.

And RAND did have a weather group then, headed by a guy named William Kellogg, who was very good. They had several very good people. They were also much involved with this project to do reconnaissance from balloons, when you had to then predict the track of the balloons and where to release them and so on. That was called, I think, Project SINBAD.[29]

So there were all of these very interesting things going on. And there also was game theory and all of the efforts to think about and use game theory.

Q1: How influential was that in the early years, game theory?

Marshall: It was fairly influential, in the sense of people saying, OK, here's this new intellectual tool or way of looking at things, and how can we adapt it and use it. People tried to do analytic work, develop ways of actually using it in warfare. But most of that never turned out. Warfare is just too complex. But the ideas, the kind of intellectual framework that it supplied was fairly influential.

And it had another effect, which was, I think, very important for RAND. I think the man who first pushed the idea, which was then taken up by some others, was named Merrill Flood. His notion was that even if you couldn't get analytic solutions to [indistinct] games, human beings have great skill at playing games, so have people play these games to try to discover what the good strategies are, even though you cannot analytically derive them.

That was done, first on a small scale, with simple games that were still nonetheless beyond analytic solution, but then, as early as the '54 time period, when I got back to RAND, there already was underway a game that was called SAW, Strategic Air War; I think that was probably the acronym for it. Alex Mood, who was a statistician was kind of the father of that. The SAW game was being used to look at the broadest kind of strategic questions as to the use of weapons and so on in a situation where the war starts as a ground war in Europe.[30]

In fact, it [SAW] was done somewhat in conjunction with this *The Next Ten Years* effort.[31] *The Next Ten Years*, of course, is the name of a paper. There was a group that was called the Strategic Objectives Committee that I and...really chaired by Charles Hitch, I guess...Hitch and Williams. And I was on it, Mood was on it...I forget who...Brodie.[32] What it was supposed to do was to reexamine, "What are our strategic objectives? What are the really big issues and questions that we at RAND ought to be addressing, or our work ought to serve as some help in addressing?"

Mood ran these games. Now, Hitch didn't quite think they were adequate, and I know he was anxious at one point about these games being pushed too much by these people, or the results of them. But the only output from the thing was this *The Next Ten Years* paper, which Hitch and Brodie and I wrote. It was circulated and people commented on it, but we were the basic drafters and not much was changed.

Q1: This was the only effort of the Strategic Objectives Committee?

Marshall: Yes, at least the only output I can remember.

Q1: And how was it received?

Marshall: I think well, but I don't know, again, how operationally effective it was.[33]

Q1: Do you think it influenced the direction and the content of the research program at RAND?

Marshall: Some, but I think not necessarily a lot.

Q1: Was there distribution outside of RAND?

Marshall: As I remember, it was only an internal RAND document. A few copies may have gotten out, but it was directed inward to RAND people.

Q1: Does *The Next Ten Years* give a good summary of the discussions within the Strategic Objectives Committee?

Marshall: Yes, I think so.

Q1: There aren't any things that you talked about in the group that aren't touched on in *The Next Ten Years*?

Marshall: I don't think so. Well, there were some things. For example, as part of this effort, I remember I was asked to try to reconstruct the size of the U.S. nuclear stockpile. It turned out to be possible because of some inadvertent disclosures at one point. Then there were things that were kind of fuzzed up, but once you knew certain base numbers… I mean when people in testifying about some new separation plant would talk about the improvement in efficiency of the barriers, and things of this sort. The actual numbers, of course, were very highly classified at that time. But this was an effort to do that. So there were things like that that didn't specifically get included in *The Next Ten Years*. And there were these various games, as I say, that were run.

But game theory as such, I think, gradually died out, weakened in effect, by the middle, late '50s. But it was very active initially, the belief that it could be useful, could eventually be applied.

Q1: What were the hopes? Ideally, what results did people want these game theory analyses to yield?

Marshall: I'm not sure I can really reconstruct that.

Q1: But there were high hopes.

Marshall: Yes, there were. Or there certainly were people who thought that it would allow us to think better about things. And you know, in a way, some of the ideas got incorporated into people's thinking. But the very direct applications were very few.

One of the things that I quickly [illegible] in the first place, in real life tended to be far too conservative, because you probably always had some information about the style of the other side's play, so to speak. He was a bit more predictable than that assumed.

Q1: What would be an example of that?

Marshall: Well, that was our interest in the Russians, in the sense that if you looked at them, understood their organizations, understood their history, you didn't have to assume that they were always going to do the most damaging, worst thing, or that they were going to randomize and so on. Real competitions really weren't like that.

Q1: I take it you weren't getting much substantial support in understanding what type of opponent we were facing from the political scientists? Or it was inadequate, their characterization of what the enemy was up to.

Marshall: Yes, right.

Q1: Was that a general feeling within RAND?

Marshall: I think it was. I think people as a whole tended to be disappointed in what it was the social scientists really could offer. And the social scientists, I think, on the one hand, were cautious about what they—When they were asked direct questions they felt they couldn't answer, something like that. Anyway, they were very cautious, and didn't feel they had any way of, I suppose, getting better at doing better. Clearly, there were some differences among them, but they tended to want to work on their own kinds of problems, in their own ways, and that's one of the reasons they initially stayed separate and stayed so long in Washington before moving eventually, under a lot of pressure, to Santa Monica. The Social Science and the Economics Departments were originally set up as a result of this meeting that took place in New York, when RAND was going to become a new, separate organization, not just a project at the Douglas Aircraft Company. And there was a discussion about, "What issues do we want to really deal with? What kind of expertise do we need? What kind of questions do we need to address?" And it was out of that that there came the notion that there ought to be political scientists, and economists, and so on.[34] But then the political scientists were very loath to get much involved in really helping. These people had a problem.

Q1: It's interesting, because there was a genuine demand for their services.

Marshall: Yes, there was.

Q1: It wasn't just icing on the cake. There was a need for this help—

Marshall: That's right.

Q1: —and it wasn't provided or it was inadequate.

Marshall: Yeah.

Q1: So they pretty much set their own research agenda—

Marshall: Yes.

Q1: —apart from the rest of you? How did they get away with that?

Marshall: I don't know. I guess by arguing, pleading. I don't know, I'm not sure of the whole thing. They did some things. I think they were not of very great quality. They started this effort looking at the Soviet military writings or some of the doctrine. I think it was very inauspicious, given the time when they actually started to do it, in the early '50s, because—and this is something that Loftus and I pushed very strongly—there was a big, big difference between what they were actually doing during that period and the doctrine. The overt doctrine tended to belittle nuclear weapons and tended to be much in the mold of their World War II operations, whereas if you saw what they were really doing, it was just very, very different.

Here was a country that was just recovering from staggering losses, damage, and so on, but that started a very big nuclear program in the middle '40s. In some ways, '45 through '50 is one of the most interesting periods with regard to the development of the Soviet military.[35] Big changes took place. Many things were started that represented really big shifts, big allocations of resources, enormous allocations of resources. There's some reason to believe that something like 50 percent of all the concrete poured in the Soviet Union for a couple of years or so in the middle and late '40s went into the nuclear program, at a time when the overt doctrine tended to pooh-pooh the importance of nuclear weapons. But obviously Stalin…

I think the real thing was that coming out of the war they saw that they were quite far behind in a number of technologies, not only the atomic bomb, but the Germans had done the V-2, their big jet aircraft, innovations in aircraft design, radar, things the Soviets were very far behind on. Then also, against the Germans they didn't need any big air defense system. After World War II they started essentially the whole air defense effort. They also started a big missile program. They took over things the Germans had underway, air defense missiles, rockets. They took over all of those.

Also, Loftus and I used to note that they had had this close-support airplane, the "Stormovik" or something, it was called, but then if you looked at the airplanes… and then the doctrine seemed to glorify that and suggest that… but they never built any, [Eds: never built] things for that close-support mission after the war was over. They were all fighter-interceptors and some bombers and so on. So at the level of the tactical aircraft, they didn't build any specific aircraft for close support. There were no new designs, no new buys of these close-support aircraft.

Q1: Now, at the time, what did you think accounted for that, the absence of close-support aircraft?

Marshall: I guess we just took it that they were focusing on air defense and these other things. They were just not oriented the way the doctrine said, that the real decisions were taken with the other things in mind.

Q1: Was the failure of the political scientists to take into account what the Soviets were actually doing, as opposed to what they were writing, due simply to their lack of access to intelligence?

Marshall: I don't think so. No, they had no inclination to do that. Loftus and I, in the end, gave up. We kept telling them, "How do you explain this? How do you explain that?" They just kept churning away, looking at the Soviets' written materials. Now, in the end, I think doctrine is actually very interesting and is a very fruitful source, but they just were in a period when it didn't provide very much insight. Later, there was this big change in the '50s to this potential doctrine of preemption and so on. And later in the '50s, of course, after Nikita Khrushchev came in as premier, there was a big switch to open endorsement of nuclear weapons. In fact, they went sort of too far in that direction and then they came back.

Outline for 29 October 1993 Talk

[Note: Italicized items were not discussed during the interview.]

Introduction

1. Purpose
   - Talk through aspects of '50s/'60s
   - Emphasis on AWM intellectual history
   - Format
     - Short talk
     - Questions from small group
2. Problem
   - Right way of integrating/treating
     - AWM intellectual history
     - Contrasting RAND intellectual framework
     - Interaction of analysis and real world

AWM Intellectual History (1950–69)

1. Probably better to focus on this period
   - '49/early '50 focus on mental health problem with Herb Goldhamer
   - Dec '69 start of work for Kissinger
   - Also Sep '53–Apr '54 teaching at University of Chicago—*some focus on Monte Carlo methods*
   - And Aug '63–Mar '65—Paris/NATO interlude
2. Treatment
   - Start in middle ~1960
     - Then ask—how did I get there
     - What happened after
3. 1960—Themes of thinking
   - Importance of macroeconomic constraints (especially in understanding the Soviets—focus on military budget/resource constraints, costing effort. Existing Soviet economic studies worthless)
   - Importance of looking at others (Soviets) within framework of:
     - Organizations as actors (shaped by history, past patterns of behavior, adaptations to social, cultural aspects of society)
     - Rational actor model not useful—Heymann study
   - Possibility of useful forecasting
     - Some aspects/past behavior patterns
     - Hence, game theory max/min not good either—always know more of others than this suggests
     - Uncertainty must be reflected
       - Others' actions not fixed, evolve, but past choices constrain
   - What people (analysts) believe may not be reality-based (reality vs. model focus)
     - NATO beliefs
     - RAND political scientists' view of allies' actions (can't afford bigger military budgets?)
4. Discussion of '50s, how this developed
   - 1950/'51/'52 not too interesting
     - Focus on measures of effectiveness—role of economists at RAND
       - Value of history
       - Herb Goldhamer
     - Targeting—>focus on intelligence and later work, but not central to

intellectual history, key in relation to Joe Loftus (learned a lot—especially importance of '45–'50 for Soviets)—failure of political scientists
- Strategic warning
  - –*How to think about others*
  - –*Irrationality of U.S. analysts*

Herman Kahn [notes in upper left margin of p. 3 of handwritten outline]
Reality focus (reinforce each other); Role of economists; Monte Carlo; What's wrong with RAND; Analysis; Models; Rationality; Treatment of uncertainty
- Important for intellectual ideas on:
  - –Long-term competition/lost later at RAND (until I revived in '69)—period of thinking about U.S. strategy vis-a-vis Soviet Union
- Value of gaming/as evolved at RAND
  - –Not based on German war games
  - –Came from game theory/analysis of strategy as games—>(Merrill Flood) idea of using game-playing skills of people to find solutions to games that were mathematically intractable
- Counterforce/preventive war (preemption) insights into others

— Things really get started in summer of 1954/return to RAND—'63
- Next Ten Years
  - –Last of long-term competition efforts/what are the big strategic moves, what is the situation—prototype of some later work (faded at RAND—why?)
  - –Use of games/SAW—criteria problem/AWM/Hirshleifer (Rational actor model problem)
- Work with Loftus
  - –Importance of '45–'50 for Soviets
  - –We can do better on forecasts
  - –Impetus/RAND failures, intelligence not trying
  - –Controversies with political scientists (look at what doing, not just writing—organizational behavior). Importance of history/trends
  - –Sovoy, leads to: AWM reading of Barnard, Simon & March/later Cyert & March (most important); (Crozier—'63 book); Central part of work until leave for Paris, Aug '63
- Continued work with Herb Goldhamer
  - –Herb's game
  - –'58 collaboration on Framework paper/Nathan Leites—*Value differences U.S./Soviet/use of judgment in ranking/assigning utility values to vectors (state)*

5. Post-Paris/Spring '65–'69

— Return to organizational behavior focus/start project at RAND
- Meet March, Cyert, Crecine
- Neustadt, Allison, May/May Group, fall '65 (Big influence)
- Application to Soviets/military programs

Carnegie school—look at others; began line of analysis Harvard Business School—>notion of strategies (very important)
[notes in middle left margin of p. 5 of handwritten outline]

— Schlesinger comes to RAND
- Summer of '62—friends
- Full-time summer '63
- After return (Paris) we both read Ardrey's *Territorial Imperative,* began discussing ethology, nonrational aspects of behavior

        –Implications for psychological/political impact of military forces, etc. ('68–'69)

- — Together (organizational behavior+ ethology)—>critique of standard deterrence theory
    - • Lead to insights on analysis of strengths/weaknesses—we/they
    - • 1966 paper on assessing strengths/military balances
- — Wikner/NetTechnical Assessment in DoD
    - • RAND contract—became involved in '67/'68 (?)—needs checking. Write papers/memos/etc.
- — Rowen as new president of RAND (late '67). Schlesinger and me as major advisers—> creation of new post—Director of Strategic Studies (strategic forces). Schlesinger appointed spring '68.
    - • Late '68 Schlesinger leaves for new administration (transition team)
    - • Early '69 AWM takes over
        - –Review of strategic program/meetings in spring/summer '69→ development of long-term competition ideas—return to early ['50s]

6. Contrasts with mainline RAND
    - — Counterforce is possible!
    - — Can do better than rational model
    - — Importance of not overestimating opponent
        - • Air defense case
    - — Organizations matter
    - — Resources matter
    - — Long-term competition theme in late '60s—shift of focus (real flash of insight for me)
7. Some comments of national level
    - — Special aspects of World War II generation
        - • Impact on '50s/Eisenhower Administration
    - — Fades in early '60s
        - • Cuba->'63/'64 shift
            - –Big strategic errors begin
            - –Focus on stability/arms control vs. competing/bankrupting (Long-term competition)
            - –Counter cost-imposing
            - –Limits on accuracy
    - — Kissinger strategic pessimism/settling for parity
        - • Contrast with Schlesinger
        - • No adjustment (query of Bill Hyland/Rodman)
    - — Long-term consequences of misestimating Soviet economy—deep sources going back to '30s/'40s—loss of belief in free markets—broader trends/ethos/importance of

1950–1969[36]

Marshall: While I joined RAND in January of '48 ['49], the real time to start this discussion, I think, is really the summer of 1950.[37] Because during the first year I spent my time, first in Washington, working with Herbert Goldhamer on this large project on mental disease, which was very interesting, but didn't contribute very much later.[38] I mean its part in my intellectual history is not very interesting. I picked up certain things, principally the critique that was emerging of U.S. thinking about how to target the limited number of nuclear weapons, and whether you ought to immediately expend all of them or have some more deliberate, politically motivated strategy for using things.

Q4: Be an interesting linkage, linking mental disease to knocking all the weapons out.

Marshall: No, this was just side conversations.

Then there are also some other years that kind of drop out in this whole period. One is when I went off to teach at the University of Chicago, September of '53 through about April of '54. And then there's a period, August '63 through March of '65, when I lived in Paris and was involved with this do-nothing organization called NATO, which had no contributing, positive features intellectually as far as I could tell.

Q3: What did you teach at Chicago?

Marshall: Mathematical statistics, the first graduate courses. Allen Wallis, who had become a friend of mine, had gone off for some big study at that time that he headed for the Ford Foundation, which was going to give a lot of money to universities, and the study was to decide where to give the money.

Q4: Was that the original purpose in you going?

Marshall: The original purpose? No, the original purpose was that I was going to complete the work for a doctorate. Instead, he persuaded me to teach his courses. And so I never finished that. Right.

Also, essentially '69 is a good year to end, because in January [December] of that year, I first went to work for Henry Kissinger. And basically while I was at RAND until January of '72, I really worked half time or more through that whole period for Kissinger on the National Security Council staff. So really the period from summer of '50, in some sense, through November of '69 was a reasonable period, with these other periods left out.

So look at 1960, then, as this intermediate point, as a way of getting at that. What were some of the themes that characterize the way I was thinking about things?

One that came, I suppose, out of my training as an economist was this importance of macroeconomic constraints, especially, as you'll see later, in understanding the Soviets. The importance of looking at the military budget, the resource constraints as a whole, led to an effort by Joseph Loftus and I to… or a proposal really, which then the Central Intelligence Agency actually adopted, I think, although maybe other people were thinking the same way, that you had to undertake this overall costing of the Soviet military effort.[39] Up until that time, the CIA had spent a lot of time looking at the economy, but a lot of it was related to industry studies—if you're going to attack the Soviet Union, what do you want to know about its industrial structure and so on—and their general economic success. But the reason you wanted to do the overall costing was that people, in looking at and trying to make forecasts of Soviet programs, did them individually, without any sense of any overall cost or resource constraints that they might have. So the idea was a way of disciplining and integrating efforts to look at the Soviet military effort as a whole. But there had been lots of things, which we will get into later, which made that always seem very important to me.

The second thing, which by that time I'd gotten into a lot, was that if you're looking at other countries, like the Soviets, the military in particular, you have to take account of the fact that what you're talking about are big, complicated organizations. You have got to move away from this notion of, "There is this thing called 'the Soviets,' that have decided X or Y," and realize that what you're talking about when you're talking about Soviet military programs is the actions or the outcome of decisions of big, complicated organizations.

Now at one level that seems obvious, but it wasn't until I'd say the late '50s that I myself had this insight into the blindingly obvious. Because almost all the discussion of these things tends to proceed in terms of, "'The Soviets' have decided, or 'the leaders' have decided X, Y, and Z." And then once you think about that, then it leads you to, "How do these organizations come to be what they are? What is their history? What is the actual organizational structure and their processes? What are their past patterns of behavior? How have they had to adapt to social-cultural aspects?"

Although it wasn't until, I guess, the early '60s that Michel Crozier wrote his book [*The Bureaucratic Phenomenon*] and I read it and was very impressed by it.[40]

Which gets you, again, beyond the organizational factors and into factors of social-cultural importance, because, as his book points out, there just are these systematic differences between French organizations, British organizations, American organizations, as to how they function, because organizations are very much involved with this issue of the relationship of subordinates to superiors. These individual cultures, right from the beginning, create those kinds of relationships and predispose people to certain attitudes about those relationships and how they ought to be structured and so on. Therefore the organizations, when they come to deal with the raw material any particular culture has provided through the nature of the culture, adapt to that, have to deal with the fact that if they want this organization to run, they have to take account of the fact that it's inhabited by Germans or French or British, and so on. In any case, you just have to move away from either this vague "The Russians decided" or "the Soviets decided" or this more formal kind of "rational actor" model.

I guess one of the reasons this became dramatically clear to Loftus and I, who I'll get to a little later, is the experience both he and I had at RAND and throughout the '50s, although he may have arrived when I was away at Chicago or maybe slightly before that.[41] He had been in the Air Force intelligence organizations, the Air Force had the chief responsibility then for following the Soviet nuclear programs, and he had headed that group and had had a lot of access to intelligence materials. He and I became close friends because of some earlier involvement. I was one of the few people at RAND who had been much involved in and more directly in intelligence things. But at any rate, we knew a lot of things which other people at RAND didn't know.

There had been a study by two or three RAND people [H. Heymann, J. DeHaven, R. Raymond, *The Soviet Intercontinental Mission Through 1957*].[42] It came out probably in '56 or maybe '57 ['54]. It was an effort to say, "Where are all the Soviet bomber bases that we don't seem to know about, or at least the people at RAND didn't know about?" They said, "The Soviets are rational planners, where would they put these bases?" They did this study which reached this very wrong conclusion that all the big bomber bases were along the Trans-Siberian railroad, but in Siberia, very far from all of the coastal areas, because they could be subject to attack and so on. This was part of the whole pattern at RAND about why you couldn't do counterforce. It was not only that we didn't know where these things were, but, in addition, they were going to be in places that would be difficult to get to and so on. Loftus and I knew that was a lot of cr…bunk. And so we used to use arguments like, "Why is March Air Force Base where it is? Why don't you think their bases are where they have been? Don't you remember that airplanes used to have short legs? If you wanted to hit somebody, where would you put your air force bases? And why would they move them? Or why would they move them and we haven't?" And all of that was totally unavailing.

Anyway, things like that disabuse you of the notion that these actions and all of these things are the result of highly rational behavior and so on. So there's that. It's very important to get into things like that.

But all of this then leads to the notion that you actually can do some useful forecasting. It's not entirely hopeless to forecast some aspects of an opponent's behavior. Also, it tells you that, of course, you don't want to use the rational model as the basis of the forecast.

And also one of the things at RAND which this leads you away from… What was very important intellectually at RAND in the early days was this notion that game theory was going to be a very important tool. Not just technically, that you could actually solve, make it a game that you could— but as a way of thinking. And in the game theory ethos of that time, the minimax strategies were the thing that you ought to follow in the face of uncertainty. Again, by '60, or even much earlier, I was very convinced that you could do a lot better than that. That you did know some things about past behavior, and you could predict in some ways the kind of probability distribution of the behaviors, much more than the gametheory view suggested.

And the other aspect is that, of course, nonetheless the future is uncertain and uncertainty has to be reflected, not artificially removed, as was true in so many forecasts, where people make their best estimate and then decide that they're going to plan against that.

The other thing, or maybe main perspective I had, was that what people, analysts believe may not be reality-based, and I've already gone into one example, this study of Soviet bomber bases that I mentioned.

Another thing that by the mid-'50s I certainly had come to believe was that most of the way in which NATO was discussed and thought about was wrong in some essential ways. There is this tendency, which may have been politically appropriate in dealing with the people who we wanted to go along with us, to treat the NATO alliance as a real alliance, rather than the situation of a major power and a bunch of protectorates, which, in my view, it was, it remained.

The first reason I came to feel that way was that I noticed—and let's get back to some of these resource issues—that some of the RAND political scientists would go to Europe and come back and have all of these explanations, beginning in the middle '50s, about why the 1952 Lisbon force goals were not being met. The political scientists claimed the reason why things the European allies had signed up to in the early '50s were not happening was because they couldn't afford that. I pointed out to them, first, that the European allies had already reached or surpassed their prewar per capita gross national product levels of output. They were all living better than they had in the '30s. How was it that they could make these big efforts and that there might… We certainly needed an explanation about why they weren't doing what they should have been doing, but it had nothing to do with the fact that they were too poor. Yet people kept using that explanation.

So I kept trying to tell the political scientists it was not a problem of economics, it was something that they should be dealing with and explaining. The only person who offered real insight about it, as far as I was concerned, was Nathan Leites, who said, "If you look at them, their belief, the Europeans' belief structure, is that there are giants, midgets, and pygmies. And that they are midgets, and they are simply not up to certain things." They're psychologically unable to act as the giants do. And then there are, of course, the… Well, no, I guess it's the other order; it's the giants, the pygmies, and the midgets. And then there are these midgets around that are still below them. And unless you see something like that, or adopt that view, which is, of course, very undiplomatic, you mislead yourself about what's probably really going on in these people all through until the very end, and probably to some extent even now they are acting in ways that fit much better the notion that they saw us as protecting them, rather than their being part of an alliance where they're all really going to fight together and going to have some real equal relationship.

Q2: Do you think the Soviet Union was a "giant" in the '50s?

Marshall: Well, it was and it wasn't. They certainly had a navy… Again, we'll get back to that. I think that's one of the other problems with the misperception of the Soviets throughout this whole thing. But they certainly had a big military program.

Q2: Did Nathan Leites think that the Europeans saw the Soviets as a "giant?"

Marshall: Yes, oh yes.

Q2: So there's one "giant" protecting them from another "giant."

Marshall: That's right.

OK, so let's go back more to looking in a bit more detail at the '50s. Again, in the period '50, '51, '52, through the summer of '53 really, it is not so interesting, I think. There were a lot of things that went on, but from this point of view of intellectual history not all of them were that interesting. For myself initially, after I joined the Economics Department at RAND, I was focused a lot on these measures of effectiveness—the "criteria problem"—which was seen as one of the major roles of the economists.[43] Even there, there were certain things that came out, I think, like the value of history in thinking about some of those issues and also, through my association with Herb Goldhamer, which continued, his insights about human behavior and how various things had to be taken into account.[44] So anyway, I worked some on that.

I worked a lot on the problem of the appropriate targets to be destroyed in the Soviet Union, how to use efficiently the small number of weapons that we had.[45] But that led to my getting very much involved in intelligence matters. Which led to this natural selection by Loftus of me as a close associate when he joined RAND. It was through him that I learned, more than I had by that time, about the Soviets and various intelligence problems. We also got into a lot of battles with, again, the political scientists at RAND, who had this focus on the Soviet military doctrine. Now I certainly think the doctrine is very important, but they absolutely refused, as far as we could tell, to look at what the Soviets actually did. So they focused on their writings, which was very unfortunate… [gap due to tape change]

…1945 to, say, the middle '50s, when they were writing, was the period where there probably was the biggest discrepancy between the written doctrine and what they were really doing. In fact, in retrospect, both Loftus and I felt that the period '45 to '50 was absolutely central in understanding the Soviets, because it was in the aftermath of the war that they undertook these massive efforts to catch up with the West and with us in military technology.[46] They pushed their nuclear program and, in fact, at the time when the doctrine was pooh-poohing nuclear weapons, they undertook this program in which they spent a very big proportion of their GNP. In fact, Loftus told me, based on some of the information he had, that something like 50 percent of all concrete poured in the Soviet Union for several years in the late '40s went into the nuclear program. This was at a time when they were rebuilding the country. And if you look at our own nuclear program, there were times when we had the nuclear program absorb an enormous fraction of some certain kind of things, like special stainless steel. There was a point in U.S. history where something like 25 or 50 percent of the stainless steel went into the nuclear program.

Q4: The conclusions that you and Loftus reached, were they largely driven by the fact that you and he had access to information which the others did not, and so you had this chance to see this graphic contrast?

Marshall: Yes, a lot of it was. Because access to these intelligence materials, like communications intelligence and so on, which at that time was very rich and detailed about many things, was very closely held. Loftus and I and maybe two or three other guys at RAND were the only people who had access to that.

Going back to this '45–'50 period, it was not only the nuclear program, but they, as we, grabbed a bunch of German scientists in aviation technology—design was where the Germans were far ahead and put out this swept-wing jet fighter —and from the ballistic missile programs. They grabbed not only the scientists, but a bunch of the missiles themselves and started shooting them at Kapustin Yar shortly after the war. They took over the people who had been—the Germans had underway two surface-to-air missiles for air defense purposes. I forget the names of them... One was "Schmetterling"—"butterfly." The other was...

Q3: "Wasserfall."

Marshall: "Wasserfall," yes, right. The Russians started out, but didn't fully create a new organization, had everything but it, of a big air defense organization. During World War II, they didn't have...they had a bunch of antiaircraft guns and so on, but they didn't have a big set of radars and so on. And they were very far behind in radar.

So you had the Russians starting a big missile program, their nuclear program, creating this air defense structure and going into that in a big way in response to our obvious strategic bombing competence. 1945–1950 was a big period of innovation, of setting the pattern for where they were going. It was very important.

Also, they started to build long-range bombers. Prior to World War II, they had some bomber enthusiasts. Most of the bombers were knocked out early in the war and never... Basically, their bombers did nothing during the war. They probably weren't even reconstituted. I don't think they had any long-range bombing capabilities at all. After the war, they went into the long-range bomber business.

So if you wanted to understand where they were going, to hell with the reading of this doctrine about, "Close-air support is the big mission for the air forces" and so on. I used to point out to the RAND political scientists that the Soviets had produced no close-support aircraft, ever; I mean specialized ones. I emphasized that they had done bombers, they had air defense fighters, and so on, and they had some medium-range bombing things, but they didn't have these "Stormovik"-type or any specialized close-support aircraft. Right.[47]

Let me say, just as an aside, these particular years—'50, '51, '52, '53—were the years when Herman Kahn and I were very close friends. We essentially spent every night, every weekend together, unless one of us was away on a trip or doing something. He and I reinforced, I guess, our notions that there were two kinds of people in the world: people who focused on reality and wanted to know how the world really functioned, and those who focused on their models or some hypotheses and barely looked at the world, or occasionally looked at the world. And we talked a lot about why it was the economists had been so successful at RAND and gradually took over much of the stuff. We talked a lot about the Monte Carlo methods and so on. We also discussed what was wrong with the RAND-type of analysis, the emphasis on the models, which we have already talked about, the emphasis on rationality in the narrow sense, how uncertainty was treated, and so on. Which finally led Herman astray, in my view, in the sense that in the middle '50s, he gave a talk that was so sensationally successful that it turned him into a kind of super briefer, almost media personality, instead of continuing to work hard on other things.[48] But he was an immensely bright guy and just a really wonderful person.[49]

Q4: Either in this portion of your speech, or your book, or in a previous one, are you going to deal in some more detail with RAND's intellectual atmosphere at this point in time? Why it recruited the kind of people it did? What that meant for its output? To what degree that was part of the mainstream of intellectual thinking at this point? To what extent it led it or trailed it in the country?

Marshall: I talked, I guess, some about that in the first thing. I'm not sure of all the criteria, but RAND, I know from my own personal case, initially treated its people fantastically and was very attractive. I know the offer that I got was 50 percent bigger than any other offer I had and, again, I only had one or two others. So it bid high for the people it wanted. It had negotiated what today would seem an outlandishly generous operation. People were to fly first class. Secondly, the notion was that people like this needed to take vacations, so they were paid double when they took a vacation day. From the beginning, they got 30 days of vacation a year or something like that. And there was a support structure in which the people actually thought that they were to help the professional staff. It was the most terrific organization to be involved in. And then the people, the collection of people were extraordinary in lots of ways…

Q6: And had essentially total freedom to choose what you did.

Marshall: Yes, right. You picked your own problems to a large extent. There was a group of people who shaped the thing, but if you had a good idea of something you wanted to work on, you could propose it and chances were that, if this was at all reasonable, you'd be allowed to do it.

Q4: In your view is that, was that the right thing to do?

Marshall: Yes.

Q4: And does that say something about how to do that sort of thing in the future?

Marshall: Absolutely. There's that, plus the fact, which I would get to later… I think the other thing that I felt I learned and took away from it came out of the program they had of having people in during the summer to work on various sorts of things. RAND, at that time, was, I think, very ahead in a lot of these areas, and therefore the people coming through were the likes of John von Neumann and all these guys. The whole notion, particularly of the summer thing, was, we have these problems; the objective is to assemble the best people in the country, not just the best people within RAND. You see, the problem with most organizations is that when they have problems to work on, their focus is on, "How can we do the best we can with the people in our organization?" But the attitude at RAND was really one that I think very much carried over from World War II. I think there is a very big impact—we'll be getting into it a little later—of World War II, when people were really serious: we had a problem, we tried get the very best people to work on it [indistinct]. So I have come away, really for the rest of my life, with the belief that what should happen is, if you have a problem, you get the very best people to work on it, and it doesn't matter whether they're in your organization or not. The thing to do is, if you have a problem, you get the best people you possible can to work on it.

Q3: But to link this back to the point where you left off, despite the fact that you have this very favorable structure, and all these best people, you still have these intellectual pathologies.

Marshall: Yes, exactly. Absolutely.

Q3: So I was curious if you could expand on the discussions you had with Herman as to what went wrong, what happened at RAND that the economists of a particular type—

Q6: If I might interject just a short point in talking about game theory, which kind of answers your question. They had guys like Thomas Schelling and like Martin Shubik, who were very articulate, very forceful, believed in their own worldviews, argued them very strongly in public. And it's hard to overturn people like that. So you had these guys who were "pathologies" arguing very strongly.

Q3: "Viruses," "viruses." But the system… In other words, was the system with regard to these "viruses" particularly weak because it was predisposed to look at the world in this way? I mean why were these people… Just force of personality? I mean Tom Schelling is hardly [indistinct].

Marshall: Yeah, right. Well, look, to answer your thing more directly: First, I'll make a general point. There is the tendency—and it's, in my view, what has largely happened to economics and, to some extent, statistics, which are the two fields I know the best—for these fields to tum inwardly upon their own models and so on, and toward mathematics and so on, and just away from reality. There seems to be some strong tendency within various fields for that sort of thing to happen.

As to why the economists were so successful, and therefore unduly narrowed the RAND approach to the problems, it was to some extent the abdication by the political scientists, and possibly to some extent it may have been that the wrong kind of political and behavioral science people were there. The so-called "Social Science" Department was primarily—I mean that's what its title was, but it really was political scientists, particularly heavily people who had been involved during the war in the propaganda and information agencies. Now Goldhamer was a sociologist, but there weren't anthropologists. Some of those were brought in, but it was part of trying to understand the Russians, much as the attempt had been made to use anthropologists to understand the Japanese during the war. It may have been that the right kind of people just didn't exist in the social science side of things, I don't know. In any case, the social scientists stayed in Washington; they preferred to be in Washington rather than Santa Monica. They had two or three people in Santa Monica. They weren't involved in the studies, the main studies. They only later migrated out there and never integrated very well with the rest of the organization. So this whole area of "criteria"—how you should think about what it is you're trying to do, how to measure it and so on—ended up in the hands of the economists.

It also was true early on that the RAND economists gained a lot of ascendancy because the engineers, who had been doing this on their own, were such idiots.

Once you looked at them [the criteria defined by the engineers] logically, they were demolished. The engineers just were no competition for the economists.

I think it's a deeper question about why it was these theories of deterrence and so on ended up the way they were, so heavily focused on essentially assuming rationality of calculations. Now there, people like Schelling may have been very important, but that was true to a large extent of Albert Wohlstetter and all these other people who were very much of the…

Q7: Yeah, you haven't mentioned Wohlstetter and Bernard Brodie and William Kaufmann, which I think are people associated with sort of the "golden age" of RAND.

Marshall: Yeah. Kaufmann was more involved. Brodie was off writing a book [*Strategy in the Missile Age*] and never got so much involved.[50] Actually, the view you mention is looking at it from the outside. They wrote these things that got published in journals, but the real guts of what was happening was in classified reports by people who did write in those areas. So I think people on the outside have a very, very distorted view about that.

Now, Albert was very important because of the inventions he made or approaches he had to how you ought to do studies. I think the most important one I mentioned in that *festschrift* thing [*On Not Confusing Ourselves*].[51] You see, again, the engineers when they started, were mathematicians who did the early big RAND system studies and tried to lay out, "Well, there are these 24 alternative designs of airplanes," or whatever it is, "and then here are these models of campaigns, and we try each one of them, we run through these calculations, and the one that does best, given all these criteria, is the one we pick." The problem early on was one wacko criterion in many cases that was not logically very good. But they had constrained themselves by creating these 24 cases and then doing that and then that's the end of the study.

Albert adopted a very different strategy, which was an incremental one. You have what people think they are already going to do, and then you design some alternatives, and you run it through a whole bunch of different calculations, and as you're going through it you begin to learn how to design new, better alternatives. And by sort of successive approximations you get better and better combinations of characteristics of a program that might be significantly better than the official listing of the initial program of these organizations, and that's what in the end you propose. You're not constrained by what you imagine or guess at the beginning as to what combinations of things are good, and you can also try a wider range of contingencies and assumptions and test for which alternative is the best. And he was very cautious in looking for an alternative that had dominance over the official, approved program. It's a much better strategy for doing studies. Also you don't make up your mind initially about what the criteria are. You start with some and add others as you think of different aspects or ways of measuring what you want.

Now, that strategy has a separate difficulty in the end, in that one of the failures, I think, on Albert's part was not to take account of the fact that what you're trying to influence are organizations. And there was the danger that in some cases you would be unrealistic about the advantages your proposal would have, because as it would really be implemented it wasn't going to have those savings. For example, one of the things that he was much involved in was various kinds of hardening, the protection of systems. You could say, "Alright, people on submarines live in so many cubic feet, why can't we have submarine levels of austerity, so to speak in protective structures on land?" So some of the things would come out being more austere and minimalist than, in fact, you could ever expect the organization [indistinct] adopted it was going to do these things, wouldn't do it to those standards. So the advantages you claimed realistically weren't going to occur.

If you were going to try to persuade an organization to do something very different, you had to have a whole implementation strategy that went along with the proposed approach; you couldn't stop with just the proposal, "Do X." You needed a whole strategy for making something like "X" actually happen. Later on—that's more a matter of the '60s—the verbiage was coined that there's a "missing chapter" in almost all the RAND studies, which is the implementation strategy, given that what you're trying to influence are these organizations.[52]

OK, the other thing that I thought was very important about RAND, particularly in the first few years, and I think it's probably true of the country as a whole, was these initial notions, which I came back to in '69, of this long-term competition. That is, that the way you really need to look at our situation, had to look at our situation, was that we were in this extended competition with the Soviets, and that we had certain strengths of our economic and technological capabilities and so on, and that we needed a strategy and policy for engaging in that competition. I think an interesting question is why that thinking kind of disappeared after the mid-'50s, and I don't know the answer to that. But it just kind of disappeared from RAND thinking. It was only later that I revived it in '69. But it was very prominent then, I mean people thinking in a more strategic vein. And it had an impact on that piece that was written called *The Next Ten Years,* written in '54, which was an attempt directed inwardly at RAND to say, this is what the situation is, this is the longer-term picture, these are the problems that people ought to be working on.[53]

Q2: Was that your piece? You did this piece?

Marshall: Right after I returned from Chicago, I was asked to join a group that Charles Hitch and John Williams, I guess, were the chief people on, but it was Hitch, Williams, Brodie, and myself, Alex Mood, James Digby, I guess, and it was called the "Strategic Objectives Committee."[54] I'm not sure what triggered it getting started.[55] I think the idea was around and I came back and I was asked to join it. [gap for tape change] ...strategic level and to provide a context within which the thinking about what specific projects people at RAND ought to work on could proceed. Partly, it was triggered, I think, by an insider perspective that I guess had come to fruition just before I left to go out to teach at Chicago. See, when I first went to RAND, and I think it was not specific to RAND, the whole notion was, if there's a war, it's going to be more or less like World War II, in the sense that the Russians invade Europe; we hold, hopefully, at the Pyrenees; ...we cripple their economy by strategic attack in the first few weeks of the war; and then it's a mobilization race between ourselves and them. We win that and we push them back and retake Europe and we do in the Soviet Union and so on. That was the image of the war.[56]

Two things happened. One, thermonuclear weapons appeared and the stockpiles were growing, and then the notion was that we've got to shift. You're going to essentially have to fight with what you've got, because they're going to hit us and we're both going to be crippled. So the only things that count, that largely are going to count, are what you have to begin with.[57]

Q2: But the shift came after this '54 publication?

Marshall: No, came before.

Q2: So it was before our thermonuclear weapons.

Marshall: Well, theirs too. Our first big test is in '52 and they have a test in '53. So the shift was already taking place. I went away to Chicago and then I came back. Partly the stimulus for this thing probably was, "Well, people haven't adjusted, even at RAND, to this; we've got to really reformulate and say how we ought to look at this whole world that we're going to have to face."

Q4: At this point, in this period of RAND's existence, given your comments about the contrast between the realities and the perceptions, were those of you who—the few of you that had insights like even Loftus, into the best we could figure out about the Soviets, did you have very different views about the risks of war than the people who did not have those insights? Or did you all differ in your opinions on the risk of war, or were you all basically in agreement, even if your perceptions were wrong in historical retrospect?

Marshall: I don't know. I'm not sure enough about the different perceptions of the risks of war. I did run two samples of people at RAND, but it was much later. One was done in the very late '50s, I think ['59].[58] I put out a questionnaire and had people fill it in to give their own estimates, but with several weighted answer questions so you could check what their estimates should have been. At least in the late '50s, the RAND average view was that there still was a very high risk of war. People were making estimates of the probability of war within the decade. The answers were, as I remember, just under 20 percent. But there was a big spread. There were some political scientists—maybe it was one of their good points—who tended to be on the low side. But then there was a big shift in the '61, '62 period, especially after the Cuban missile crisis.

Q2: Wasn't that shift after Sputnik and the Gaither Committee and all that?

Marshall: No, I don't think that had a lot of effect. I think the height of concern about war really was probably earlier, in the mid-'50s.

Q2: So nobody got sucked into the "missile gap" argument at RAND?

Marshall: No, that's not right. I think most people did. Again, there were a few of us, like Loftus and I, who had better connections in the intelligence world and really had a different view of that.

Q7: Wasn't it the Air Force intelligence community that was concluding that there was a "missile gap", and there were other elements in the intelligence community, like CIA, that were much more skeptical?

Marshall: It's a complicated picture. The Air Force had the very high estimates. I forget the year, but, like, they were estimating the number of Soviet missiles at 500, or the Soviets were going to have 500 missiles. The Army and the Navy essentially said they had one or two or none. The CIA, in its official position, came out kind of cutting the difference. If you talked to the CIA analysts, they were with the Army and the Navy. But officially, as an organization, it didn't take that view.[59]

Q7: When you were looking at your scenarios and strategic options, "this is how the war might go," there was, I believe, at least until 1954, some serious discussions in senior circles, that is up to the point in Project Solarium, of preventive war—

Marshall: Yes.

Q7: —against the Soviet Union—

Marshall: Yeah.

Q7: —and the so-called strategy of "massive retaliation" in the eyes of the Air Force was really massive preemption. To me, that would have had a major bearing and effect on your judgments about how war would have played out. Was that factored in? I think that would be interesting.

Marshall: I think, in the first place, there were several people at RAND, and including some of the visitors like von Neumann, who were very much pushing the notion of preemptive war.[60] In fact, there are some interesting documents, one by this man John Williams, called *Stalking the Tiger* [*Hunting the Tiger (And Other Aspects of the Active Life)*]. But on the whole, the RAND people were very much against it.[61] For reasons that I mentioned earlier, there were a bunch of people that were absolutely persuaded that we didn't know where the hell the targets were, so you couldn't do it. Then there were a whole bunch of others who, I think, felt for whatever reason, whether political or emotional, that it just was not right to even consider it, and so multiplied arguments, like, you couldn't do it. Loftus and I felt you probably could do it. In fact, later, around '59 or '60, we had sort of convinced Kaufmann that the option was feasible. In fact, he was one of the people who did become kind of converted to our views, and he undertook this study of counterforce.[62] I think, until really the early '60s, that probably the U.S. had a real option.

But it didn't, for a variety of reasons, get much consideration at RAND. The Kaufmann thing happened very late, and was not—During most of the period, people sort of either, as I say, had this view that it was just impossible, we didn't know enough...

Q7: Another interesting thing, I think especially in terms of what we're seeing now, if memory serves, didn't Schelling write, at least in the public literature, a paper during the early '50s in which he reported a very high probability that you would see large-scale proliferation over, say, the next 15 years? Did that ever come into play at RAND in terms of your thinking?

Marshall: People thought some about proliferation, but I think it wasn't anything like that early. I suppose very early people thought it was going to happen fairly quickly, but I don't remember a lot of attention paid to that.

Q4: I sort of took you off the trail during this last section. You had just said something about reaching the conclusion that the Russians had a number of thermonuclear weapons, and you started building towards this image of how the war might have evolved, changed dramatically. Is that sort of, was that a shift in thinking at RAND that was sort of universal and sort of quickly adopted, or was it a struggle for people to come to the notion or even to think about the war in very…

Marshall: I think the answer is yes and no. I think most people refer to that… Although, going back to the reasons for this Strategic Objectives Committee, my conjecture is that it says something about [indistinct] that the Strategic Objectives Committee was motivated by the fact that some people felt that the organization as a whole wasn't really adjusting, adapting focus.

OK, let me pick up again. One other thing that came out of the '50s is this matter of the value of gaming or gaming as a tool. In an interesting way, it's not based on any look at the military history of war-gaming or the value that the Germans placed on it and so on. But it really, in an odd way, I think, at RAND came from this game-theory effort that had such an emphasis in the early '50s. Because it quickly became apparent that, except for the simplest games, you just couldn't analytically do very much with…to analyze games.

There was a man named Merrill Flood, who proposed that, "Human beings have great skills at playing games. If they play them frequently enough, they do devise good strategies. If what we're looking for are good strategies, why don't we just formulate games and play them, and make use of these skills that people have?" And so that was done, with progressively more and more complicated games.

By this period around '54, there was a game that had been put together by this fellow Alex Mood, who was a statistician, and possibly some other people. The game I think was called "SAW," "Strategic Air War" game or something like that.[63] It was used as part of this Strategic Objectives effort to look at large-scale nuclear exchanges and so on.[64] I remember that one of the things that happened, one of the few times I remember anything like this, was Hitch corning to me just prior to one of the meetings of the Strategic Objectives Committee to make sure I was going to vote right when this issue of what role in the final report should be claimed by the results of these games. Because, again, they were kind of crazy. The criteria were mis-specified initially, and a man named Jack Hirshleifer and I developed a more complex measure of worth of the outcome. Initially, these mathematicians specified the way you measure the outcome of the game was where the front line was in the middle of Europe—where what was happening in the war was both the U.S. and the Soviet Union were being destroyed. So the early plays didn't make a lot of sense. You had people trading half of the U.S. GNP for a few miles of movement in the front line. So at any rate…

Q2: What was your better measure of effectiveness?

Marshall: We got the U.S. GNP and other things in the score and gave it a big weight and… Right. Look, my life seems to me…that it's kind of doing the obvious. Seems to be something that is important. Because it's just amazing what people do. No reasonable person is going to actually do this, but you have these people who have this game, and they decide this fairly simple criteria and away they go.

Q4: Are you at some point in this program going to devote some time and energy to this issue about very bright people coming up with new ways to look at problems? And what happens when you get off on this pathological…

Marshall: I don't know. But I think that one of the reasons this gaming thing took off at RAND, and later there were some interesting games—Herb Goldhamer ran one of the first of these political-military games I know about…[65]

Another item I had down here was counterforce, which we have already discussed. And the preemption business, which was discussed a lot at RAND. Although the thing that was striking to me—let me say one more point about it—was, for a place like RAND, how much the positions people took were determined on emotion. Because most of the resistance to it was just that it was seen as… I don't know whether it was immoral or this and that… Yet it seemed to me that the logic, if you're just going to talk about strategy in a very cold-blooded way, not that that's necessarily the final answer, but the logic of the people who favored some kind of preemption was just enormous…given the views other people had. If you have some big technical advantage, and you can deprive the other guy of it, and you think it's central to the whole situation in the world, you have to have more than, "We've never done it before" as an argument. Anyway…

Q2: What do you attribute that to?

Marshall: …it was striking how logically weak the oppositions were. Most people were against it, but didn't have really well thought-out strategic reasons for not doing it.

Q2: We used them in Japan, so why is that—Is it because we used them in Japan that…

Marshall: No…I don't know.

Q4: To what extent, given your experience subsequently, do you think that really mirrored also the way administrations think?

Marshall: Clearly, people have been very reluctant to use these things. There's no doubt about that. But…

Q7: Not just restricted to nuclear weapons. The same thing is true…

Marshall: The other thing I suppose is interesting about that, although I think it sets in later in the '60s, when the main problem was this sense that, in the end, the Russians were going to win. That certainly was like Henry Kissinger's view, to some extent. There wasn't perhaps as much of that in the '50s, but if you really took a long-term perspective, you might have even redoubled the arguments for preemption. But it was striking how poor the arguments against preemption were. Bernard Brodie wrote a paper on the subject, and I guess one ought to go back and reread that, but most of the argument against preemption seemed to me very, very weak.

OK. We've already covered some of this. Let me go back to the work with Loftus, which really is the central matter, I think, for much of this period. I told you about how he comes to RAND probably around '53, and when I had gotten back after the six months or so teaching at Chicago. I covered the fact that we early on had come to agreement that '45 to '50 is very important for an understanding of the Soviets. We were also convinced, as I mentioned, early on that we could do better than the RAND people were doing on forecasting.

See, one of the things that was going on at RAND at that time was that the intelligence community did not make longer-term estimates themselves. There were a few things, but they were very limited. The RAND people who were looking at the future, designing programs for the future, therefore were put in a position of making up their own forecasts. Loftus and I looked at the things that they were producing and thought that they were just awful, either because they had this failure of this rationality model that we mentioned earlier... You had the political scientists, who by that time were, as I said, focused on reading what the Soviets were writing and not noticing what they were doing, and they were not a basis for anything. But then you had people probably over-inventive on the side of the Soviets. An example of which was—I think one of the things that, at least at RAND, very much set back analyses in the air defense area—And something you have to remember is that in the '50s, the U.S. had a very vigorous program to build up its air defenses. In fact, I think that by the end of the '50s, we were spending more on air defenses than we were on the strategic forces.[66] But as the RAND people got into this in the middle '50s or so... And by the way, they did for a while have a very interesting behavioral scientist, a psychologist named John Kennedy, who developed some very good simulation stuff, which then spun off into that...

Q2: SDC?

Marshall: ...SDC [System Development Corporation], which did some very interesting things.[67]

But the people, in doing their analyses, were super-inventive on the part of the Russians. I mean, at the radar perimeter, they had the Russians' bombers coming in with thousands of decoys and cruise missiles. In fact, cruise missiles initially were largely invented as bomber decoys. That's the way our thinking was. It was only in probably the early '60s that people like Al Latter and company said, "Why don't we put bombs on these things?" which was a heresy in the eyes of the Air Force. Anyway, they made the problem so hard that you just couldn't defend the country. You always kept getting the answer that it was nearly impossible. It wasn't clear that that was really ever true.[68]

Anyway, Loftus and I didn't think much of what they were doing, so we undertook our own project, probably after another thing I worked on a lot, but which was not that intellectually important—work on strategic warning.[69] We started this project called "Sovoy," In which we offered to start making projections for the rest of RAND.[70]

Q4: Of the Soviets?

Marshall: On the Soviets.

Q2: Using classified information?

Marshall: Yeah.

Q2: I mean using everything you had.

Marshall: Yes, right. In fact, we had created a special arrangement with the intelligence people at the analytic level to give us much more material than they otherwise would, because of our own connections with these guys, especially Loftus's. Then we got to this problem of, "How are we going to do this?" Then you get to the things that I was mentioning earlier. Loftus and I already believed that you need to know all about the history of these organizations and where they come from. It's an important fact, for example, that the Soviet missile people [the Strategic Rocket Forces], when they were formed, took over one of the existing academies—What do you think it was? It was the artillery academy. So you've got to know about these guys, where they're corning from and so on. And then I have this, as I said, blindingly obvious insight of, "They're organizations. That's what we're trying to predict the behavior of." Then you get into, "What the hell do we know about organizations?"

So I start to read—this all starts about the spring of '57, probably, or '56 sometime—Chester Barnard, Herbert Simon [*Administrative Behavior*], Simon and James March [*Organizations*]. Later, when it comes out, Richard Cyert and March [*Behavioral Theory of the Firm*], which was probably the most influential book for me. And then later, when Crozier comes out in '63 with his book, I read that.[71]

Q3: *The Bureaucratic Phenomenon.*

Marshall: Yeah, right. And this really was the central part of the stuff I worked on, except for when I went away to the Gaither Committee and so on. But until I left for Paris, my work was on how do we do a better job of understanding the Soviets, predicting them, thinking about them [indistinct], with this matter of, "We've got to understand the allocation of resources, what their constraints are, what the long-term costs are of the level of effort they put into the military," and so on.

That reminds me of another thing I did around the late '50s. I did a lot of work on the accuracy of our cost estimates of our own programs at various stages. You find out the early estimates were pretty lousy.[72]

One other thing is this collaboration with Herb about deterrence and war-fighting, which looked at the contrasting utility functions of ourselves and the Russians at various stages of the exchanges that may go on.[73]

So then I go away to Paris. Pleasant living, no intellectual content, and learn more about how abysmal organizations can be, how whole organizations can exist and do practically nothing with lots of people paid a lot of money.

Q3: Were you sent there or...?

Marshall: Yes. What happened was Secretary of Defense Robert McNamara came to an agreement with Dirk Stikker, who was then the Secretary General of NATO. It was in '65 that there was going to be a five-year review process of the programs within NATO, or there was then. McNamara persuaded Stikker that, "Look, this thing is corning up in '65. What we should do is have some studies done so that when we come to this review, we have some basis for deciding what's good, what's bad."

Q4: Revolutionary concept.

Marshall: So he approached RAND to put together a team of people to go over. It was headed by a guy from RAND who was then the head of the Economics Department, Burt Klein, and included me, Fred Hoffman, and Oleg Hoeffding, and Peter Szanton, who was in the government at the time, came and that was it. We were to do these collaborative studies with the other countries.[74] I worked with a succession of French colonels who, for a variety of reasons, were the head of the thing. The study that I did looked at the actual stockpiles of ammunition and other things, like tanks, that were held as replacements for attrition or for expenditure in war. There was a set of standards that people were supposed to have, so many days of supply and so on and so forth. This was an across-theboard study of, "What did people actually have?" Right.

Q3: I was just wondering if you volunteered for this or were…

Marshall: No. I was asked because by that time I had done some studies on NATO problems.

When I returned in the spring of '65, the thing I then proposed was that I organize a program of study at RAND on organizational behavior and how our knowledge of that could be used in a wide variety of things, including this business of understanding the Russians.[75] I was allowed to do that, to use not only my own time, but to get three or four other people to work with me. Because of that, I first met Jim March of Carnegie-Mellon University, and from him got leads to some students who might be recruited. Pat Crecine came for a year because of that. I went to Harvard and talked with Dick Neustadt. I was told I ought to do that because of some of the stuff he was doing. Out of that came the creation of the so-called "May Group" chaired by Professor Ernest May, which began to meet probably in the fall of '65, to talk not only about organizational behavior, but this whole business of how do you understand the behavior of governments, governmental organizations. Graham Allison was his research assistant, or some associate of his, and was the rapporteur of the group. And out of that in some ways came the book *Essence of Decision*.[76] When I started to look at this business about organizational behavior, I looked around and located this cluster of people at Carnegie that was intellectually attracted to it. Then I went and talked to the people at the Harvard Business School to see what they were doing about trying to understand organizational behavior. It was there that I came upon these notions of strategy that they were developing. And both of those things had, I think, a really big impact on what I did later.

The other thing that happened about that time—some of it predates my going off to Paris—was that James Schlesinger had come to RAND to spend a summer in '62, I believe, and we became acquainted and friends. Then he came back as a full-time person just about the time I was leaving for Paris [summer 1963]. But after I returned, I spent a lot of time talking with him. One of the things that the two of us read, about somewhere in that period, was Robert Ardrey's *The Territorial Imperative*.[77] And I got very interested in, "Here's another way of thinking about human beings and how they behave." Still later, and not so much based on Ardrey's stuff, but on the work of other people, I became further interested in these nonrational aspects of behavior. And I arrived at this notion that maybe RAND is not a rational animal.

Q4: That's it right there.

Marshall: And there are interesting things to be said about that. Also, Schlesinger and I began to talk, probably in the '68–'69 period, about the implications of all of that stuff that Ardrey and others were doing by that time. I'm thinking about the psychological and political impact of military forces.

So it seems to me, intellectually, the organizational, behavioral stuff, plus Crozier's connection of the organizational stuff to culture, and this ethology, led me to this low opinion or critique of the standard deterrence theories, the whole rational actor view. And, of course, I obviously, from the '50s, was already a little skeptical about rationality as a way of thinking about what people do or predicting what they do.

Then, I don't understand exactly now why I wrote that '66 paper on assessing military power [*Problems of Estimating Military Power*].[78] I was encouraged to do it by Bill Kaufmann for some political science association meeting here in Washington [the paper says, "American Political Science Meetings in New York, September 6–9, 1966"]. 1966 seems early to me, at least in terms of my memory, because—maybe it happened earlier than I thought—Fred Wikner, who I don't know if some of you know, visited RAND to obtain analytic support for his work. There had been created here in the Defense Department a Net Technical Assessment function. He came to RAND, but my memory of it is that this was probably '67, '68 [not '66, when *Problems of Estimating Military Power* was written]. And I started to do a little work for him on a contract about, "How do you do net technical assessments? How would you really think about that?"[79] Anyway, the chronology here may be wrong, and something that one needs to check.

Q2: He came out to RAND in '67, '68 and asked you to start working on this?

Marshall: Yeah.

Q2: So it was created in '64, '65?

Marshall: I don't know. That's something that needs to be checked. I don't know what the chronology is. My guess is the earliest it probably was created was '66.[80] But I don't know whether one of the reasons for writing that...paper on assessing the military balance [*Problems of Estimating Military Power*] was to support Wikner. I just can't remember why I did that. I guess something I need to do is to go back and look at the papers I wrote for Wikner.

Also something that happened about this time is Harry Rowen became the president of RAND, sometime in '67. Jim Schlesinger and I used to spend a lot of time with him because both of us thought there were a lot of problems at RAND and that maybe he was going to fix them.[81] It didn't happen that way, but... One of the things that did was lead to this creation of a new post of Director of Strategic Studies—that is, dealing with strategic forces—that Schlesinger took over in probably the spring of '68 [1967].[82]

Then Schlesinger left RAND in very late '68 to rejoin some of the transition teams in the turnover that was taking place to the Nixon Administration, and then left permanently in January of '69. Rowen asked me to take over the position of Director of Strategic Studies, which led to my doing a lot of thinking about the strategic program. Out of that came this thing that I wrote on the long-term competition [*Long-Term Competition with the Soviets: A Framework for Strategic Analysis*].[83]

I guess I was moved to do that because when I looked at what people were working on at RAND with respect to strategic forces, it just seemed kind of a mishmash of all kinds of things. So I tried to address the question, "What if you really had some sort of program that had some intellectual strategy to it? What should it look like?" And so I addressed the question of, "How should you think about strategic forces?" Then the things that came back to me were, first, these ideas from the early '50s of, you're in an extended competition with the other side. Then, from the Harvard Business School stuff that I had been spending a lot of time on since the middle '60s, comes this notion, you should have a strategy for this competition. And one of the things you want to focus on are your strengths and weaknesses, his strengths and weaknesses. One of your objectives is to drive him out of business. There may be some risks in doing some things too hastily, and you want to look at that. And then, what you're talking about are organizations, and they have all kinds of built-in vulnerabilities, problems that they can't correct very easily, and so there may be things you can exploit.

I just had this sense that this is a terrific new way of looking at these things. At one level, it all seems to me very obvious—yes, you should look at things like that. And it's odd, as I say, that some of the people at RAND in the early years had some of these views although they weren't integrating them with these ideas on the behavior of organizations and so on. It's just odd to me, as I think about it, that this kind of perspective just disappeared from the middle '50s until the late '60s.

And so that's what I have to say on that, I think. We can take it up later. But let me just run over a couple of things.

What about the contrast with the mainline thinking at RAND? One contrast is, and we've talked about it already: I believed counterforce was possible. During much of the time, people at RAND didn't.[84] Even the shooting down of the U-2 didn't really make people feel that it was entirely possible, although it shifted the balance of argument.

Q3: Could you expand on that a little bit, the shooting down of the U-2 and its effect?

Marshall: It made public the fact that you had this reconnaissance thing. And, in fact, I remember Alain Enthoven stopping me in the parking lot and saying something like, "Now I understand why you've been…"[85]

Most of the rest of my outline we've already talked about. Let me just end with some comments about the national level or the more general things. It's something I've talked a lot about. I think World War II and the kind of generation of people who came out of it, who played an important role, already had some important level of participation in it, are very important. It had a big impact, I think, on the character of the thinking of the people at RAND. It had an impact on those who had come from that the World War II experience, somebody like Hitch, for example, or people in the Eisenhower Administration—I think there were a lot of the same people. I think it needs to be articulated and understood, but I just have the feeling that people who went through that in some responsible position just came away with a sense of, life is serious, you make decisions, if you have a problem you think about it, and then you decide what you're going to do.

The reason I raise that is I think that in the Eisenhower period, when you had this big shift taking place that we talked about earlier, because of the growth of the nuclear stockpiles on both sides, there was a whole succession of special study groups and so on. It seems to me symptomatic that soon after a problem was surfaced, a serious group of people was put together to look at it and they came up with suggestions. The suggestions were not always followed, but the issues were addressed.

There was also in the Eisenhower period this whole notion of an extended competition. For example, the question of, "Should we overspend?" There was a longer-term perspective, particularly in the early years. I think, given the performance of the U.S. government from the early '60s on, the work in the '50s is impressive to me. Maybe if you look back in detail it wouldn't come out that way, but that's the way I feel.

It's certainly different in my view, than in the late '60s, early '70s, when you have this final kind of coming to terms, in a certain sense, with strategic parity. Here we were, we had an overall strategy relative to the Russians, particularly in our alliance arrangements and so on, that in some ways depended on an advantage in strategic weaponry. Parity happens, and there is no reconsideration of, "Now we have a different situation, should we do something differently? Do we now have to have a different balance in conventional forces?" Never even addressed. And when the people were writing Henry Kissinger's memoirs on this period, I asked them, Peter Rodman and Bill Hyland—and I was there part of that time—"Was this issue ever raised?" Their answer was, basically, they saw no record of it.[86]

Now it seems to me, in addition, of course, that another period that deserves a lot of study is the period around '63, '64, where in the aftermath of the Cuban missile crisis I think there was just a big shift. I don't know whether it was that people found that experience on reflection or even at the time so terrifying that they never wanted to be anywhere as near as close that they thought they might be to real war, but I think there was a big shift. Kennedy came in on the argument that Eisenhower wasn't doing enough, that we've got to up the budget, we've got to compete more effectively, and so on, and all of a sudden there's this switch and we're into arms control, settling for parity and even, to some extent in the administration, but certainly later in the Congress, there were these regulations, stipulations, limits on the accuracy of our weapons so that we don't threaten the Soviet silos.[87] There was a whole set of decisions which are, in my view, anti-strategic or represented a loss of nerve. And so there was a period in there when we were, in some sense, allowing the Soviets to impose costs on us by the increases they were making in the accuracy of their weapons and retrofit programs and so on, increases in hardness and some other things, efforts to begin looking at global missiles and so on, while we were abstaining from imposing similar costs on them as "unfair," "destabilizing," and all that horseshit.

So I think—I'm never going to do that—but I keep proposing to people that they really look at that, these special aspects of the people, the generation from World

War II, this kind of critical period in the early '60s, when I think there was a kind of flip-flop, switch…loss of nerve, really.

Then there was also, when you get to Kissinger—it isn't that he's to be picked on—this wide class of people with this pessimism about the U.S. competing with the Russians over the long term, which is one of the fruits of this absolutely awful effort by not just the CIA, but almost the whole of academia, on the Russian economy—its problems, the size of the military burden, and all of those things. I think it's really true that Henry saw himself, with what he thought of as kind of an idiotic populace, making the best deal with this power that was going to dominate inevitably. The thing that differentiated Schlesinger from him was…

In fact, I remember in a *Time* article [11 Feb '74] there was a little box in the middle of the article which had one of the most significant statements by Schlesinger—but most people don't pay any attention to it—that he saw himself as a "revivalist," getting people's spirits up.[88] Why the hell should we lose? He certainly didn't think that we were going to necessarily lose. But there certainly was a lot of that.

And I think you can't understand why these analyses of the Soviet economy and other things were so lousy unless you go back to the '30s and the '40s and the generation of people who came along and dominated much of the economics profession and certainly did not believe in free markets.

Q2: How would you weigh that in relation to this sense that there was a sort of lost faith in America and American ideals after Vietnam? Is it your sense that—Which one do you think was most important in this kind of fatalism or pessimism about being able to compete with the Soviets—is it the loss of faith in America, or these lousy estimates on the Soviets, that they could build, build, build and it didn't cost them anything?

Marshall: I don't know. I was never affected very much by this, the Vietnam thing. I'm not a good person to judge how others felt about that. But I don't think…I paid next to no attention. But I don't have a good feel for it. It certainly shook up people. I don't know. I'm not a good judge of that.

Q7: But you did write a paper. There was a short paper involved, a rebuttal to somebody, maybe Kaufmann or somebody else, talking about Vietnam as a military failure, a failure of military organizations to adapt.

Marshall: Yeah, right.

Q7: …a RAND paper.

Q3: There was a Bob Komer paper.[89]

Q7: Maybe it was a response to Komer.

Marshall: Yeah, right.

Q7: …you had taken issue with Komer…

Q4: I'd like to come back to your point about the economists.

Marshall: Yeah.

Q4: Let's see if I got this right. Some of what happened in the '60s, was perhaps determined by the fact that the economists who were beginning to predominate at that point in time were really trained in the Depression period with the notion that the free enterprise system doesn't work. Planned systems do work. Therefore the Russians, because they have a planned system will work better than... Is that kind of...

Marshall: Yeah, that's right. See, it's a goddam scandal, I think, the way that G. Warren Nutter was treated. When you read his book now—it's true he spoke more vigorously—but the book itself is the mildest statement of ... poses the question, What reason is there to believe that in the longer term the centrally managed economy will grow faster than the free economy?[90] He put a lot of information on the table that suggested there wasn't any reason. But he was seen as an absolute ideologue and really excluded from first-rate universities. He just wouldn't be considered for Yale or places like that. And for that reason, this guy ended up at the University of Virginia. It's these other guys who are the ideologues, the whole bunch of them, largely, and even at RAND, somebody as sensible as Hitch. I think it was Allen Wallis who was telling me about how the criticism that was made of the RAND economics effort on the Soviet Union...which was a whole bunch of people, Abe Bergson-trained people, violently...treated any criticism as being ideological, when the fact was here were these guys with all their few suppositions....

One of the things I guess I forgot to comment on: Here Loftus and I were trying to get this work done, do this projection. We would come to RAND economists analyzing the Soviet economy and ask them questions. In the first place, they had no answers, or didn't even want to work on any serious problems. And they claimed, "Oh, we can't make answers to questions like that." I guess that led me, along with other times, to have this view that poor answers to good questions are a lot more valuable than good answers to poor questions. These, we thought, were the good questions that people really ought to be working on. But you couldn't get any of those guys to even address them. They had their things. These people were not only worthless on the few questions, but had over time, I think, perpetrated this view of the Soviets as constantly growing, a very inflated view of their GNP, while simultaneously underestimating the size of the resources going into the military.

See, it was explained to them, "How was it that in this somewhat smaller economy, let's say half ours, they could spend six percent of their GNP and have a military establishment which in many ways was competitive with ours?" We were spending five percent of our GNP. Well, they would say we don't have complex economic arguments about how...though they were more efficient in those industries of that sort.

Q2: Capture economies of scale.

Q4: In reference to your comments on Kissinger: To what extent were Kissinger's views a reflection of his predictions on the Russians, as opposed to his sort of pessimistic views about the Western side...

Marshall: Combination of both. Certainly pessimism about us and...

Q7: Especially about the Congress.

Marshall: But he also made this argument that they're this, in some sense, young, vigorous, expansionist society, with faith in what they're doing.

Q3: I have one quick note on this subject of the economists and a comment on your talk overall. In talking to people of your age, who reflect some of these views, have some of these views on the Soviet economy...which is not so much that they went through the '30s and saw the failure of market economies, but that they went through the '30s and saw what the conventional economists had said about the Soviet economy and how poor it was...

Marshall: Yeah.

Q3: ...and then they saw the expansion of World War II and these vast amounts of military production of the Soviets. And then the view was, to give my father's version, "All the goddam economists were wrong. They said the Soviet economy was worthless, wasn't worth anything, and here were these 40,000 tanks that came out."

Marshall: Right.

Q3: So it was essentially that that economy, in fact, had triumphed. That's the footnote.

The general comment is, I think this was great. This talk is great for one primary reason: The conventional view of the relationship of the RAND analytical community to the American defense establishment was that you have a bunch of military officers who, for a variety of reasons, were pathological and ignore reality. And yet a bunch of brilliant, objective scientists at RAND who [indistinct] perceived reality, and you proceed to set the situation straight. What you portray is a picture of analysis as a social phenomenon, in which you have communities of people at work on various problems. And reality and truth in these social communities may be [indistinct] a generation. The Great Depression is one of them. World War II is another event. The explosion in thermonuclear power is another event. And the Cuban missile crisis is the fourth. There's obviously others, but those seem to be the four events over a period of maybe 30, 40 years. And once every 10, 15, 20 years something really comes and knocks in on these communities and kind of forces them to sort of look out [indistinct] and reorient them slightly. But then after these shocks, they...something happens and then they behave in terms of their product in ways which are determined by the nature of their internal social dynamics more than it is by... This is not a perfect explanation of what [indistinct], but it tells you more what's going on than the model of, analytical scientist looking at a fact or reality and analyzing it and coming to an objective [indistinct].

Marshall: Yeah.

Q3: As I listen to the whole of your presentation, I think you're not only kind of setting the record straight with regard to the '50s and '60s, but you're also presenting here a fairly coherent view of how it is collective intellectual activity takes place in the context of these [indistinct] states. It's that there is a social theory of intellectual output.

Marshall: Yeah.

Q3: Not an unimpressive accomplishment. But, the thing to do is to take—OK, is that right or is that wrong? OK, I think that's the view which you have used to structure. You might want to go back and think about, "OK now, is that absolutely right, or how would we go about thinking about whether or not that has captured the whole picture or some important part of some other thing?" [Indistinct] But as it is, I think this is fantastic.

Q2: What do you want to do with the book? What do you want... You said you want this to be an intellectual history. But what do you want beyond that? What do you want it to do?

Marshall: I don't know if I really want it to do anything. I don't have any great goals for that.

Q2: I'm not talking about what kind of sales you want to make...

Q4: The last chapter, I think, remains to be written, and it's the question that [indistinct], given that you capture that, what does that enable you to say about the future? Don't trust societies except when you can shake them?

Marshall: Yeah, in a way, this thing today is a little too down on RAND because I think it really was a fantastic place, and lots of good things do happen, and it was a place at which people thought about some things where they certainly were on the right line of many things, like this whole business of hardening and protection. I mean the Air Force was just crazy. As Herman Kahn used to say, hardening is like, supposing a man just hates aluminum...

Q3: They hate the color gray...

Marshall: Yes, right.

Q3: ...and you come up with this great new stuff which is twice as strong and half the weight, and you say, "I just created...," and they say, "But it's gray." And then you paint it blue. But they say, "It's still gray underneath."

Marshall: That's exactly right. You know, that's the way they felt about hardening. Putting one of these pretty, almost bird-like things [bombers] into a shelter *underground!* "Jesus Christ!" "What's that for?"

So there's things like that or—which I didn't talk about—the kind of questions that people were raising about the targeting policy, particularly in the early years at RAND, and which Brodie joined in.[91] Right, I mean it had very little impact on the notion that...

Q3: That was the thing that you did more of in the talk at Harvard, what is the impact of what's going on at RAND on the [indistinct].

Marshall: Yeah, right. But you're right, I mean about certainly this kind of, like, social processes of analysis.

Q7: It struck me... I guess I have two observations—what you said about the social phenomena of the organization, and what you said about these flashes of insight of the obvious. And I think the preceding discussion provides a good explanation for why the obvious, or why innovation, whether it's military or otherwise, is often very reluctantly accepted until the evidence is almost overwhelming, because "It doesn't fit my model" or "It doesn't fit my..."

Q3: From the standpoint of a card-carrying academic, let me tell you, that's how it works in the academic world—very slow.

Q7: ...the second is... You alluded to, but didn't go into much detail on the work that you did with respect to intelligence and strategic warning.

Marshall: Yeah.

Q7: And yet I think it would be interesting for a lot of people to hear more about that, and to make the link, as evidently you're going to do with Graham Allison, with what Roberta Wohlstetter eventually produced [*Pearl Harbor: Warning and Decision*].[92]

Marshall: Well...

Q7: Because if you talk about the World War II generation and what made it...

Marshall: Yeah.

Q7: ...Pearl Harbor...

Q3: She does have that acknowledgment...[93]

Marshall: I forget exactly whether I proposed the topic to her, or she had several options and I said, "Pick that one, there are a lot of things to be done there."

Q7: I mean that really speaks to the psyche of the generation that comes out of World War II in terms of avoiding another Pearl Harbor.

Marshall: That was very big, there's no doubt about that. And that's why you have, I think, all of this focus on withstanding a first strike. Again, you see, almost everything that came out of RAND was premised on the notion that we just were not going to start the war. And that's why this other, small cluster of guys said, "Wait a minute. Here's the situation. Well, why wouldn't we strike first? And here's about the time we ought to be doing it. Don't wait until it's too late." So everything was really focused on the notion that the other side was probably going to get in the first blow. And then you have to design against that.

Q7: You were also involved in the Gaither Committee, which is a direct reaction to what is perceived to be going on with the Soviets, and their ability to attack with little or no warning, and then Albert Wohlstetter writes his paper ["The Delicate Balance of Terror"]...[94]

Marshall: The main trigger for the Gaither Committee, even though it didn't turn out to be quite that...was the calculations that were being done by this Net Evaluation Subcommittee of the NSC that Eisenhower had set up in the early years of his administration.[95] With the rapidly rising stockpiles and the yields of weapons, you had to face the question, What are you going to do about this? So the stimulus was, really, how to defend the country. And the offensive side got...

Q7: Civil defenses...

Marshall: Yeah. So it was active defenses... Actually, Jerry Wiesner headed a group that was part of the Gaither Committee. One of these things was looking at ballistic missile defenses.[96] He turned "traitor" later, I mean left that... But all kinds of crazy things were going on. There was a set of people, also up at MIT, earlier that was very against thermonuclear weapons development.

Q7: Herbert York, wasn't he Ike's adviser at the time? He certainly wrote a lot of papers after the fact that he was against the development of the thermonuclear weapon.

Q4: In subsequent lectures, chapters, you're going to deal ·with intelligence and your role?

Marshall: Well, I'm going to get on to the '70s, and the '80s, and so on, so I mean the '70s... Yes.

Q4: About an hour and a half ago, you made a comment, you sort of said, there is a sentence something like, "Oh, and then there was my involvement in warning, but that wasn't particularly intellectually rewarding," or something.

Marshall: It didn't do a lot for me, intellectually, I don't think, as against these other things.

Q4: There is a path, that when you get into the '70s, I think you ought to go back to the '50s and '60s and bring that...

Marshall: Alright.

Q4: ...and make that a full picture, because I think there is a lot that you can say about, well, indications and warning, for example.

Q3: There is one aspect, which you did highlight with regard to its effect on you, which does have to do with intelligence, which is the one sentence that I remember, which is that "COMINT was very rich in those days."

Marshall: Yes.

Q3: And there is some stuff in the open literature about Soviet weaknesses in encryption practices and stuff like that. Now, obviously it's very difficult or very sensitive to think about how you do that.

Marshall: Yes. We're not going to do much writing on that.

[Indistinct; multiple voices]

Q3: But yet your access to that form of intelligence played a very major role in...

Marshall: Oh yes, absolutely. It all really happened because of my targeting work, initially. Then in the summer of '52, the guy who ran Air Force Intelligence in Europe [Gen. Millard "Poppy" Lewis] approached RAND and wanted some work done on various topics, and one of them was strategic warning. Hitch asked me and Jim Digby to work on that. I went and lived for five months in Wiesbaden, I guess, at that time. First, we wrote some stuff and had to convince some people in Washington before we spent four or five months in Wiesbaden working on that.[97] Then this, whole intelligence thing became a very big part of what I did, because I had some comparative advantage in that.

But there were other things that I did. Harvey DeWeerd and I did some work in the middle '50s on tactical nuclear weapons and civilian casualties in Western Europe and what might happen and what you might do about that, our own targeting policy.[98]

Q7: Was the lack of intellectual reward a function of your knowledge of that special intelligence and the feeling of, there wasn't that much to be concerned about in terms of a surprise attack?

Marshall: No, no. I think it was just that it wasn't until Loftus came there, and we really got into this business of how do you think about problems of forecasting and understanding what the Russians are really doing, that, from the longer-run point of view, I feel that there's something interesting to talk about, because it then leads to all these other things. The strategic warning thing didn't have that. There's some insights into how crazy some of our own arrangements could be, but that's just part of 60,000 instances of where less-than-optimal behavior is underway.

Q2: What happened to Loftus? You don't mention him when you come back from Paris. Did he leave RAND, or…?

Marshall: In the middle '60s, he retired on a kind of medical disability. I think it was a complicated thing. I think partly psychological, partly physical. I used to see him occasionally when I went out to visit him. But he was a terrific guy, really.

Q7: But even in later years, when you'd keep reaching back from the office, you keep reaching back to Loftus because he had knowledge of special sources…

Marshall: That's right.

Q7: …and special operations…

Marshall: Yeah, right.

Q7: You're trying to get guys to go tap Loftus's memory.

Marshall: He really was a fantastic guy. He was a fellow who had been a Notre Dame football player. He spent the war in PT boats, about 18 months circling Shemya Island.

Q4: Eighteen months going around Shemya in a PT boat?

Marshall: Right. And after the war, he went to Johns Hopkins, I think as an economics PhD. And for some reason, he got very early interested in nuclear things. Then, maybe in '48 or '49, I don't know, he went into intelligence and headed up the Air Force… It was earlier than that, because he was on board when the first nuclear weapon went off, the Russian one. So he probably joined that in about '48, and he was the head of the Air Force's whole effort on following the Soviet nuclear program.[99] Then, as I say, he was there till he came to RAND about… he might have come to RAND just before I left for Chicago.[100]

Q2: Did he come to RAND as an Air Force officer?

Marshall: No, he was a civilian.

Q7: There was an effort that you were interested in at one point in the '70s, where it was the Soviet control of nuclear weapons—

Marshall: Yeah, right.

Q7: —and some of the knowledge that Loftus had was [indistinct].

Marshall: Yeah, right. Also he and I, in the middle '50s, when both of us, again, knew that counterforce clearly was far more feasible than anybody at RAND had thought about, he and I, but particularly he, went back and looked at, did some very interesting work on· World War II counterforce efforts against the German V-Is and V-2s—the efforts were called Operation Crossbow—and looked for the lessons you could learn about, OK, if you go for counterforce, how do you really have to do it? Let's look at a case where people really were serious and did this.[101]

Loftus died about a year ago.

Q2: Andy, how do you deal with, in your intellectual construct… I wouldn't ask it now, except while, since we're on the subject of organizational behavior, how do you deal with individuals? How do you deal with dominant personalities? How do you deal with the impact of somebody like a Marshal Nikolai Ogarkov on the Soviet military, the Soviet military writ large? Do you just include that as part of your organizational paradigm, or those are all excursions?

Marshall: I don't know. I guess that's another factor. Clearly, there are individuals that matter at various points in time. I don't know, I guess you just have to be aware of that, and try to deal with that. But I don't have anything very systematic…

Q2: Well, we've always asked the intelligence community, "Well, what can you tell us about"—pick one, Chervov, or whoever—and they always invariably come back and go, "Beats the shit out of me. He drinks scotch, likes jazz…" I mean…

Marshall: Oh yeah, well that. Yeah, actually I got into some of that for Henry Kissinger. Henry was very brilliant and liked to know about these key individuals, particularly at the political level. Again, the big failure of the intelligence community to really know… They had some capabilities in that area, but they were very snooty about it. You know, "That's not our top-line product. Why does Henry… why are you so interested in that?" You know, kind of hand's length, holding their nose, "If you really want it…," you know. "Why don't you read our top-line (worthless) estimate?"

Q2: But you don't seem to…

Marshall: I don't feel I have anything special to say. Obviously some individuals matter enormously. At RAND, I think, at the beginning of it, individuals mattered tremendously. The leadership was the key people. They just had a big impact on the kind of organization of RAND.

Q2: One of the things that comes to mind is this Dead Hand system that first John Hines told me about—

Marshall: Yeah, right.

Q2: —then finally Bruce Blair blurbs it out in *The Washington Post* [*New York Times*].[102] But, you know, in large part the Soviet general officers that John talked to said that came about primarily because, as a result of this couple of exercises where General Secretary Leonid Brezhnev and Prime Minister Alexei Kosygin said, "Look, we are not going to initiate a nuclear strike." That the Soviet General Staff said, "OK fine, we'll bypass the military and political leadership. We'll set up this system that automatically does it." Clearly, that's a major… If there wasn't Brezhnev or Kosygin or somebody else in there they'd go, "Yeah, no problem. Last guy that stands is Russian. That's all we care about." Major impact on—But we're just so lousy at it that it's not even… Approach it episodically as best you can?

Marshall: Yeah, right. Anyway, I don't think I have anything special to say on that.

Q3: Well, except in this special context of the RAND business, the system of the key men. You talked a little bit about it, and when I went through some of your oral histories out at RAND… Because Frank Collbohm's role…

Marshall: Yes.

Q3: …why were these special conditions created and preserved over a long period at RAND?

Marshall: Yes, that's right.

Q3: And if you just read about Collbohm, it seems like he gets into a screaming fit with the Air Force just about every day in order to safeguard the RAND people from…

Marshall: And most of the RAND people were not at all grateful. They thought he was kind of dumb, a dumb engineer.

[Indistinct; multiple voices]

I mean the initial sort of notion of what this place should be like and how people should be treated is truly extraordinary.

Q3: If it hadn't been for him…

Marshall: That's right, it probably wouldn't have happened. Well, with the possible exception that General "Hap" Arnold [Commanding General of the Army Air Forces]…[gap for tape change] …I think it came out in one of those innovations meetings that we had.

Q3: Right.

Marshall: Arnold's view.

Q3: [Indistinct] Which in World War II gives him his vision of technology and the development of his relationship [indistinct].

Marshall: Yeah, that's right. [Indistinct] individuals who [indistinct] had a very long-term perspective [indistinct].

Q3: This is something that it might be interesting to track down, see whether or not, either at RAND or in Collbohm's family there are records that would indicate how he thought about setting up RAND as a special place.

Marshall: Yeah, right.

Q7: The parallel—well, this is not an exact parallel, but—is J. Robert Oppenheimer and the scientists at Los Alamos, General Leslie Groves tends to push military norms on these people. I don't know if that ever went through his mind…

Q3: I'm not sure. I don't know much about it. I think it's also something of an oversimplified picture. I mean it's Leslie Groves who picks Oppenheimer. Oppenheimer's not the leading candidate—it's James Conant. Conant is "Mr. American Establishment Scientist," the man with all the credentials. And it's clear from what Groves thought at the time, "I know Conant, I know this other…Ernest Lawrence, and I've never heard of Oppenheimer." [Indistinct]…he's the only person who can make this whole program run. It's Oppenheimer who always had very nice things to say about Groves.

[Indistinct; multiple voices]

Q7: …it was two other things. It was, number one, his ability to manage, which I guess the other two, Lawrence and Conant, had also demonstrated. But there were a lot of frictions, too, between what Groves wanted to do with those people and what Oppenheimer wouldn't let him do with them.

Q3: Yeah, true.

Q7: When you were talking about earlier how to organize the book, are you looking for something to organize the entire book?

Marshall: No, I think we have an idea about the organization of the book as a whole. It's how in each of these chapters to deal with this intellectual history, and then there's what was happening in the organization, in the immediate organization, and what's happening maybe more broadly [indistinct].

Q2: Organizing themes…

Marshall: …how to do that.

Q3: Well, there are at least three of them stressed: Andy's intellectual development; did it alter the analytical community at RAND; and then what the hell is going on in the rest of the world? Well, the real world, in terms of what people actually do.

Marshall: Yeah, right.

Q3: And that's not such a [indistinct] thing to do.

Marshall: Now there will be later sessions on the '70s, '80s, early '90s, I guess—I don't know where to cut it off.

Q3: From a literary point of view, the way you did it today was effective. You say, RAND in some sense reaches its culminating point of glory sometime around 1960. And then, "What was it like?" And then going back, "How did we get there?" From a story point of view it works.

Q7: Schlesinger once had a short comment—or at least it was attributed to him—he said, "RAND was a dung heap, but sometimes in dung heaps roses grow."

Q3: Unfortunately, there are lots of dung heaps. And roses.

Marshall: OK. Well, thank you.

Endnotes

1 Eds: This chapter includes two sections. The first section covers the early 1950s and is excerpted from an interview focused on the book project conducted on 16 September 1993. The second section covers the entire period from 1950 to 1969 and is an interview conducted on 29 October 1993. The outline was prepared for the second interview.

2 Eds: Alexis de Tocqueville, *Democracy in America*, ed. Eduardo Nolla, trans. James T. Schleifer, (Indianapolis: Liberty Fund, 2010) I: 655–656.

3 Eds: Ernest R. May, John D. Steinbruner, and Thomas W. Wolfe, *History of the Strategic Arms Competition, 1945-1972, Part I,* Alfred Goldberg, ed., (Office of the Secretary of Defense Historical Office, March 1981).

4 Eds: David Alan Rosenberg, "American Atomic Strategy and the Hydrogen Bomb Decision," *The Journal of American History*, 66, no. 1 (June 1979): 65–87.

5 Eds: Thomas B. Cochran, William M. Arkin, Robert S. Norris, and Milton M. Hoenig, *Nuclear Weapons Databook: Volume II, U.S. Nuclear Warhead Production* (Cambridge: Ballinger, 1987), 16.

6 Eds: Thomas B. Cochran and Robert Standish Norris, *Russian/Soviet Nuclear Warhead Production*, NWD 93-1 (Washington: National Resources Defense Council, 1993), 20.

7 Eds: Basic National Security Policy, NSC 162/2, 30 Oct. 1953, in *Foreign Relations of the United States, 1952–1954, National Security Affairs*, Vol. 2, Part 1, 578-82; John Foster Dulles, "The Evolution of Foreign Policy," Speech before the Council on Foreign Relations, New York (12 Jan 1954) in *Department of State Bulletin*, 30, no. 761 (25 Jan 1954): 107–110; Lawrence Freedman, *The Evolution of Nuclear Strategy*, 2nd ed. (New York: St. Martin's Press, 1989), 76–90.

8 Eds: Robert S. Norris and Hans M. Kristensen, "Global Nuclear Weapons Inventories, 1945-2010," *Bulletin of the Atomic Scientists* (July/Aug. 2010): 81.

9 Eds: Lorenz M. Luthi, *The Sino-Soviet Split: Cold War in the Communist World* (Princeton: Princeton University Press, 2008); Chen Jian, *Mao's China and the Cold War* (Chapel Hill: University of North Carolina Press, 2001); Michael S. Gerson, *The Sino-Soviet Border Conflict: Deterrence, Escalation, and the Threat of Nuclear War in 1969*, CRM D0022974.A2 (Arlington: CNA, 2010), 6–9.

10 Eds: General Maxwell D. Taylor, *The Uncertain Trumpet* (New York: Harper, 1960).

11 Eds: Freedman, *Evolution of Nuclear Strategy*, 227–244.

12 Excerpted from an interview conducted 16 September 1993. See also "The Development of Strategic Thinking at the RAND Corporation, 1948-1963: An Economist's View," transcript by James Digby and Joan Goldhamer, 2 Aug 1985; Transcript of interview of Marshall by Alfred Goldberg and Maurice Matloff (Historical Office, Office of the Secretary of Defense), 1 Jun 1992, 1–9.

13 Herbert Goldhamer and Andrew W. Marshall, *The Frequency of Mental Disease: Long-Term Trends and Present Status*, R-157 (Santa Monica: RAND Corp., July 1949), published as the book *Psychosis and Civilization: Two Studies in the Frequency of Mental Disease* (Glencoe: Free Press, 1953).

14 Andrew W. Marshall, "Forward" to Herbert Goldhamer, *The 1951 Korean Armistice Conference: A Personal Memoir*, P-7869 (Santa Monica: RAND Corp., 1994), xi–xii.

15 For background on the "criteria problem," see Charles J. Hitch and Roland N. McKean, *The Economics of Defense in the Nuclear Age* (Cambridge: Harvard University Press, 1960), 158–181; R.N. McKean, "Criteria," in Edward S. Quade, ed., *Analysis for Military Decisions* (Chicago: Rand McNally & Co., 1967), 81–91. For examples of Marshall's work in this area, see A.A. Alchian, G.D. Bodenhorn, S. Enke, C.J. Hitch, J. Hirshleifer, and A.W. Marshall, *What is the Best System?* D-860 (Santa Monica: RAND Corp., 4 Jan 1951); A. Alchian, J. Hirshleifer and A. Marshall, *Sound-Offs on Pay-offs*, D-892 (Santa Monica: RAND Corp., 20 Jul7 1950); A. Alchian, J. Hirshleifer, and A. Marshall, *Report of Cost Team Plus Panel*, D-829 (Santa Monica: RAND Corp., 20 Sep 1950); J. Hirshleifer and A. Marshall, *Proposed Payoff Criterion for MSA* [Multiple Strike Analysis], D-817 (Santa Monica: RAND Corp., 2 Oct 1950); J. Hirshleifer and A. Marshall, *Revised "Best Guess" Salvage Value Assumption for MSA,* D-820 (Santa Monica:

RAND Corp., 5 Oct 1950); J. Hirshleifer and A. Marshall, *A Proposed Salvage Value Procedure for MSA,* RM-476 (Santa Monica: RAND Corp., 20 Oct 1950); A.W. Marshall and Stephen Enke, *The Subsitution Value of U-235 in Two Assumed Strategic Campaigns (U)*, D(L)-923 (Santa Monica: RAND Corp., 20 Apr 1951), SECRET//RESTRICTED DATA (on the subject of this report, see also Stephen Enke, "Defense Economics: Applying Economic Criteria," *American Economic Review,* 55, no. 2 (May 1965): 418–420; and Stephen Enke, "Some Economic Aspects of Fissionable Material," *Quarterly Journal of Economics,* 68, no. 2 (May 1954): 217–232; S. Enke, E.W. Lee, and A.W. Marshall, *The Deployment of Local Defenses Suggested to Barlow by Economics,* D-935 (Santa Monica: RAND Corp., 11 May 1951).

A.W. Marshall, On the Accuracy of Immediate Bomb Damage Assessment with Radar Observation of the Bomb Fall *(U)*, D-989 (Santa Monica: RAND Corp., 6 Aug 1951), SECRET//RESTRICTED DATA; A.W. Marshall, *Coverage and Separation Distance of Point Targets (U)*, D-999 (Santa Monica: RAND Corp., 14 Aug 1951), SECRET//RESTRICTED DATA; A.W. Marshall, *The Estimation of Parameters in a Physical Vulnerability Model*, D(L)-1059 (Santa Monica: RAND Corp., 24 Oct 1951); A.W. Marshall, *Probability, Coverage, and Multi-element Targets*, D(L)-1142 (Santa Monica: RAND Corp., 15 Jan 1952); Marc Peter, Jr. and Andrew Marshall, *A Re-examination of Hiroshima-Nagasaki Damage Data*, RM-820 (Santa Monica: RAND Corp., 1 May 1952), FOUO.

A.W. Marshall and J.F. Digby, *Indications of Imminent Hostilities (U)*, D-1477 (Santa Monica: RAND Corp., 29 Dec 1952), CONFIDENTIAL; A.W. Marshall and J. F. Digby, *The Military Value of Advanced Warning of Hostilities and Its Implication for Intelligence* Indicators *(U)*, SM-14 (Santa Monica: RAND Corp., 20 Jul 1953), TOP SECRET (classification probably has been downgraded to CONFIDENTIAL or the document has been declassified). Digby discusses this work briefly in "Contributions of RAND to Strategy in the 1950s," in Andrew W. Marshall, J.J . Martin, and Henry S. Rowen, eds., *On Not Confusing Ourselves: Essays on National Security Strategy in Honor of Albert and Roberta Wohlstetter* (Boulder: Westview Press, 1991), 22. A short summary of SM-14 appears in Merton E. Davies and William R. Harris, *RAND's Role in the Evolution of Balloon and Satellite Observation Systems and Related Space Technology,* R-3692-RC (Santa Monica: RAND Corp., Sep 1988), 49; J.E. Loftus and A.W. Marshall, *The Limitations and Potentialities of the U.S. Strategic Warning System (U)*, D(L)-4944, TS-1710, Classified (this paper is referred to in Joseph E. Loftus and Andrew W. Marshall, *RAND Research on the Soviet Military,* D-4943-PR (Santa Monica: RAND Corp., 21 Feb 1958); J.E. Loftus and A.W. Marshall, *Strategic Warning: A Proposed Collection Concept (U)*, D-6048 (Santa Monica: RAND Corp., 20 Feb 1959), Classified; J.E. Loftus and A.W. Marshall, *The Structure of the U.S. Warning System (U)*, D-6064 (Santa Monica: RAND Corp., 26 Feb 1959), Classified; J.E. Loftus and A.W. Marshall, *Soviet Passive Detection Capability (U)*, D-6300 (Santa Monica: RAND Corp., 15 May 1959), Classified; Russell T. Nichols, *Economics Division Progress Report,* D-3740 (Santa Monica: RAND Corp., 15 Jul 1956), 7; T.E. Greene and J. Vogel, *Engineering Division Activity Report, First Half of 1956,* D-3750 (Santa Monica: RAND Corp., 15 Jul 1956), 46; Russell T. Nichols, *Economic Analysis Department Progress Report,* January 15, 1957, D-4052 (Santa Monica: RAND Corp., 7 Jan 1957); Russell T. Nichols, *Economic Analysis Department Progress Report,* D-4398 (Santa Monica: RAND Corp., 15 Jul 1957), 16–17, 19.

16 B. Brodie, C.J. Hitch, and A.W. Marshall, *The Next Ten Years,* D-2700 (Santa Monica: RAND Corp., 30 Dec 1954).

17 A.W. Marshall and H.A. DeWeerd, *Weapons Limitations and the Defense of Western Europe*, D-3785 (Santa Monica: RAND Corp., 27 Jul 1956); A.W. Marshall and H.A. DeWeerd, *Weapons Limitations, Nuclear Sharing, and Graduated Deterrence—A Report on a Trip to Washington, SHAPE, USAFE, and the United Kingdom, October 29–November 29, 1956*, D(L)-4037-PR (Santa Monica: RAND Corp., 26 Dec 1956). Another DeWeerd-Marshall paper, *Nuclear Weapons Limitations and NATO*, P-1013, 16 Jan 1957, is listed in *Social Science Division, Social Science Division Progress Report, June 30, 1957*, D-4396-PR (Santa Monica: RAND Corp., 1 Jul 1957).

18 R.L. Belzer, W.W. Kellogg, A.W. Marshall, and J.S. Thompson, *Preliminary Report of the Reconnaissance Committee (U)*, D-1138 (Santa Monica: RA ND Corp., 1952), SECRET//

RESTRICTED DATA; R.L. Belzer, W.W. Kellogg, A.W. Marshall, and J.S. Thompson, *The Monte Carlo Method Applied to a Comparison of Three Strategic Reconnaissance Strategies,* RM-811-2; *Appendix*, RM-811-3 (Santa Monica: RAND Corp., 16 Apr 1952), FOUO; H. Kahn and A.W. Marshall, *A Simplified Study of Reconnaissance in Strategic Bombing Campaigns,* RM-787 (Santa Monica: RAND Corp., 6 Mar 1952); A.W. Marshall, *RAND Systems Analysis Course: Material for Workshop—A Study of Alternative Reconnaissance Strategies in Strategic Bombing Campaigns,* D(L)-2559 (Santa Monica: RAND Corp., 3 Nov 1954).

19 J.E. Loftus and A.W. Marshall, *Research On Soviet Military Programs (U)*, D(L)-6500 (Santa Monica: RAND Corp., 26 Jun 1959), SECRET; J.E. Loftus and A.W. Marshall, *On Improving Midterm Estimates of Soviet Military Capabilities for Use in U.S. Military Planning (U)*, RM- 2892-PR (Santa Monica: RAND Corp., Aug 1962), SECRET; J.E. Loftus and A.W. Marshall, *Forecasting Soviet Force Structure: The Importance of Bureaucratic and Budgetary Constraints (U)*, RM-3612-PR (Santa Monica: RAND Corp., Jul 1963), SECRET; Andrew Marshall, "Propositions from Case Study on Forecasting Soviet Force Structure," 20 Jul 1970, in Hans Heymann, Jr., rapporteur, *Bureaucracy and National Security Policy: A Conference Report* (Draft), R-644-PR/ARPA (Santa Monica: RAND Corp., Feb 1971), 46–47; A.W. Marshall, "Studies of Bureaucratic Decision making," 14 Jan 1966, 2–3; Andrew W. Marshall, "Arms Competitions: The Status of Analysis," in Uwe Nerlich, ed., *The Western Panacea: Constraining Soviet Power through Negotiation,* Volume II of *Soviet Power and Western Negotiating Policies* (Cambridge: Ballinger Publishing Co., 1983), 6.

20 Report to the President by the Security Resources Panel of the Office of Defense Mobilization Science Advisory Committee, "Deterrence and Survival in the Nuclear Age," 7 Nov 1957 (TOP SECRET; Declassified 10 Jan 1973), in Department of State, *Foreign Relations of the United States 1955-1957, Vol. 19: National Security Policy* (Washington: CPO, 1990), 639–661. While on the staff of the Gaither Committee, Marshall was a member of a "small group of people with systems analysis-analytical kinds of skills that worked as consultants and experts to people in the other panels and did various kinds of tasks." Transcript of Marshall interview by Goldberg and Matloff, 1 Jun 1992, 8. An organization chart for the committee shows an "Evaluation Quantitative Assessments" group headed by Robert C. Prim (Bell Telephone Labs) and Stanley J. Lawwill (Strategic Air Command).

Herbert Goldhamer and Andrew W. Marshall, with the assistance of Nathan Leites, *The Deterrence and Strategy of Total War, 1959-1961: A Method of Analysis,* RM-2301 (Santa Monica: RAND Corp., 30 Apr 1959). An earlier version of this paper is Herbert Goldhamer and Andrew W. Marshall, *The Deterrence and Strategy of Total War, 1959-1960 (U)*, D-5614 (Santa Monica: RAND Corp., 1 Oct 1958), Classified.

A.W. Marshall and J.W. Tukey, *A Stringent Inspection Scheme Directed Against the Danger of Surprise Attack (U)*, SAW/180, 2 Nov 1958, SECRET; Memo, Andrew W. Marshall, Vincent V. McRae, and Albert Wohlstetter to Mr. Foster, subj: The "Accident" Problem, 19 Nov 1958, with attachments, "Part I: The Problem of Nuclear Weapon Accidents and Other Causes of Accidental War"; "Part II: Some Comments on Specific Statements by Eastern Experts," SAC/00/34, File 9T.2 US Experts b. Inter-office Memos SAW/00/1-50, Meetings-Talks-Conferences, Box 99, Lot no. 58D133, Records Relating to Disarmament, 1942-1962, 1953-62 Period, General Records of the Department of State, Record Group 59, National Archives]; Summary Minutes, Five Power Working Group Meeting on Systems of Control, 22 Oct 1958 (CONFIDENTIAL; Declassified 4 Oct 1994), SAW/FPC/4, File 9T.1 k. Five Power Conversations SAW/FPC/1-11, Box 98, Lot 58D133, RG 59, National Archives. Marshall was a member of the Instruments of Detection and Control Systems Croups of the U.S. delegation.

For discussions of the Surprise Attack Conference, see Bernhard C. Bechhoefer, *Postwar Negotiations for Arms Control* (Washington: Brookings Institution, 1961), 464–487; Johan J. Holst, "Strategic Arms Control and Stability: A Retrospective Look," in Holst and William Schneider, Jr., *Why ABM? Policy Issues in the Missile Defense Controversy* (New York: Pergamon Press, 1969), 245–284; and Memo, Rear Adm. Paul L. Dudley, USN to the Joint Chiefs of Staff, subj: Report on the Conference on the Problem of Surprise Attack, Geneva, Switzerland, 30 Dec 1958 (SECRET//NOFORN; Declassified 4 Oct 1994), with attachment, "A Summary of the Preparatory and Conference Activities of the U.S. Delegation to the Conference on the Problem of Surprise Attack" I File 9T.4 r. Proposed Basis for Examination of Possible Systems for Observation and Inspection of Naval Force, Box 109, Lot no. 580133, RG 59, National Archives.

21 A.W. Marshall and J.E . Loftus, *Latest AFCIN Missile Estimates for the 1965 War Game (U)*, D-6242 (Santa Monica: RAND Corp., 28 Apr 1959), SECRET//RESTRICTED DATA; A.W. Marshall and J.E. Loftus, *Further information on 1965 War Game Assumptions: Missile Reliability and Accuracy and Exchange of Intelligence (U)*, D-6375 (Santa Monica: RAND Corp., 1 Jun 1959), Classified; Bruce L.R. Smith, *The RAND Corporation: Case Study of a Nonprofit Advisory Corporation* (Cambridge: Harvard University Press, 1966), 170–171; Fred Kaplan, *The Wizards of Armageddon* (New York: Simon and Schuster, 1983), 216–217; Memo, J.F. Digby [no recipient] [no subject], 11 Feb 1959 (SECRET; Declassified 11 Dec 1985), RAND Archives/J. Digby Files/Box: Strategic Studies/F: SOFS 1959; "Agenda: Steering Committee Meeting, Strategic Offensive Force Project," 17–18 Feb 1959, RAND Archives/J. Digby Files/ Box: Administrative/Career; Memo, J.F. Digby and H.S. Rowen to SOFS Project Leaders, subj: Some US and SU 1963 Posture Discussions, M-1425, 3 Mar 1959 (SECRET; Declassified 12 Dec 1985); Memo, J.F. Digby to SOFS Distribution List, subj: Review of Talks for SOFS Steering Committee Meeting, M-3862, 18 Jun 1959, RAND Archives/J. Digby Files/ Box: Chronological Files; Memos: Outgoing (Defense), 1949–1951 & 1959–1960; List of assorted SOFS-related documents provided by Walter Nelson (Head, Classified Information Services, RAND Corp.), 25 Feb 1994.

22 Social Science Division [Victor Hunt, W. Phillips Davison, Paul Kecskemeti, Irving Janis, Jean Hungerford, Eric Willenz, Roberta Wohlstetter, and Joseph Goldsen], *The Warning of Target Populations in Air War: An Appendix of Working Papers,* RM-275 (Santa Monica: RAND Corp., Nov 1949).

23 Brodie joined RAND as a full-time senior staff member in Aug 1951. Barry H. Steiner, *Bernard Brodie and the Foundations of American Nuclear Strategy,* (Lawrence: University Press of Kansas, 1991), 252.

24 Ibid., 78–111. Brodie was special assistant to the Air Force Chief of Staff, Aug–Sep 1950, and special assistant to the Air Force Chief of Staff, acting as RAND Corp. consultant, Nov–Jun 1951 (252).

25 Bernard Brodie, "Must We Shoot from the Hip?" 4 Sep 1951, reprinted in Marc Trachtenberg, ed., *The Development of American Strategic Thought: Writings on Strategy, 1945-1951* (New York: Garland Publishing, Inc., 1987), 235–272.

26 Russell D. Buhite and Wm. Christopher Hamel, "War for Peace: The Question of an American Preventive War against the Soviet Union, 1945-1955," *Diplomatic History* 1, 14, no. 3 (Summer 1990): 367–384; Marc Trachtenberg, "A 'Wasting Asset': American Strategy and the Shifting Nuclear Balance, 1949-1954," *International Security,* 13, no. 3 (Winter 1988/89): 5–49; John Williams, *In Response to Paxon's Memo on the Probability of War,* D-1513 (Santa Monica: RAND Corp., 3 Feb 1953); Williams, *Regarding the Preventive-War Doctrine,* D(L)-1777 (Santa Monica: RAND Corp., 27 Jul 1953); Memo, Williams to Bernard Brodie, subj: The Next Nine Years, 11 Months, M-5372, 25 Oct 1954; Memo, Williams to Brodie, subj: The Subject, M-5598, 5 Nov 1954, RAND Archives, Digby Files/Box: Strategic Studies/File: SOC, 1954-1955; Clay Blair, Jr., "The Passing of a Great Mind," *Life*, 42, no. 8 (27 Feb 1957): 96; Steve J. Heims, *John von Neumann and Norbert Wiener: From Mathematics to the Technologies of Life and Death* (Cambridge: MIT Press, 1980), 484; Norman Macrae, *John von Neumann* (New York: Pantheon Books, 1992), 27–28, 332.

27 John Williams, *Hunting the Tiger (and Other Aspects of the Active Life) (U)*, S-16 (Santa Monica: RAND Corp., 26 Mar 1954), SECRET.

28 On von Neumann's general interest in weather modification, see *John von Neumann,* 16–17.

29 RAND's Role in the Evolution of Balloon and Satellite Observation Systems and Related Space Technology, 19–23, 39–43, 59–61, 74–77, 94.

30 *Rules for an Aggregated War Game,* RM-1046-2 (Santa Monica: RAND Corp., 3 Feb 1954; revised 27 Jul 1954).

31 Minutes of Strategic Objectives Committee [1960 Committee], 10 May 1954, RAND Archives/J. Digby Files/Box: Strategic Studies/File: SOC, 1954-1955.

32 Minutes of Strategic Objectives Committee (1954—30 Apr, 10 May, 17 May, 24 May, 4 Jun, 25 Jun, 9 Jul, 23 Jul, 6 Aug, 3 Sep, 2 Dec; 1955—21 Jan, 4 Feb, 14 Feb, 7 Mar, 12 Apr, 25 Apr, 7 Jul) (Declassified 13 Dec 1985) RAND Archives/J. Digby Files/Box: Strategic Studies/ File: SOC, 1954-1955. Members of the SOC included Bernard Brodie, James Digby, Charles Hitch (first chairman), Victor Hunt, Arnold Kramish, James Lipp (second chairman), Andrew Marshall, Alex Mood, John Williams (third chairman), and, toward the end, Herman Kahn and Malcolm Hoag.

33 For criticisms of *The Next Ten Years*, see Memo, J.D. Williams to B. Brodie, C.J. Hitch, and A.W. Marshall, subj: Unkind Comments on *the Next Ten Years*, M-4419, 2 Sep 1954, RAND Archives/J. Digby Files/Box: Strategic Studies/File: SOC, 1954-1955]; Joseph E. Loftus, *Two Minutes on The Next Ten Years,* D(L)-2726 (Santa Monica: RAND Corp., 14 Feb 1955).

34 *The RAND Corporation: Case Study of a Nonprofit Advisory Corporation,* 60–65; *The Wizards of Armageddon,* 67–72. According to Bruce Smith (*The RAND Corporation*), "It was not until 1956 that the bulk of the social scientists finally moved from RAND's Washington office to the main headquarters in Santa Monica, California," 64.

35 "Arms Competitions: The Status of Analysis," 8–10; Ernest R. May, John D. Steinbruner, and Thomas W. Wolfe *History of the Strategic Arms Competition, 1945-1972,* Alfred Goldberg, ed. (Washington: Office of the Secretary of Defense Historical Office, Mar 1981) (TOP SECRET//RESTRICTED DATA/NOFORN; Sanitized), Part I, 81–103, 242–333; Part II, 824; Steven J. Zaloga, *Target America: The Soviet Union and the Strategic Arms Race, 1945-1964* (Novato: Presidio Press, 1993).

36 Interview conducted on 29 October 1993. See also "The Development of Strategic Thinking at the RAND Corporation, 1948-1963: An Economist's View," Transcript of Marshall interview by James Digby and Joan Goldhamer, 2 Aug 1985; Transcript of Marshall interview by Alfred Goldberg and Maurice Matloff (Historical Office, Office of the Secretary of Defense, 1 Jun 1992), 1–13.

37 To help devise strategies for the efficient and effective use of nuclear weapons, the Air Force in 1946 created an independent organization—designated Project RAND and located within the Douglas Aircraft Company—to conduct research on intercontinental air warfare. By 1948, the arrangement was proving unworkable. Existing and prospective RAND employees were leery of working for a subsidiary of an aircraft manufacturer rather than a truly independent organization, and other aerospace firms began to protest that the existence of RAND gave Douglas an unfair advantage in the competition for government contracts. Consequently, RAND officials created an independent nonprofit corporation, the RAND Corporation, to house Project RAND and its assets. The RAND research agenda was broad, but at its core was an effort to "make a major scientific attack on the whole theory of warfare, in the broadest sense of the words." RAND quickly became one of the nation's leading centers for research on nuclear strategy and policy. It also conducted pioneering work in rocketry and space systems, digital computing, and artificial intelligence. Some of its most influential work was in applied mathematics and economics. RAND analysts made advances in game theory, modeling and simulation, cost analysis, and Monte Carlo methods as they sought to improve military decision-making. By the mid-1960s, RAND comprised over 1,100 employees and conducted more than $23 million of funded research annually. *Project RAND Second Annual Report,* RA-15075 (1 Mar 1948), 3, quoted in Andrew David May, "The RAND Corporation and the Dynamics of American Strategic Thought, 1946-1962" (PhD diss., Emory University, 1998) 23. David R. Jardini, "Out of the Blue Yonder: The RAND Corporation's Diversification into Social Welfare Research, 1946-1968," (Ph.D. diss., Carnegie Mellon University, 1996), 9–11.

38 Herbert Goldhamer and Andrew W. Marshall, *The Frequency of Mental Disease: Long-Term Trends and Present Status,* R-157 (Santa Monica: RAND Corp., Jul 1949), published as the book *Psychosis and Civilization: Two Studies in the Frequency of Mental Disease* (Glencoe: Free Press, 1953).

39 The following RAND papers may be related to this effort: Andrew W. Marshall and Joseph

E. Loftus, *Soviet Military Expenditures, 1955-1963: An Official Summary View (U)*, D-6365 (Santa Monica: RAND Corp., 27 May 1959), SECRET; Andrew W. Marshall and Joseph E. Loftus, *CIA Preliminary Estimates of Soviet Military Expenditures, 1947-1959 (U)*, D(L)-7085 (Santa Monica: RAND Corp., 8 Jan 1960), SECRET.

40 Michel Crozier, *The Bureaucratic Phenomenon* (Chicago: University of Chicago Press, 1964).

41 According to Fred Kaplan, Loftus joined RAND in late 1954, working in the Washington office until the end of that year. Kaplan, *The Wizards of Armageddon* (New York: Simon and Schuster, 1983), 207, 209, 212.

42 Hans Heymann, Jr., James C. DeHaven, and Richard C. Raymond, *The Soviet Intercontinental Mission through 1957 (U)*, R-281 (Santa Monica: RAND Corp., 1 Aug 1954), CONFIDENTIAL; James C. DeHaven, *The Soviet Strategic Base Problem*, RM-1302 (Santa Monica: RAND Corp., 16 Aug 1954), FOUO.

43 For background on the "criteria problem," see Charles J. Hitch and Roland N. McKean, *The Economics of Defense in the Nuclear Age* (Cambridge: Harvard University Press, 1960), 158–181; R.N. McKean, "Criteria," in Edward S. Quade, ed., *Analysis for Military Decisions* (Chicago: Rand McNally & Co., 1967), 81–91. For examples of Marshall's work in this area, see A.A. Alchian, G.D. Bodenhorn, S. Enke, C.J. Hitch, J. Hirshleifer, and A.W. Marshall, *What Is the Best System?* D-860 (Santa Monica: RAND Corp., 4 Jan 1951); A. Alchian, J. Hirshleifer, and A. Marshall, *Sound-offs on Pay-offs*, D-892 (Santa Monica: RAND Corp., 20 Jul 1950); A. Alchian, J. Hirshleifer, and A. Marshall, *Report of Cost Team Plus Panel*, D-829 (Santa Monica: RAND Corp., 20 Sep 1950); J. Hirshleifer and A. Marshall, *Proposed Payoff Criterion for MSA* [Multiple Strike Analysis], D-817 (Santa Monica: RAND Corp., 2 Oct 1950); J. Hirshleifer and A. Marshall, *Revised "Best Guess" Salvage Value Assumption for MSA*, D-820 (Santa Monica: RAND Corp., 5 Oct 1950); J. Hirshleifer and A. Marshall, *A Proposed Salvage Value Procedure for MSA*, RM-476 (Santa Monica: RAND Corp., 20 Oct 1950); A.W. Marshall and Stephen Enke, *The Substitution Value of U-235 in Two Assumed Strategic Campaigns (U)*, D(L)-923 (Santa Monica: RAND Corp., 20 Apr 1951), SECRET//RESTRICTED DATA (on the subject of this report, see also Stephen Enke, "Defense Economics: Applying Economic Criteria," *American Economic Review*, 55, no. 2 (May 1965): 418–420; and Stephen Enke, "Some Economic Aspects of Fissionable Material," *Quarterly Journal of Economics*, 68, no. 2 (May 1954): 217–232; S. Enke, E.W. Lee, and A.W. Marshall, *The Deployment of Local Defenses Suggested to Barlow by Economics*, D-935 (Santa Monica: RAND Corp., 11 May 1951).

44 Herbert Goldhamer, *Human Factors in Systems Analysis*, RM-388 (Santa Monica: RAND Corp., 15 Apr 1950).

45 A.W. Marshall, *On the Accuracy of Immediate Bomb Damage Assessment with Radar Observation of the Bomb Fall (U)*, D-989 (Santa Monica: RAND Corp., 6 Aug 1951), SECRET//RESTRICTED DATA; A.W. Marshall, *Coverage and Separation Distance of Point Targets (U)*, D-999 (Santa Monica: RAND Corp., 14 Aug 1951), SECRET//RESTRICTED DATA; A.W. Marshall, *The Estimation of Parameters in a Physical Vulnerability Model*, D(L)-1059 (Santa Monica: RAND Corp., 24 Oct 1951); A.W. Marshall, *Probability, Coverage, and Multi-element Targets*, D(L)-1142 (Santa Monica: RAND Corp., 15 Jan 1952); Marc Peter, Jr. and Andrew Marshall, *A Re examination of Hiroshima-Nagasaki Damage Data*, RM-820 (Santa Monica: RAND Corp., 1 May 1952), FOUO.

46 Andrew W. Marshall, "Arms Competitions: The Status of Analysis," in Uwe Nerlich, ed., *The Western Panacea: Constraining Soviet Power through Negotiation, Volume II of Soviet Power and Western Negotiating Policies* (Cambridge: Ballinger Publishing Co., 1983), 8–10; Ernest R. May, John D. Steinbruner, and Thomas W. Wolfe, *History of the Strategic Arms Competition, 1945-1972*, Alfred Goldberg, ed. (Washington: Office of the Secretary of Defense Historical Office, Mar 1981) (TOP SECRET//RESTRICTED DATA//NOFORN; Sanitized), Part I, 81–103, 242–333; Part II, 824; Steven J. Zaloga, *Target America: The Soviet Union and the Strategic Arms Race, 1945-1964* (Novato: Presidio Press, 1993).

47 "The Development of Strategic Thinking at the RAND Corporation, 1948-1963: An Economist's View," 24–25; Raymond L. Garthoff, *Soviet Strategy in the Nuclear Age* (New York: Frederick A. Praeger Publishers, 1958), 170–195.

48 H. Kahn and I. Mann, *Techniques of Systems Analysis,* RM-1829-1 (Santa Monica: RAND Corp., 3 Dec 1956; revised Jun 1957).
49 Andrew W. Marshall, Eulogy at Herman Kahn memorial service, National Defense University, 15 Jul 1983.
50 Bernard Brodie, *Strategy in the Missile Age* (Princeton: Princeton University Press, 1959).
51 Andrew W. Marshall, "Strategy as a Profession for Future Generations," in Marshall, J.J. Martin, and Henry S. Rowen, eds., *On Not Confusing Ourselves: Essays on National Security Strategy in Honor of Albert and Roberta Wohlstetter* (Boulder: Westview Press, 1991), 305–306; A.J. Wohlstetter, F.S. Hoffman, R.J. Lutz, and H.S. Rowen, *Selection and Use of Strategic Air Bases,* R-266 (Santa Monica: RAND Corp., Apr 1954; second printing Jun 1962).
52 On the "missing chapter," see C. Wolf, Jr., "Working Group I—Improving Policy Analysis: Introductory Paper," 17 Jul 1970, A.W. Marshall, "Notes on Improving Policy Analysis: Attachment to Introductory Paper (Working Group I)," 17 Jul 1970, and "Working Group I: Improving Policy Analysis: Summary of Findings," 23 Jul 1970, in Hans Heymann, Jr., rapporteur, *Bureaucracy and National Security Policy: A Conference Report* (Draft), R-644-PR/ARPA (Santa Monica: RAND Corp., Feb 1971), 57–58, 61–63.
53 B. Brodie, CJ. Hitch and A.W. Marshall, *The Next Ten Years,* 0-2700 (Santa Monica: RAND Corp., 30 Dec 1954).
54 Minutes of Strategic Objectives Committee (1954—30 Apr, 10 May, 17 May, 24 May, 4 Jun, 25 Jun, 9 Jul, 23 Jul, 6 Aug, 3 Sep, 2 Dec; 1955—21 Jan, 4 Feb, 14 Feb, 7 Mar, 12 Apr, 25 Apr, 7 Jul) (Declassified 13 Dec 1985), RAND Archives/J. Digby Files/Box: Strategic Studies/ File: SOC, 1954-1955. Members of the SOC included Bernard Brodie, James Digby, Charles Hitch (first chairman), Victor Hunt, Arnold Kramish, James Lipp (second chairman), Andrew Marshall, Alex Mood, John Williams (third chairman), and, toward the end, Herman Kahn and Malcolm Hoag.
55 On the origins of the SOC, see "The Development of Strategic Thinking at the RAND Corporation, 1948-1963: A Radar Engineer's Point of View," Transcript of interview of James Digby by Joan Goldhamer, 2 Jul 1985, 18–20.
56 R.T. Nichols, *STRAP: Summary of Economic Panel Discussions,* RM-257 (Santa Monica: RAND Corp., Oct 1949); Steven T. Ross, *American War Plans, 1945-1950* (New York: Garland Publishing, Inc., 1988); Melvyn P. Leffler, *A Preponderance of Power: National Security, the Truman Administration, and the Cold War* (Stanford: Stanford University Press, 1992); David Alan Rosenberg, "Toward Armageddon: The Foundations of United States Nuclear Strategy," PhD dissertation, University of Chicago, 1983.
57 C.J. Hitch and W.M. Capron, *Implications of Potential Weapon Developments for Strategic Bombing and air Defense: A Preliminary Study,* RM-868 (Santa Monica: RAND Corp., 10 Jul 1952).
58 Memo, H. Goldhamer, 0. Helmer, A. Marshall, T. Schelling, and M. Weiner to SOFS [Strategic Offensive Forces Study] and Friends, subj: Results from Attached Questionnaire (U), M-3157, 15 May 1959, CONFIDENTIAL. Some of the results of the survey are cited by Marshall in Daniel Ellsberg and Paul Kecskemeti, rapporteurs, *Minutes of the Monterey Seminar on Active and Passive Defense, III: Tile Spectrum of Central War Moves,* D-774-1 (Santa Monica: RAND Corp., 24 Oct 1960) (SECRET; Declassified 22 Dec 1975), 5–6, RAND Archives/J. Digby Files/Box: National Security Seminar; Publications/Correspondence, 1960-1961. Another questionnaire is mentioned in Memo, Olaf Helmer and Marshall to National Security Seminar Participants, subj: Attached Questionnaire, M-4857, 30 Aug 1960, RAND Archives/J. Digby Files/Box: National Security Seminar, Publications/Correspondence, 1960 1961.
59 The following National Intelligence Estimates, among others, were declassified in full or in part in late 1994 and can be obtained from the Office of the Information and Privacy Coordinator at the CIA: SNIE 11-10-57, The Soviet ICBM Program, 10 Dec 1957; NIE 11-5-58, Soviet Capabilities in Guided Missiles and Space Vehicles, 19 Aug 1958; NIE 11-8-59, Soviet Capabilities for Strategic Attack Through Mid-1964, 9 Feb 1960; NIE 11-5-60, Soviet Capabilities in Guided Missiles and Space Vehicles, 3 May 1960; NIE 11-8-60, Soviet Capabilities for Long-Range Attack Through Mid- 1965, 1 Aug 1960; NIE 11-8-61, Soviet Capabilities for Long-Range Attack, 7 Jun 1961.

60 On von Neumann's views, see Clay Blair, Jr., "The Passing of a Great Mind," *Life*, 42, no. 8 (27 Feb 1957): 96; Steve J. Heims, *John von Neumann and Norbert Wiener: From Mathematics to the Technologies of Life and Death* (Cambridge: MIT Press, 1980), 484; Norman Macrae, *John von Neumann* (New York: Pantheon Books, 1992), 27–28, 332.

61 During 1953–1954, John Williams and Bernard Brodie debated preventive war, with Williams for and Brodie against. See Williams, *In Response to Paxson's Memo on the Probability of war*, D-1513 (Santa Monica: RAND Corp., 3 Feb 1953); Brodie, *A Commentary on Preventative-War Doctrine*, D-1715 (Santa Monica: RAND Corp., 11 Jun 1953), reprinted in Marc Trachtenberg, ed., *The Development of American Strategic Thought: Writings on Strategy, 1952-1960*, Vol. 1 (New York: Garland Publishing, Inc., 1988), 131–148; Williams, *Regarding the Preventative War Doctrine*, D(L)-1777 (Santa Monica: RAND Corp., 27 Jul 1953); Williams, *Hunting the Tiger (and Other Aspects of the Active Life) (U)*, S-16 (Santa Monica: RAND Corp., 26 Mar 1954), SECRET; Memo, Brodie to Williams, subj: Reply to Your M-5110, M-5185, 13 Oct 1954; Memo, Williams to Brodie, subj: The Next Nine Years, 11 Months, M-5372, 25 Oct 1954; Memo, Brodie to Williams, subj: A Moratorium on Similes, M-5484, 1 Nov 1954; Memo, Williams to Brodie, subj: The Subject, M-5598, 5 Nov 1954; Memo, Brodie to Williams, subj: A Parting Plea, M-5665, 9 Nov 1954. See also Memo, Victor Hunt to Williams, Brodie, and SOC, subj: The Dangers of Bi-polar Discussions, M-5615, 8 Nov 1954; Memo, Brodie to Hunt, subj: The Dangers of Bi-polar Discussions Circumvented, M-5717, 11 Nov 1954 [All memos are from RAND Archives/J, Digby Files/Box: Strategic Studies/File: SOC, 1954-1955]. Brodie included a discussion of preventive war in *Strategy in the Missile Age* (227–241).

62 W.W. Kaufmann, *Counterforce IV: Summary Briefing*, D(L)-7880-PR (Santa Monica: RAND Corp., 6 Sep 1960); Ltr, William W. Kaufmann to Lt. Col. D.F. Martin, Office, Assistant for Coordination, Deputy Chief of Staff, Plans and Programs, HQ USAF, 7 Jul 1960, with attachment, "Informal RAND Reaction to Briefing on Counterforce," 7 Jul 1960 (SECRET; Declassified 4 May 1977); Memo, J.F. Digby to The Research Council, subj: Report on War Fighting Capability Project, 21 Nov 1960; Memo, J.F. Digby to F.R. Collbohm, subj: Information for War Fighting Capability Study, 23 Nov 1960, RAND Archives/J, Digby Files/Box: Strategic Studies/F: Air Defense, Strategic Studies, 1960.

63 *Rules for an Aggregated War Game*, RM-1046-2 (Santa Monica: RAND Corp., 3 Feb 1954; revised 27 Jul 1954).

64 SOC [1960 Committee] minutes, 10 May 1954.

65 Herbert Goldhamer and Hans Speier, "Some Observations on Political Gaming," *World Politics*, 12, no. 1 (Oct 1959): 71–83; H. Goldhamer, *Toward a Cold War Game*, D(L)-2607 (Santa Monica: RAND Corp., 22 Oct 1954); H. Goldhamer, *Summary of Cold War Game Activities in the Social Science Division* [First Round], 0-2850 (Santa Monica: RAND Corp., 12 Apr 1955); P. Kecskemeti, *War Games and Political Games*, D-2849 (Santa Monica: RAND Corp., 15 Apr 1955); P. Kecskemeti, *Summary of Cold War Game Activities in the Social Science Division: May Experiment* [Second Round], D-2975-RC (Santa Monica: RAND Corp., 20 Jun 1955); H. Goldhamer, *The Political Exercise: A Summary of the Social Science Division's work on Political Gaming, with Special Reference to the Third Exercise*, D-3164-RC (Santa Monica: RAND Corp., 1 Sep 1955); E. Schnitzer, *Third Political Exercise: Summary and Documents*, D-3639 (Santa Monica: RAND Corp., 1 Sep 1955); E. Schnitzer, *Fourth RAND Political Exercise: Summary and Documents*, D-3163 (Santa Monica: RAND Corp., 25 May 1956); J. Goldsen, *The Political Exercise: An Assessment of the Fourth Round*, D-3640-RC (Santa Monica: RAND Corp., 30 May 1956).

66 "In the 1950s, the United States spent about two-thirds as much on air defenses as on nuclear offensive forces, some $200 billion in ten years (in today's dollars). In 1960, the United States actually allocated more to its defenses against Soviet nuclear attack than to its nuclear offensive forces." Fred Charles lkle, "Nuclear Strategy: Can There Be a Happy Ending?" *Foreign Affairs*, 63, no. 4 (Spring 1985): 813.

67 Garry D. Brewer and Martin Shubik, *The War Game: A Critique of Military Problem Solving*, (Cambridge: Harvard University Press, 1979), 111–114.

68 A.L. Latter, R. Latter, F. Collbohm, W.B. Graham, A.W. Marshall, and E.A. Martinelli, *Defense Against Bombers (U),* D-10130 (Santa Monica: RAND Corp., Jun 1962), SECRET// RESTRICTED DATA; A.L. Latter, James Digby, W.B. Graham, R. Latter, A.W. Marshall, and E.A. Martinelli, *Report of the RAND Ad Hoc Group on Defense (U),* RM-3237-PR (Santa Monica: RAND Corp., Aug 1962), SECRET//RESTRICTED DATA; A.L. Latter, R. Latter, F. Collbohm, W.B. Graham, E.A. Martinelli, and A.W. Marshall, *Report of the RAND Ad Hoc Group on Defense (U),* D-10055 (Santa Monica: RAND Corp., Jun 1962), SECRET// RESTRICTED DATA.

69 The mid-50s study effort on strategic warning was called Project HAYLOFT (RAND Project 4028: Indicators of Soviet Intentions). It was done for the Directorate of Intelligence, Office of the Deputy Chief of Staff for Operations, HQ USAF. The HAYLOFT study group was led by Joseph Loftus and included Marshall, Robert Belzer, Richard Raymond, Nicholas Rescher, Pat Keith, and Roberta Wohlstetter. The work of Loftus and Marshall is documented in J.E. Loftus and A.W. Marshall, *The Limitations and Potentialities of the U.S. Strategic Warning System (U),* D-(L)-4944, TS-1710, Classified. This paper is referred to in Joseph E. Loftus and Andrew W. Marshall, *RAND Research on the Soviet Military,* D-4943-PR (Santa Monica: RAND Corp., 21 Feb 1958); J.E. Loftus and A.W. Marshall, *Strategic Warning: A Proposed Collection Concept (U),* D-6048 (Santa Monica: RAND Corp., 20 Feb 1959), Classified; J.E. Loftus and A.W. Marshall, *The Structure of the U.S. Warning System (U),*D-6064 (Santa Monica: RAND Corp., 26 Feb 1959), Classified; J.E. Loftus and A.W. Marshall, *Soviet Passive Detection Capability (U),* D-6300 (Santa Monica: RAND Corp., 15 May 1959), Classified; Russell T. Nichols, *Economics Diversion Progress Report,* D-3740 (Santa Monica: RAND Corp., 15 Jul 1956), 7; T.E. Greene and J. Vogel, *Engineering Division Activity Report, First Half of 1956,* D-3750 (Santa Monica: RAND Corp., 15 Jul 1956), 46; Russell T. Nichols, *Economic Analysis Department Progress Report, January 15, 1957,* D-4052 (Santa Monica: RAND Corp., 7 Jan 1957); Russell T. Nichols, *Economic Analysis Department Progress Report,* D-4398 (Santa Monica: RAND Corp., 15 Jul 1957), 16–17, 19.

70 J.E. Loftus and A.W. Marshall, *Research on Soviet Military Programs (U),* D (L)-6500 (Santa Monica: RAND Corp., 26 Jun 1959), SECRET; J.E. Loftus and A.W. Marshall, *On Improving Midterm Estimates of Soviet Military Capabilities for Use in U.S. Military Planning (U),* RM-2892-PR (Santa Monica: RAND Corp., Aug 1962), SECRET; J.E. Loftus and A.W. Marshall, *Forecasting Soviet Force Structure: The Importance of Bureaucratic and Budgetary Constraints (U),* RM-3612-PR (Santa Monica: RAND Corp., Jul 1963), SECRET; Andrew Marshall, "Propositions from Case Study on Forecasting Soviet Force Structure," 20 Jul 1970, in *Bureaucracy and National Security Policy: A Conference Report,* 46–47.

71 Chester L Barnard, *The Functions of the Executive* (Cambridge: Harvard University Press, 1938); Herbert A. Simon, *Administrative Behavior: A Study of Decision-Making Processes in Administrative Organizations,* second edition (New York: Free Press, 1965; first published in 1945); James G. March and Herbert A. Simon, *Organizations* (New York: John Wiley & Sons, 1958); Richard M. Cyert and James C. March, *Behavioral Theory of the Firm* (Englewood Cliffs: Prentice Hall, 1963).

72 A.W. Marshall and Eugene Brussell, *On the Accuracy of Cost Estimates: A Preface,* D-5014 (Santa Monica: RAND Corp., 10 Mar 1958); A.W. Marshall and W.H. Meckling, *Predictability of the Costs, Time, and Success of Development,* P-1821 (Santa Monica: RAND Corp., 14 Oct 1959; revised 11 Dec 1959), in *The Rate and Direction of Inventive Activity—Economic and Social Factors: A Conference of the Universities-National Bureau Committee for Economic Research and the committee on Economic Growth for the Social Science Research Council* (Princeton: Princeton University Press, 1962), 461–475.

73 Herbert Goldhamer and Andrew W. Marshall, with the assistance of Nathan Leites, *The Deterrence and Strategy of Total War, 1959-1961: A Method of Analysis,* RM-2301 (Santa Monica: RAND Corp., 30 Apr 1959). An earlier version of this paper is Herbert Goldhamer and Andrew W. Marshall, *The Deterrence and Strategy of Total War,1959-1960 (U),* D-5614 (Santa Monica: RAND Corp., 1 Oct 1958), Classified.

74 B. Klein, G. Schubert, O. Hoeffding, F. Hoffman, A. Marshall, *The NATO Force Planning Ex-*

ercise: Scope, Aims, and Organization (U), D (L)-12548-ISA (Santa Monica: RAND Corp., 15 Jun 1964), Classified; Section on NATO Force Planning Exercise in "Some RAND Products," in "RAND Research Results 11" notebook, 16 Sep 1969, RAND Archives/Robert Specht Files; Ltr, James F. Digby to Henry S. Rowen, Deputy Assistant Secretary of Defense for International Security Affairs (Planning and NSC Office), 6 Dec 1963, with encl, B.H. Klein, "Progress Report on RAND Participation in the Stikker Exercise *(U),* 14 Nov 1963 (CONFIDENTIAL; Declassified 23 Dec 1975), RAND Archives/J. Digby Files/Box: Chronological Files, Memos: Outgoing (Defense), 1949–1951 and 1959–1960.

75 Albert P. Williams, Jr., *The Record and Future Plans for RPN 1412: Decisionmaking Processes in Military Organizations,* WN-7043-PR (Santa Monica: RAND Corp., Sep 1970).

76 Graham T. Allison, *Essence of Decision: Explaining the Cuban Missile Crisis* (Boston: Little, Brown & Co., 1971).

77 Robert Ardrey, *The Territorial Imperative: A Personal Inquiry into the Animal Origins of Property and Nations* (New York: Atheneum, 1966).

78 A.W. Marshall, *Problems of Estimating Military Power,* P-3417 (Santa Monica: RAND Corp., Aug 1966).

79 A.W. Marshall, *Comparisons,* R&D *Strategy, and Policy Issues (U),* WN-7630-DDRE (Santa Monica: RAND Corp., Oct 1971), SECRET.

80 According to Don R. Harris (a member of the NSC Net Evaluation Subcommittee), the Net Technical Assessment staff in DDR&E was created in 1966. MR, Harris, subj: The Current Status of Net Assessment, 12 Mar 1973, in Capt. James W. Lucas, "Net Assessment and the Threat."

81 Memo, J.R. Schlesinger to H.S. Rowen, subj: Difficulties in and a Mechanism for Partial RAND Restructuring by DoD Program Packages, 30 Jan 1967; Memo, Draft, Andrew Marshall [to Rowen?], [no subject], 31 Jan 1967; Memo, Andrew Marshall to Harry Rowen, subj: Possible Focus for RAND work in NonDefense Areas, M-2799, 13 Apr 1967; Memo, A.W. Marshall and S.G. Winter, Jr. to H.S. Rowen, subj: A RAND Department of "Management Sciences—The Case in Brief," M-8668, 29 Dec 1967, RAND Archives.

82 Schlesinger was RAND's Director of Strategic Studies from 1967–1969. History Staff, Central Intelligence Agency, *Directors and Deputy Directors of Central Intelligence* (Washington: CIA, 1989), 21.

83 A.W. Marshall, *Long-Term Competition with the Soviets: A Framework for Strategic Analysis (U),* R-862-PR (Santa Monica: RAND Corp., Apr 1972). See also A.W. Marshall, *Long Term Competition with the Soviets: A Framework for Strategic Analysis,* JN-21542-PR (Santa Monica: RAND Corp., 15 Apr 1971) (CONFIDENTIAL; Declassified 7 May 1980).

84 *The Wizards of Armageddon,* 204–219; Gregg Herken, *Counsels of War* (New York: Alfred A. Knopf, 1985), 79–81; Marc Trachtenberg, *History and Strategy* (Princeton: Princeton University Press, 1991), 27–30.

85 "Most civilian analysts at Rand simply weren't allowed to see secret materials, and so exaggerated the difficulty of finding Soviet military targets. That all changed in May of 1960, when the Soviets shot down one of the CIA's U-2 spy planes on its way to photograph a missile testing center at Plesetsk. Not long afterward, Marshall remembers, a Rand analyst stopped him one day in the parking lot and said, 'Now I understand what you and Loftus were talking about.'" Thomas Powers, "Choosing a Strategy for World War III," *The Atlantic Monthly,* 250, no. 5 (Nov 1982): 91.

86 Eds: This conversation was with Helmut Sonnenfeldt and William Hyland in late 1980/early 1981.

87 On efforts to limit improvements in the delivery accuracies of U.S. ballistic missiles, see Alton Frye, *A Responsible Congress: The Politics of National Security* (New York: McGraw-Hill, 1975), 68–83; and Donald MacKenzie, *Inventing Accuracy: A Historical Sociology of Nuclear Missile Guidance* (Cambridge: MIT Press, 1990), 268–270.

88 "As Secretary of Defense, [Schlesinger] recognizes that his function is to provide both intellectual and administrative leadership: 'You cannot control a department of this size; you have to guide it,' he said last week. I'm really a revivalist.'" "Watching Birds and Budgets," *Time,* 103, no. 6 (11 Feb 1974), 16.

89 R.W. Komer, *Bureaucracy Does Its Thing: The Impact of Institutional Constraints on US/GVN Performance in the Vietnam War*, WN-7244-ARPA (Santa Monica: RAND Corp., Feb 1971), FOUO.

90 G. Warren Nutter, *Some Observations on Soviet Industrial Growth*, Occasional Paper 55 (New York: National Bureau of Economic Research, 1957); G. Warren Nutter, *Growth of Industrial Production in the Soviet Union* (Princeton: Princeton University Press, 1962).

91 *The Wizards of Armageddon*, 204–219; *History and Strategy*, 3–46; Barry H. Steiner, *Bernard Brodie and the Foundations of American Nuclear Strategy* (Lawrence: University Press of Kansas, 1991), passim.

92 Roberta Wohlstetter, *Pearl Harbor: Warning and Decision* (Stanford: Stanford University Press, 1962).

93 "The initial stimulus for the book came from my friend, Andrew W. Marshall. For constant encouragement and advice through five years of research I am especially grateful to him and to Bernard Brodie and Harvey DeWeerd." Ibid., xi. Wohlstetter's work for the book was done as part of Project HAYLOFT. *Economics Division Progress Report*, 7; Social Science Division, *Social Science Division Progress Report, June 30, 1956,* D-3730 (Santa Monica: RAND Corp., 11 Jul 1956), 4.

94 Report to the President by the Security Resources Panel of the Office of Defense Mobilization Science Advisory Committee, "Deterrence and Survival in the Nuclear Age," 7 Nov 1957 (TOP SECRET; Declassified 10 Jan 1973), in Department of State, *Foreign Relations of the United States, 1955-1957, Vol. 19: National Security Policy* (Washington: GPO, 1990), 639–661. While on the staff of the Gaither Committee, Marshall was a member of a "small group [of] people with systems analysis-analytical kinds of skills [that] worked as consultants and experts to people in [the] other panels [and did] various kinds of tasks." Transcript of interview of Marshall by Goldberg and Matloff, 1 Jun 1992, 8. An organization chart for the committee shows an "Evaluation Quantitative Assessments" group headed by Robert C. Prim (Bell Telephone Labs) and Stanley J. Lawwill (Strategic Air Command).
Albert Wohlstetter, "The Delicate Balance of Terror," *Foreign Affairs*, 37, no. 2 (Jan 1959): 211–234.

95 Morton H. Halperin, "The Gaither Committee and the Policy Process," *World Politics*, 13, no. 3 (Apr 1961): 360–384; *The Wizards of Armageddon*, 125–127; Net Evaluation Subcommittee Folders, 1955-1964, Box 66, Records of the National Security Council, Record Group 273, National Archives.

96 Security Resources Panel of the Office of Defense Mobilization Science Advisory Committee, *Active Defense and SAC Vulnerability*, 27 Nov 1957 (TOP SECRET//RESTRICTED DATA; Declassified 11 Sep 1992), Declassified Documents Collection, 1993/001293.

97 A.W. Marshall and J.F. Digby, *Indications of Imminent Hostilities (U)*, D-1477 (Santa Monica: RAND Corp., 29 Dec 1952), CONFIDENTIAL; A.W. Marshall and J.F. Digby, *The Military Value of Advanced Warning of Hostilities and its Implications for Intelligence indicators (U)*, SM-14 (Santa Monica: RAND Corp., 20 Jul 1953), TOP SECRET (classification probably has been downgraded to CONFIDENTIAL or the document has been declassified). Digby discusses this work briefly in "Contributions of RAND to Strategy in the 1950s," in *On Not Confusing Ourselves: Essays On National Security Strategy in Honor of Albert and Roberta Wohlstetter*, 22. A short summary of SM-14 appears in Merton E. Davies and William R. Harris, *RAND's Role in the Evolution of Balloon and Satellite Observation Systems and Related Space Technology*, R-3692-RC (Santa Monica: RAND Corp., Sep 1988), 49.

98 A.W. Marshall and H.A. DeWeerd, *Weapons Limitations and the Defense of Western Europe*, D-3785 (Santa Monica: RAND Corp., 27 Jul 1956); A.W. Marshall and H.A. DeWeerd, *Weapons Limitations, Nuclear Sharing, and Graduated Deterrence—A Report on a Trip to Washington, SHAPE, USAFE, and the United Kingdom, October 29–November 29, 1956*, D(L)-4037-PR (Santa Monica: RAND Corp., 26 Dec 1956) [KC]. Another DeWeerd-Marshall paper, *Nuclear Weapons Limitations and NATO*, P-1013, 16 Jan 1957 is listed in Social Science Division, *Social Science Division Progress Report, June 30 1957*, D-4396-PR (Santa Monica: RAND Corp., 1 Jul 1957).

99 According to Kaplan, Loftus joined the Target Programs office, Air Target Division, Air Force Intelligence in June 1950. *The Wizards of Armageddon,* 209. As of May 1949, Loftus was Director of the Teaching Institute at American University. Joseph E. Loftus, "Strategy, Economics, and the Bomb," *The Scientific Monthly,* 68, no. 5 (May 1949): 310–320.

100 As noted earlier, Kaplan (*The Wizards of Armageddon)* says Loftus joined RAND in late 1954.

101 J.E. Loftus, *The CROSSBOW Campaign Against the V-2 Missile Threat,* D-6823 (Santa Monica: RAND Corp., 30 Sep 1959). A Sep 1960 RAND memo refers to a related AWM paper "in preparation," *Lessons from the Crossbow Campaign for Contemporary Considerations,* D-6824. Memo, E.S. Quade to Concept Group, National Security Seminar, subj: Seventh Concept Group Meeting, Wednesday, September 28, 9:00 A.M., Main Conference Room, M-5338, 21 Sep 1960, with attachment, "Publications of Possible Interest to the Concept Group, National Security Seminar," 21 Sep 1960, RAND Archives/J. Digby Files/Box: National Security Seminar, Publications/Correspondence, 1960–1961.

102 William J. Broad, "Russia Has 'Doomsday' Machine, U.S. Expert Says," *The New York Times,* 8 Oct 1993, A6; Bruce G. Blair, "Russia's Doomsday Machine," *The New York Times,* 8 Oct 1993, A35.

3 The Nixon-Ford Administrations: 1969–1976[1]

Introduction

When President Richard Nixon entered office in 1969, the United States military was deeply embroiled in the war in Vietnam. Although the totals had declined slightly from their peak the previous year, there were still more than 535,000 U.S. troops in South Vietnam and little prospect of victory.[2] The Tet Offensive in 1968, during which Communist forces suffered 37,000 casualties, was a tactical failure but a strategic victory for the North Vietnamese. It seriously set back U.S. efforts to pacify the countryside, terrified the civilian population of South Vietnam, and inflicted more U.S. casualties than the American public was willing to tolerate. The war was sparking growing domestic protests and squeezing defense budgets. The diversion of troops and equipment from Europe to Vietnam and the morale and readiness problems resulting from years of military operations also reduced the deterrent value of U.S. forces against the Soviet Union. Civilian leaders in the United States adopted a new approach—Vietnamization—that sought to decrease U.S. involvement by improving and modernizing South Vietnam's forces so they could assume a greater share of the burden and allow U.S. forces to withdraw.[3] Vietnamization was adopted as official U.S. policy after a National Security Council review in the spring of 1969.[4]

Vietnamization represented what later became known as the Nixon Doctrine. During an impromptu press conference in Guam in July 1969, Nixon outlined his views that U.S. allies should contribute more manpower for their own defense. The United States would continue to honor its treaty commitments, including providing military and economic assistance, and would furnish a nuclear umbrella to its allies, but would encourage and expect its allies to assume primary responsibility for their own defense. Nixon explained, "[W]e must avoid the kind of policy that will make countries…so dependent upon us that we are dragged into conflicts such as the one we have in Vietnam."[5]

While the United States was heavily engaged in Vietnam, the number of Soviet strategic missiles had grown substantially. By 1969, Soviet nuclear warheads totaled more than 10,500—far fewer than the 27,500 warheads in the U.S. arsenal but a nearly ten-fold increase from 1960.[6] Nixon wanted to change the character of America's relationship with the Soviet Union from confrontation to negotiation, and he sought to initiate talks on strategic arms. A prerequisite for serious strategic arms limitation talks, however, was the deployment of an antiballistic missile (ABM) system. The Soviets had already begun deploying a system around Moscow; Nixon believed that if the United States did not have its own system, it would have less leverage during the arms talks. In 1969, the Senate approved funding for an ABM system by one vote.[7]

The approval paved the way for the Strategic Arms Limitation Talks (SALT) in November 1969. The first SALT Treaty was signed in May 1972, in Moscow. The treaty limited the number of ABM launchers to 200 for each nation, although the number was halved two years later. The systems would be limited to the defense of the national capitals and one missile site in each country. The treaty also halted construction of additional fixed land-based intercontinental ballistic missile (ICBM) launchers and limited the number of modern ballistic missile submarines and submarine-launched ballistic missiles (SLBMs). Among the more controversial aspects of the treaty were the provisions that allowed the Soviet Union more ICBMs and SLBMs than the United States and the omission of any mention of Multiple Independently Targeted Reentry Vehicles (MIRVs). U.S. critics of the treaty feared that once the Soviets equipped all of their missiles with MIRVs, their numerical advantage would increase substantially, enabling greater flexibility and a strategic advantage.[8]

The emergence of nuclear parity between the superpowers coincided with a decline in U.S. conventional military capabilities. The war in Vietnam weakened Congressional support for military spending, which impacted force structure. Between 1969 and 1976, the U.S. defense budget decreased by nearly 29 percent in constant dollars, while active duty endstrength declined almost 40 percent. The number of tactical fighter squadrons in the Air Force fell from 93 to 69; the number of ships in the Navy dropped from 885 to 536; the Marine Corps shrunk from 310,000 to 192,000 marines, and the Army contracted from 1,512,000 to 779,000 soldiers. The cuts restricted the U.S. capacity to project conventional military power.[9]

In Western Europe, the United States sought to rationalize NATO's defense investments by convincing European allies to buy U.S. weapons in areas where American technology was superior, while the Americans would purchase European weapons when they were better. Several European states agreed to buy U.S. F-16 fighters and the United States agreed to purchase the Roland II, a Franco-German short-range air defense system. The United States also pushed NATO to approve a Long-Range Defense Concept that envisioned a capability to defend Western Europe without quickly resorting to nuclear weapons. The resultant document, approved in May 1975, placed a "major emphasis" on conventional capabilities but noted that those capabilities would require only a "modest annual increase in real terms in defense expenditure" because "NATO has already achieved a large measure of success in this regard." In the end, the improvements in NATO defense capabilities were limited.[10]

Although the SALT Treaty was intended to ease the nuclear competition with the Soviet Union, the Soviets were no longer the only nuclear-armed U.S. competitor. China detonated its first nuclear weapon in 1964 and its first thermonuclear weapon in 1967. By 1969, China possessed approximately 50 nuclear warheads.[11] The Nixon Administration undertook a number of reviews beginning in 1969 to identify options for dealing with the potential Chinese threat as it decreased the presence of U.S. troops in Asia. Nixon, however, began to doubt the policy of isolating China over the long term. Skirmishes between Soviet and Chinese forces along the Sino-Soviet border in March 1969 gave Nixon an opportunity to exploit the growing Sino-Soviet split. After another two years of debate within the administration, Nixon announced on 15 July 1971 that he would visit China in 1972. In the communique issued at the end of Nixon's visit, the United States and China agreed to begin moving toward normal diplomatic relations and not to seek hegemony in the Asia-Pacific region.[12]

In the Middle East, Egypt and Syria, still smarting from their disastrous defeat at the hands of the Israelis in the 1967 Six-Day War, launched a simultaneous attack against Israel on 6 October 1973. The Israelis were caught by surprise and suffered heavy losses, but committed mobilized personnel and equipment to battle as rapidly as they reported for duty. After 8 days of fighting, the Israelis had stabilized the Syrian front and turned their attention to the Sinai. By 24 October, they had moved three armored divisions across the Suez Canal and cut off the Egyptian 3rd Army from the remainder of the Egyptian forces.[13]

Early in the conflict, the U.S. national security establishment came to view the October War as a potential battle laboratory. It presented a unique opportunity to collect information on the performance of modern weapons and the effectiveness of tactics, techniques, and procedures. When Israel realized that it was running out of critical military supplies, particularly munitions, and turned to the United States for help, Secretary of State Henry Kissinger demanded a quid pro quo. In exchange for providing Israel with additional ammunition, fighter aircraft, and other war materials, the United States would have the opportunity to study the war up close shortly after it ended.[14]

Outline for 14 December 1993 Talk

[Note: Italicized items were not discussed during the interview.]

Introduction
1. Sep '69–Jan '81
2. Period in government to cover
    — Important antecedents
        • Intelligence involvement
        • Net assessment precursors
        • Long-term competition
        • Previous work in government
        • Bureaucratic strategy

Chronology Overview
1. Sep '69–Jan '72
    — RAND/Kissinger
        • Intelligence—White House
            –Flow
            –Quality
            –Bio
        • Net assessment ('70)
        • Intelligence reform
    — Consulting with Schlesinger
2. Jan '72–Nov '73: NSC
    — Intelligence
        • NSC Intelligence Committee
        • Case studies
        • Organizational behavior experience
    — Net assessment: NSSM 183, NSSM 186 start up
    — Advice to Schlesinger/DCI Nov '71—
3. Nov '73–Jan '81
    — Schlesinger/SecDef/Start of Net Assessment
    — *Rumsfeld*
    — *Brown*

Antecedents
1. Intelligence
    — '52: Strategic warning/Digby—five months in Germany
    — '55–'56: Strategic warning—Alaska, Japan, Hawaii
    — '56: CIA consultant
    — '65: Bross, NIPE
2. Net assessment
    — '68/'69 (?): RAND papers for Wikner
3. Long-term competition/Strategic planning, '69/'70
    — '65–'68: Harvard Business School
4. Previous government connections
    — '52: Five-month stint in Germany
    — '57: Gaither Committee
    — '58: Surprise attack talks, four months
    — '61: Nuclear accidents (three to four months)
    — '62: Cuban missile crisis aftermath
    — '63–'65: Paris, delegation to Norway

- '66/'67: Consultant to Rowen at OMB (with Schlesinger)
- '65–'72: DCI/NIPE

5. Bureaucratic strategy
   - RAND/reflections on impact of studies/ideas
     - Proposal for RAND sales department
     - Letters/memos more impact than reports/timing/tailoring
   - Organizational behavior analysis
     - Timing
     - Adviser to high power
     - *Crisis*
       - *Budget trends*
     - Changing minds—framing problems/vocabulary (Goldhamer). Leads to choice of diagnosis, not therapy. Revulsion against hubris of systems analysis/how much is enough(?), reflections on

Stories

1. Sep '69–Jan '72
   - Meeting with Kissinger/last vacation
     - Motives of Kissinger
   - Second meeting, Nov—tasks
   - Dec—China analysis
   - Jan—getting started on first task
     - Interviews
   - Quality of intelligence—missile forces; leads to SS-9 study—exemplar
2. Net assessment story—delegation from Blue Ribbon Commission—[______?]
   - Bio/Panama coup—structure
   - Summer of '70 – Kissinger request
     - Back pocket program if Soviets were bad
       - Wayne Smith; AWM, Schlesinger (OMB), Bill Kaufmann, Fred Hoffman, Charlie Herzfeld, Larry Lynn
       - [Prior review?] of NSSM 3—strategy—AWM/Schlesinger—first net assessment/long-term competition framework/asymmetries/trends/organizational behavior
   - Dec '70—Nixon—>Kissinger/Shultz: reform intelligence community/25% cut
     - Smith/Schlesinger put in charge; AWM is Kissinger's representative/most work done by OMB
     - Report in Mar '71
     - Final report in Sep 71; AWM drafts directive/memos; Haig/Smith broker (insertion of Net Assessment paragraph); issued Nov '71
     - Negotiation to come on board—do so, Jan '72
     - Tweedy/Helms reactions/role of DCI/Intelligence Community staff, Air Force general
3. Jan '72–Oct '73 at NSC
4. Appointed to CNO Executive Panel (spring)
   - Focus on intelligence
     - Laird response—slows Net Assessment
     - Pick Pickett—advice of R. Christensen—reasons for choice—reform of organization—need strategy
     - (Robin Pirie (Pat Parker))
     - NSC Intelligence Committee—report card/grades
       - Tweedy as [?]
       - Case studies: crisis—insights; Yugoslavia—insights; special case of Oct '73 war

5. *Request for [_______?]; background; reaction; insights*
   - Reaction of CIA
     - Promises/poor response
     - Why so [PO'ed?]—change in Agency, '50s vs. '70s
6. Schlesinger as DCI
   - Note to/context of
   - Trample stairways at 4:00 pm
   - Scared middle management/Proctor
   - Big failures
     - Chinese missile program
     - Soviet defense burden
     - Psychological profiles
     - Probability estimates/multiple outcomes
     - RAND work on reproducible scoring systems—insights
     - *No training for analysis*
     - No customer analysis/interviews—results
   - Start of Net Assessment
     - Slow/delayed until Spring '73
     - Working out scheme (role of DoD [_____?]
       - NSC tasking
       - DoD (other) performance
       - Experience of first/Parker
     - Schlesinger negotiation—request soon after arrival at DoD
7. Schlesinger as SecDef—3 to 4 page paper—advice when he took over
   - Reflections on Kissinger's motives/interest
   - SS-19 fiasco—Haig/Duckett mistakes in estimate of volume
   - *Insight into submarine program (saw first reports/meet with skipper)*
   - Raising Soviet submarine project/visit/security—Schlesinger view of
   - Summing up of period
     - Madness of arms control seen close up
     - *Kissinger contrast of Chinese and Soviets*
     - Awful state of intelligence community/analytic weakness—*lack of management skill/*final disenchantment
8. DoD
   - Schlesinger period: Nov '73–Nov '75
     - Start with three areas
       - NATO/Warsaw Pact
       - Strategic
       - Maritime
     - Three tasks
       - Assessments—>SecDef
       - Stimulate Services (net assessment)
       - Research program/tools and methods
     - Intermediary with Intelligence Community/CIA
       - Somalia (story)
       - Military burden
     - *Visit to Marines—role for Marines*
   - *Rumsfeld: Nov '75–Jan '77*
     - *Lots of requests*
       - *Strategic/missiles*
       - *Deterrence*
     - *Short assessments*

- *Rumsfeld questions on Navy*
- *Initial efforts at strategic planning (July '76), Roche*

— *Brown period*

- *Best, most productive period*
  - –*Developed analysis effort—final shape of assessments*
  - –*Brown as reader/action-oriented*
- *Special success*
  - –*C3I*
  - –*ASW assessment*
  - –*Power projection*

1969–1975[15]

Marshall: I targeted to talk about the period from September of '69 through January of '81. I'm not sure we are going to make it. I think there may be too much. At any rate, my notes run out a little bit before then. It's really the period of the first chunk of my time in government, up essentially to the end of the Carter Administration.

I thought also that I would go back and say a bit about some important antecedents to this. Now, we've covered some of this already, but not in, as I got through it, as much detail, or things had been left out in the earlier discussion, because they weren't, in some sense, intellectually that important.

There's a bunch of antecedents in my relations with the intelligence community, or the intelligence area. There is this kind of precursor work that I did on net assessment while at RAND, and then there is this long-term competition set of ideas. Also it might be worth talking a little bit about my previous involvement in government [indistinct] stems [indistinct] RAND. A number of these things have an impact on the bureaucratic strategy I adopted.

So the period, I think, has something like the following breakup to it: From September of '69 to January of '72, I was really at RAND, but working a lot for Henry Kissinger when he was Assistant to the President for National Security Affairs. Really about half time, I think. Also, because James Schlesinger had left RAND in, I suppose, January of '69 to join the Nixon Administration, I was also doing consulting with Schlesinger at the Office of Management and Budget. And so, you want to cover some of those kinds of things.

Then there's the period of January [April] '72 through November [October] of '73, when I was actually on the National Security Council staff, and we need to talk about that. Now, simultaneously with that, Schlesinger left OMB [Eds: Office of Management and Budget], went to the Atomic Energy Commission, and then he became the Director of Central Intelligence. And I spent a fair amount of time talking with him or wrote things for him during that period.

And then, there's the period of November '73 to January '81, here in Defense, with Schlesinger, Donald Rumsfeld, and Harold Brown.

So I guess I will just talk about some of this and about particularly these antecedents, just to kind of list some things.

The intelligence world: As I mentioned earlier, I had gotten drawn into that by the work I was doing at RAND quite early because a lot of it was connected with targeting the Soviet economy. I got very much involved in that.

But probably the most important thing was in '52 there had been a request to RAND from the man, his name was General Millard "Poppy" Lewis, who was the chief intelligence officer to USAF in Wiesbaden. Among the topics he wanted worked on was strategic warning. Jim Digby and I were asked to work on that. We ended up, I ended up spending five months in Germany, in Wiesbaden working on the strategic warning problem.[16]

Then in '55, '56, Joseph Loftus and I, again, were working on the strategic warning problem, again, for the Air Force. But by that time, the Air Force had created a worldwide warning system where there were particular nodes. One was in Alaska, which we went to and stayed a week or so at that. Another was in Japan, and we spent time there. Also, because of the nature of the information available then, we spent a lot of time out in the field in various communications intelligence collection centers. And also spent a couple of days, I think, in Hawaii. At any rate, I was very much involved in the COMINT world; I mean the strategic warning problem.[17]

Then, in '56 [1958], I became a Central Intelligence Agency consultant, which lasted through to the time I joined the government.

Q2: You were no longer at RAND?

Marshall: Yeah I was at RAND.

Q2: But you were going to work for the CIA?

Marshall: Yes, as a consultant, right. Then in '57 through maybe '63, with Loftus, I ran this Sovoy project that we talked about before.[18]

But probably more significant from the connection to the intelligence community was that when I returned from France, in the spring of '65, I was asked to come and talk with John Bross, who I had not known at all before. I don't know quite why he wanted to see me. But, at any rate, we talked, I think probably based on some of the things that Loftus and I had published about how to improve estimates and forecasts. Bross had been set up by John McCone [Director of Central Intelligence, 1961–1965] with something called the office of National Intelligence Programs Evaluation, the NIPE staff, which was the predecessor of the Intelligence Community staff, eventually.[19] Bross had been an Office of Strategic Services officer in World War II. Really one of the "princes of the realm" within the CIA establishment and clandestine services. He had probably started with a lot of money, had made a lot of money on Wall Street, married one of the Marshall Fields daughters. Anyway, he liked me.

There were very few consultants for the NIPE staff. It was quite small, and there were just two or three consultants, I think. So I came to learn a lot about the top-management kind of problems from his perspective and advising Richard Helms [DCI, 1966–1973], who he was a close friend of. That connection continued, again, also up until the time I joined the government.

So from the early '50s on, I had been pretty involved in intelligence, particularly this connection with Bross was from this period. At any rate, it opened up a lot of issues and problems to me.

On the net assessment thing, I think I've already mentioned the papers I wrote for Fred Wikner, and probably we will have to go back to the RAND efforts and locate some papers that I wrote then, Research Memoranda and so on. Wikner had set up the Net Technical Assessment effort in the Defense Department. It was probably set up in '67, so somehow during '68, '69, I spent time working on some problems for Wikner, and I can't quite remember exactly what I wrote. But then that's something we can search in the RAND library.[20]

Then, in addition, with regard to my previous governmental kind of connections…in '57, I was a member of the Gaither Committee staff, and involved with that from about August through December.[21] In '58 [Aug–Dec], I was involved in the preparation and as a member of the delegation that went to Geneva in connection with the Surprise Attack talks with the Russians.[22] Then in '61, I was on a special subgroup of the McCloy arms control study that was set up right after the Kennedy Administration took over, that looked at nuclear accidents and what you could do to prevent that. My work for that group went back over the whole history of the whole series of close calls we had.[23]

Q2: Who was that for?

Marshall: It was run out of the State Department, at the time, but the other people on the thing were—well, I guess Paul Nitze came over to testify—but people like Marv Stem, Thomas Schelling, me… I forget some of the other people now, but there were maybe six, eight of us.[24]

Q2: Was that Policy Planning?

Marshall: Could have been.

Q2: It wasn't intelligence, it wasn't the Bureau of Intelligence and Research?

Marshall: No, no.

Then in the immediate aftermath of the Cuban missile crisis, I was involved in some of the efforts to follow what was continuing to go on, but also this issue of what had the Russians maybe learned from the thing and so on. Then…

Q2: That was for the CIA?

Marshall: No, that was for Harry Rowen, a Defense Department thing, because Rowen had come to Washington, along with Charles Hitch and several others from RAND at the beginning of the Kennedy Administration.[25]

From August of '63 through about March of '65, I lived in Paris and was part of a delegation to NATO.[26] In the course of that, Rowen had me go head a delegation, make a couple of visits to Norway, because the Norwegians had approached the U.S., wanting kind of bilateral discussions about the defense of Norway.[27] In their view, if you looked at AFNorth [Eds: Allied Forces Northern Europe], and who was supposed to come to help them, the U.S. was not very well represented in AFNorth at all. Yet the only people they had any faith in who were ever going to come and help them was the U.S. So they wanted to open direct discussions about what was really going to happen, as against purported NATO plans for what was going to happen. So I did that.

And then, in the period around probably '66, '67 [1965/66], Schlesinger and I were consultants to Rowen when he was at OMB [Bureau of the Budget].[28] Some of it was related to defense, but a lot of it was related to this effort that President Lyndon Johnson started to try to spread the wonderful cost-effectiveness analysis to the rest of the government.[29] Anyway, Rowen asked us to help on that.

I've already mentioned this consulting to Bross from '65 through '72.

The other kind of antecedent thing that I think is worth talking a little bit about is the kind of bureaucratic strategy I adopted when I finally did actually get in government, which, on reflection, seems to me to reflect the impact of my interest in organizational behavior and studies, things I have been thinking about since the late '50s. Based on all of that, I had the view that organizations were pretty hard to change, and to influence very much, and that timing was very important.

It also, even when I was at RAND, had led me to feel that the RAND way of conveying the results of studies was very poor. In the first place, the studies themselves were often not very tailored to the organizations that you were really trying to change. That, in fact, coming out of some of the work that we had done in the middle '60s on organizations, led to this proposal that all RAND studies have an implementation section where the issue of how we would actually make something happen ought to be addressed. Since, practically speaking, no studies ever had that, I guess Charlie Wolf coined a phrase for saying that was the "missing chapter" in all the studies.

Also it seemed to me that very often the pattern was that the study would be completed, and then the results would be briefed, but be very badly timed with respect to the acceptability or responsiveness of the organization, and that actually RAND ought to set up a kind of sales department. That the strategy for RAND ought to be that there ought to be people who focused on the customers and exploited past work, no matter what its date. Then, the opportunities for really inserting the ideas arose.[30]

It also seemed to me, in my own experience, that letters and memos rather than the study reports were much more effective than the kind of standard briefings, the kind of rigmarole that usually went with it.

Anyway, I came into government with the view that the only way to really affect things was probably either to be a close adviser of somebody who had sufficient power to actually try to change things—but even then people overestimate what even people who are supposed to be in charge can do—and that the alternative was that you had to change people's minds. And that you could do that, in part, by changing the way in which they looked at problems, framed problems. Also, I guess, through Herb Goldhamer I had come to appreciate the importance of vocabulary in changes in that.

All of this led to the choice with respect to this whole net assessment effort of diagnosis, not therapy, that immediately recommending what to do was not going to work very well. Secondly, that people ought to think out their own solutions.

Also, it's worth mentioning that I had been struck by what happened to Alain Enthoven [Assistant Secretary of Defense for Systems Analysis] and the whole systems analysis business. One of the ironies is that when the systems analysis stuff was being thought through and argued through back in the '50s, I mean right from the beginning, after considerable discussion the whole notion was that for systems analysis, there was a class of issues that it could address and there were larger, more important issues that it couldn't. That, in particular, this notion of how much was enough, or how much you ought to spend on, say, the strategic bombing effort, warfare thing was something that it couldn't deal with. That people basically had to assume that there were decisions that were going to be made on some much more general level of broad allocations and that the systems analysis function was to, within already generally established budgets, to help in spending it in cost-effective ways.[31]

When Enthoven came to record the history of his years in the Pentagon, as I say, it's very ironic that he should pick the title *How Much Is Enough?* for the book with K. Wayne Smith, because that's precisely what was decided, after long argument, was not the kind of problem you could actually address.[32] But I guess in response to Secretary of Defense Robert McNamara's demands of whatever, or Enthoven's hubris, that's where it ended up, trying to settle those kinds of questions.

OK. So then let's go to the beginning of September '69 through '72, which was this period I was working for Kissinger.

Q2: This notion that notes and memos are more effective than briefings—how do you square that with, I think, your notion that people don't read anymore? I mean, do you give them stuff and then...

Marshall: I don't think they will read long reports. Well, some people obviously do. I mean...

Q2: Harold Brown.

Marshall: ...or Schlesinger. But the ideal thing is probably to sit down and just talk to them. But I think that sending them a letter that's fairly short and makes the simple points and so on, and is tailored to the occasion, to some problem they might have and so on, is just a hell of a lot better than some report that comes in the door.

Q4: I am most intrigued by the consumer focus here. Your point that you almost do studies, you stockpile them and then you wait until the opportunity is there to do something useful.

Marshall: Well, you may brief them at the time for whatever good that is, but yes. So that what a place like RAND tends to offer is all of the past studies, not just the one which is just coming off the press.

Q4: I don't want to get you off the path of your book, so let me just put a topic on the table. I think that in this chapter it would probably be very useful to put a couple of paragraphs or a couple of pages to what was happening to the intelligence organizations that you were starting to deal with at that point in time in their own evolution. I mean now we are in the 1970s, we are 20 years in the agencies' formation, you have senior managers who came from the old school, etc., etc., etc.

Marshall: Well, yes. We'll get to that, I think, anyway.

Q4: Because it influences both how you affect them and how effective they were able to...

Marshall: Well, as you know, I think the CIA of the '50s and early '60s is very, very different from the '70s, much to the discredit of the '70s.

Q4: You could raise a broader issue that whenever you have a new organization, CIA, you saw it at a certain time, in other...you get these particularly high-quality people and then they gradually...

Marshall: And I think I mentioned that Bross himself was very concerned about that. I remember him talking to me, talking on the people he already saw in the late '60s. His view was that these people didn't know anybody else but people in this agency, and that the people who originally formed it or who were recruited in the '50s, throughout the '50s, had a lot of other associations, they knew other people in the United States and elsewhere. By the late '60s, you had the effect of this career design choice that the Agency made of trying, for security reasons, to take people right out of universities and have a one-career lifetime. There were people that spent their whole time in there. The result is that they don't know anybody else, I mean in the truly professional way, lack breadth of experience, and so on, contacts.

OK. So let's start in September of '69. The reason I start the thing there is that in September '69, just before I was going to go off on a long trip to Europe with my wife Mary, Kissinger called up and said he wanted me to come to see him. I went to see him on my way to Europe, I stopped in Washington a couple days and it was then that he told me that he wanted me to come and do what was then, he thought, a two-month study to look at the flow of intelligence into the White House. He wanted that done because President Richard Nixon was extremely dissatisfied with the intelligence coming into the White House. Henry himself was very, very negative in the way he talked about the thing.[33] Evidently, fairly early in the Administration, Nixon and Attorney General John Mitchell had taken some of the early National Intelligence Estimates and read them closely, and thought they were abominable in terms of the writing, the logic, everything. Henry also was very down on the political analyses. He thought it was worse than that available in the newspapers, which was lousy. So I agreed to do that, but told him I was going on this vacation, and...[gap for tape change]

I did that. The pattern of my life at that point was that at least every couple of years, I would go to France, spend six weeks touring around, having a good time. So I probably didn't get back until late in October. And I had these other responsibilities at RAND. I got a couple of nagging phone calls from Al Haig, Kissinger's military assistant, asking when was I going to show up and so on. I finally did show up, and that was essentially the first week in December of '69.

I had a long discussion with Kissinger and it was decided that I would do two studies. One was this study he originally talked about, which was on the flow of intelligence into the White House—Where did it come from? How did people decide what they were going to send? What could be done to fix that and help that? The second study was to be a kind of quality report card on intelligence, and after some discussion, the area that was to be focused on, was that of the strategic missile forces of the Russians. I began work on those.

The only other thing that is worth mentioning is that in addition to that, I did a third little study. While I was there—I was there from essentially the first part of December through the first week of May of '70—a coup occurred in Panama. One of the things that happened immediately was that questions were asked about, "Who the hell are these guys? What do we know about them?" The answer turned out to be not a hell of a lot. So Kissinger wanted to know why was that, why didn't they know more? What was the actual state of the biographic intelligence? So I looked at that and indeed drafted a kind of guide book about what…you really want to know about these people. I did read a bunch of things and they were poor, unless you were thinking about entertaining these people and wanted to know whether they like scotch or bourbon, and things like that. Not much information there.

So I went to work on that thing, although I was also sidetracked on a couple of things, like the Russians had just threatened the Chinese and I got involved with, you know, "What do you think about that? What do we do?" and so on.[34] At any rate…

Q2: Are you going to say a few words about your studies?

Marshall: Yes, right. What I did on the first one about the flow was, I went over to the Situation Room, which was the focal point for stuff coming in, to find out, Where did the stuff come from? Then I went and found out, but through interviewing people, talking with them, how each of the organizations decided what they would send. I suppose the most interesting part of that was, say, the contrast between the National Security Agency and the CIA. NSA perhaps was favored by the fact that traditionally the head of the Sit Room staff came from NSA, but nonetheless, every couple of months, they sat down with that fellow and a couple of other people, I guess, and went over what were the topics, the areas that seemed of most interest. Then internally they put out a kind of directive of, "These are the topics of which material is most wanted," and they changed it every couple of months. CIA, in contrast, made no effort to ascertain the areas of most interest, but, judged by what was in *The New York Times* to determine the topics. I don't know whether they felt very alienated from this administration or what it was, but they just made no effort. The other thing was that they had this product, this President's Daily…report…

Q2: Brief? PDB? We called it PDB. We called it President's Daily Brief.

Marshall: Right, which they sent down every day. By the time I had arrived on the scene, it was very clear that Nixon was not reading it. So the question was, "Why the hell are you guys sending it? And how do you decide what you put in there?" So one of the things that I did was to read in the very early part… I got copies, for like the first six months of the administration, of that document. Very quickly the NSC staff had begun to generate a competing daily analysis. So I read all of those. Also, since Nixon was a person who, if he read anything, almost invariably made marginal notes, you could tell that fairly quickly he had stopped reading the CIA product and continued to read from the other one. I also did a statistical analysis of the degree to which these two things covered the same topics. The NSC one always had more topics, and the overlap between the two, where they covered the same topics and so on, was only about 60 percent. It didn't have so many more that that's the reason why there's more. But at any rate, the point was that CIA just was not customer-oriented at all. And they also refused to give up. They understood that they weren't being read, but it was seen as their premier product and they weren't going to give it up.[35]

Q2: Who was doing the NSC one, was it coming out of the Sit Room, or off the NSC staff?

Marshall: It came out of the NSC staff. I'm sure there were some Sit Room pieces, it was assembled in the Sit Room, but it was prepared by the regional people just from that one.

Q4: Did you try to do any similar sort of analysis of the flow of information going from the Sit Room to NSA?

Marshall: No, no. So I wrote a report describing the whole thing to Kissinger. And, I guess, I made a number of suggestions to him. But there wasn't a hell of a lot that could be done about that. I told him that as far as I could tell, whatever was there was really arriving, and that the analysis was awfully poor, which he knew, and that…

At any rate, I don't remember that there was anything particularly striking that I could recommend to him on the quality of the intelligence, which I looked at the forces, went over what was available, gave it relatively low marks. Of course the NIEs were part of that, it was not only the daily or weekly sort of things. I remember that one part of the report dealt with what seemed to me the decline of the old key staff in the Office of National Estimates, which when it started out… the whole notion of it was that it was to be made up of people largely from the outside. It would be some people from the intelligence community that would become parts of the Office of National Estimates, but that retired military guys, or ambassadors, people from the business world, or academia would be in it. By the early '70s, '69, it had become the last stepping stage of retirement. People's careers would end within the CIA, not just within the community as a whole and that it was enough. There was still like a general, maybe somebody from State, but everybody else was really from CIA. So it had lost its original character, didn't on the whole do that great, I mean with analysis.[36] And if Henry wanted to change something, …the quality of things, I mean that's one thing that needed to be changed, focused on.

The other thing that I suggested, which led to a follow-on study, was the notion of picking some part of the Soviet missile program, and, in some sense, by spending a lot of time directing and advising and so on, to get an exemplary study done. This led to this special study on the SS-9 and its antecedents. The study was aimed at trying to understand, "Why does this thing look the way it does? What's the whole history? What is the design bureau?" and so on, everything about it. Also, there were some reasons to believe that the successive generations of Soviet missiles had been designed to reduce the number of men involved in their day-to-day functioning and so on. The objective was to do a study that was both historical and had a bureaucratic cast to it and so on. That study was done and turned out to be moderately good, not great, but…

Q2: And that was your study? I mean you did this study?

Marshall: No, I supervised it. After sending in the report in May, I agreed to do that. So then over the course of the next nine months to a year I supervised that.[37]

Q2: Who was the principal…was this just CIA or the whole intelligence community?

Marshall: This was CIA.

Q2: CIA. And can you say who the CIA person was that was working primarily on that?

Marshall: I don't remember the names. There were some people... But it was a fairly large effort.

Then what happened was, while I was continuing to do that, Kissinger in the summer of '70, asked Wayne Smith, who had just come on board [indistinct] taken Larry Lynn's place—this is the Larry Lynn who went off to the University of Chicago, who had been in Systems Analysis prior—to do a private, separate study of program augmentations that might be undertaken if the Soviets misbehaved, proved recalcitrant, and so on. So Smith assembled a group to do that. That group consisted of myself, Schlesinger, who was still at OMB, Bill Kaufmann, Fred Hoffman, Charlie Herzfeld, and Larry Lynn, who was coming back and doing that. That's the group that I remember. I think Herzfeld did some stuff on research and development programs. Kaufmann, I think, did something on tactical forces. Hoffman did something, I think, on strategic forces. And as the general, broad background to all of this, Schlesinger and I did what was, in effect, a first net assessment. Herzfeld very much argued that we should do that. That is, give the broad background of, here's what the situation with respect to military balances are, here you have the long-term trends, and so on. Actually, I did most of it because Schlesinger really didn't have the time to do that.

Let me mention one thing going back. When I was writing up this thing on the quality of the intelligence, a delegation of three people, whose names I can't remember, came to see me from Defense. These were people who were on the staff of this Blue Ribbon Panel. They were in the midst of an argument which they finally resolved in favor of recommending this creation of the net assessment activity in Defense that was to report directly to the Secretary.[38] That was made to Secretary of Defense Melvin Laird in the summer of '70, but he never did anything about it. They came to lobby me to include in my recommendations to Kissinger the reestablishment of the net assessment activity at the NSC, which I did not include.[39] It didn't seem to me that it had to do with the quality-of-intelligence issue. They had an argument for doing that, which was reasonable in some ways, but anyway, I didn't pick that up.

The idea of that came from a panel member who—I tracked back afterwards, and, again, I can't think of the guy's name—had been the chief executive officer of TRW.[40] Anyway, he was on the Blue Ribbon Panel, and he was the person within that whole committee that argued very much that what the Secretary of Defense needed was a net assessment group. In fact, under his scheme of things, it was a net assessment/strategic planning activity that on an organization chart was drawn as a box to the side with a line directly to the Secretary. He believed that what was needed was some way of pulling together some comprehensive picture of what the state of affairs was and where things were going, what the biggest issues were, and so on. As I say, I later had a talk with him in which I was trying to understand what motivated him.

At any rate, I did this net assessment thing. In fact, I have a copy of it someplace here.

Q2: Did you call it that, "net assessment?"

Marshall: Yeah.

Q2: Is it still classified?

Marshall: Yeah.

Q2: Getting about to the point where we might want to think about declassifying that.

Marshall: Well, I don't know. Again, this whole effort was done very quietly, as an unadvertised thing privately for Kissinger, to provide him with initiatives that he could, the way he put it, have in his hip pocket.

Q2: Well, you probably wouldn't want to, or you may not want to declassify that part, but the net assessment part in some sort of way.

Marshall: One could look at it, but...

The next thing that happened, as we were just, I think, bringing this effort to a close, because it started in the summer, was that in December of '70, Wayne Smith reported that Nixon had met with Kissinger and George Shultz, who was then the head of OMB, and kind of pointed at the both of them and said that he wanted a study undertaken of the reform of the intelligence community, that he wanted to make major changes and also wanted to cut the budget 25 percent. I heard later that one of the things that happened was that he had, I guess while flying out of Washington on Air Force One, flown over the CIA building and looked down and saw those huge parking lots and said, "What in the hell are all these people doing? I don't see anything coming out of this that's worthwhile." I don't know that that was the whole thing. There were many motives. But at any rate, Shultz put Schlesinger on this and Smith was asked by Kissinger to form up this effort that was going to look at this thing.

Smith didn't have the time, so he asked me, when I'm still at RAND, to be Henry's guy on this thing. Of course Schlesinger owned this cell within OMB that followed the intelligence community, and also had Bill Kaufmann working for him as a consultant. So most of the work, or a good deal of at least pulling stuff together, was done by the OMB people. But they were so in the pocket of the intelligence community that they didn't in the end have much to contribute about what you ought to do about these guys.

There was an initial report in March of '71 and a final report in September, maybe a little earlier, probably August of '71.[41] Then some decisions were made and, I don't know who did some initial drafting, but in any case, in the middle of September, I undertook to draft the final kind of decision memoranda, the letters and memos to Helms and to this one and to that one, Laird, etc.

It all ended in being issued in early November of '71.[42] Schlesinger by this time was off at the Atomic Energy Commission. So I wrote the thing and then Haig and Smith brokered it around, just to see it got acceptability. It was they, somehow, who inserted a couple of paragraphs about this establishment of a net assessment activity at the NSC and started to lobby me to take that, which I did in January [April] of '72. I talked to... well, Bross had been replaced by Bronson Tweedy, another Establishment figure, also from the clandestine services, who left and ran the Audubon Society for a while.[43]

I talked to Tweedy and so I kept him informed as this study went on. But it became very clear that Helms really, wasn't seeking any change, didn't want the added responsibilities that changes would entail, because under this thing, the DCI was given a lot of new responsibilities, or old responsibilities which had never been exercised. And the DCI was then given this Intelligence Community staff to improve his ability to manage the whole budgeting and direction of the things...

Q4: Would it also be fair to say that Helms could see himself ending up in a fight with the Secretary of Defense?[44]

Marshall: Yeah, right.

Q4: Frequently non-political.

Marshall: Right. Finally, after spending a lot of time while still at RAND, half in, half out of the government, I came to the NSC in January [April] of '72 for real as a government official.

Q2: Now, did you report to Henry?

Marshall: Yes.

Q2: You directly reported to him?

Marshall: Yes, right.

Q2: Where were you located? Where was your office?

Marshall: In the Old Executive Office Building. In fact, initially, I had an office that Admiral Robert Welander, the guy who was cashiered, well, anyway, kicked out of the NSC because he had this naval enlisted man who had been...

Q2: Reporting stuff back to Admiral Tom Moorer in the Joint Chiefs of Staff Chairman's office.

Marshall: Yes, right, right. Then I changed, exchanged offices with John Lehman, who was down in, I forget our final number, but it was like the third floor, like 304 or something like that.

Q4: This was originally meant to be a temporary assignment for you, you were going to be here for just a couple of years.

Marshall: Yes, right.

Q4: And you find yourself getting ensnared in the Washington scene.

Marshall: Yes.

Q4: How you managed to maintain your sort of "terrorist" role of upsetting large bureaucracies over a period of a number of years. So maybe we could explore this in this book. Or maybe your friends and associates would like to explore it.

Marshall: Yes, yes. OK. So I came on board and it was just me initially. Then I got a secretary via NSA, Carol Doheny, and then Lexi Crow, and then I recruited Chip Pickett, and then later Robin Pirie, Joan Hunerwadel. So there were, I guess, just six of us.

Q4: Actually, it was just five, because Carol left. Pickett arrived in early July, and Carol was not there [indistinct].

Marshall: We really had two kinds of functions. One of which, on the intelligence side, really took up by far the most time, because while the thing was called the Net Assessment Group, I think the reason the thing was created was they wanted someone to monitor this implementation of this intelligence reorganization. Also, this reorganization had created an NSC Intelligence Committee that Henry was to chair, and he needed somebody to staff that and to work that.

And then we were to start up this net assessment function, which turned out to be difficult, because while Laird had never done anything about this summer 1970 recommendation that he create a net assessment activity, the morning after this memo came out from Nixon, he created a Director of Net Assessment, who was given no staff. Wikner was appointed to it, and the whole idea was that he, in that role, was part of the Secretary's own office staff, and that task forces would be set up and there would be no permanent staff of any sort, it would just be ad hoc'd and so on. In fact, as far as I know, they did absolutely nothing; I mean they didn't do any assessments. The one thing that got started, in fact—I guess by then Wikner had left—was done, not in the Secretary's office, but under this Net Technical Assessment group.

Another thing I would mention is, why did I choose Chip Pickett, or how did I have that happen? Well…In the first place, my notion was, what is the nature of the problem here? It is to have these organizations change, and therefore I want somebody who is interested in management, organizations, how to make them function better. So I went to my friends in the Harvard Business School and said, "Do you have any recommendations?" I talked to Joe Bower and I talked to C. Roland Christensen, who were the two people who I knew best.

I had met Bower and then, through him, Christensen, and I had involved the both of them in a special panel or group that I had put together, probably in '67, when Ivan Selin was the head of Systems Analysis [Deputy Assistant Secretary of Defense (Strategic Programs), OSD Systems Analysis]. Selin had seen some of the stuff that we had done on organizations of the Russians and so on, Russian military, and he wanted a special look at that. So I recruited Jim March, Christensen, Bower, and myself to look at what did we really know about the Russian military organizations, how it was put together, how they functioned. We met several times in Washington and then wrote a report about what we felt was new, what could be done in a war, and so on.[45]

Q2: When was that?

Marshall: '67, I believe.

I can't remember whether Bower had any recommendation for the NSC Net Assessment Group, but Christensen had two. One he mentioned, a person who was already in Washington, who I think was in the government, but might want some other position. I talked to him, he was a very engaging fellow, but really wasn't interested. He wanted to leave government and make some money and so on. The second recommendation was George [Eds: "Chip"] Pickett, who had the advantage of being in the Army, an intelligence officer already, who was about to graduate from Harvard Business School, and who might be the first. So I talked to George Pickett and decided I wanted George Pickett.

So then, I guess the main thing to talk about… I mean, again, with so much of this it's hard to say that it has much to do with intellectual development.

Q2: What's that?

Marshall: I mean if this is an intellectual history…

Q2: No, I think it is. I think it is a collection of people and why…

Marshall: No, no, I understand that. I understand.

Q4: Tell me another thing. I kind of like the shaping issue, sort of what was going on. At that point in time Bower was teaching this course in business policy, and Bower, who started out as— He wrote his dissertation on financial analysis and capital investment in some industries, and what he did was find out that the bureaucratic interpretation of that phenomena was better than a financial interpretation, and wrote a book on it. The other thing is that he had the draft chapters from the *Essence of Decision,* decision-making during the Cuban missile crisis.[46] And that was assigned reading in his course. It turned out that the vast majority of the students in the class really didn't like those chapters. They didn't find that it was very relevant as a framework for analysis. There were some who were absolutely captivated by it. So there really was sort of an accidental— The point is that, even the business world at this point was beginning to capture and teach their students the whole issue of how administrative behavior and what it meant, and how one had to think about it, more than just simply [indistinct].

Marshall: Well, Bower had been one of the participants in that…

Q4: The Ernest May study.

Marshall: …the May Group that I talked about last time. Right.

Q2: You want to say a few words about why you picked Robin Pirie? How?

Marshall: I picked him, I guess, on the advice of Pat Parker, who had been in Systems Analysis with him and told me Pirie had come a cropper of the Admiral, I guess, Hyman Rickover, and would be available. I interviewed him. I think it was kind of Parker's recommendation, early availability, and so on.

Q2: So he was still in the Navy then?

Marshall: Yes, oh yes.

Q2: I mean, he didn't get out as a result of this problem with Rickover or anything?

Marshall: No, no. One of the things I found when I got to the NSC was that while they said I could have so many people, it was also clear that I had to get some people for free, which people like the two of them were. The other people…

Q4: They had to come out of your budget.

Marshall: …were on the payroll.

Q4: There's another point that is sort of a subtext. You made a statement many times in the past that when new intellectual disciplines are developed, they sort of are formed and certainly begin to see their formation in groups, in academia. And if the agencies were doing their job, what you would do would be to penetrate, and

I don't mean that in the negative sense, penetrate the academic community and find out who was beginning to think about these subjects long before they were written about. And the issue of organizational behavior, administrative behavior was one of those aspects.

Q2: Was there a specific problem that you saw putting Pirie on, the way you saw putting Pickett kind of on this organizational…

Marshall: No. I thought that Pirie would be the person who would spend most of his time on the assessment activity, while Pickett and I would be the main people working on the intelligence business. Then, as I have already mentioned, the net assessment thing took a long time to germinate, because we had to just overcome a lot of bureaucratic obstacles, because Laird just didn't want to do anything to cooperate in any way. I think the most interesting stuff that came out of the work that Pickett and I were doing was in connection with this NSC Intelligence Committee. One of its functions was to provide feedback and appraisal of the performance of the intelligence community in meeting the requirements of the top-level people. So we undertook several case studies, one set of which had dealt with crises that had occurred, for which we were trying to understand the adequacy of the intelligence community support to decision-makers in the crisis periods. Out of that came one of the biggest lessons I felt that I learned, which was that when we set these task forces to work to look over the past—Chip kind of shepherded most of these, we put them to work under somebody—the first version of the report always was, "All things considered, the intelligence community…"

Q2: …did a damn good job.

Marshall: Yes, right. Then you had to go back and forth and finally you learn that things weren't so hot. I just came away with the feeling that there was an enormous gulf that separated the people in the intelligence community from the top-level people, certainly from Kissinger and Nixon. The two groups of people had a totally different view of the world and what the nature of the crisis was, what the stakes were. And so, if you put these people to work to try to grade the thing, they came up with these good grades, and it took some time for them to even see that maybe things had not gone as well as they thought. Later, when I used to give some talks to mid-term training sessions of the Agency people, I used to tell them that I thought the situation was a little bit like: Here's a chessboard, and there are a couple of players at the board, and they have a set of advisers. The problem with the advisers is that they don't know what the game is about, what the moves of the pieces are.[47]

I think in several cases it was very clear that for Kissinger and Nixon, how things came out mattered more than it did for some people who seemed to feel that people are dying, we have to end this thing as soon as possible, almost any terms will do; whereas people like Kissinger and Nixon saw crises as opportunities, as well as problems that had arisen that had to be terminated as soon as possible. Also, they had goals, like U.S. prestige and so on, that these other people had very little interest in, as far as I could see. There was even a case, I forget which of the ones, probably the India-Pak crisis, or something like that, where the question was asked, and the reply back from the CIA was so poor—in fact I talked to the CIA people later about it. Whereas people like Nixon and Kissinger cared a lot about the details of the outcome and worried about long-term consequences, these other people thought there were no long-term consequences, so that the question that had been posed was one that their response didn't answer at all. So you just have this difference that has to be taken account of and maybe you correct.

The other thing I remember was we took a look at Yugoslavia, to look at the political and economic analysis, and convened a set of people. I remember Nathan Leites was involved in looking at some of it, and I got Joe Kircheimer involved in comparing the analysis of the Wall Street banks with the intelligence community's analyses. Nathan's view, I remember, was, "Why don't they translate and circulate the best articles from German and Swiss newspapers? They're much better than anything they do." On the economic side, the comparison was very much to the detriment of the intelligence community.[48] Again, you come away with the view that there is just a lot of expertise and knowledge out in the world, and a lot of it is better, more relevant than stuff these guys produce. And they ought to be in the business of mining that more than they are.

Then we had the great good fortune that while we were there at the NSC we had another big crisis. The Yom Kippur War occurred. One of the things that we had found when we tried to find out, in fact, what various estimates people had even read, or taken any account of, in doing the earlier cases— When the war started, I set Pickett to work, and he started interviewing people while the war was going on, to try to understand the relationship, and the performance of the intelligence community. Pickett wrote a very thick, really terrific study. I think two copies were sent out, one to Kissinger, who was then at State, and one to Schlesinger over at the Defense Department. I have no idea what has ever happened to any of those.

Q4: The Net Assessment staff were, however, banned from talking to the Defense Intelligence Agency.

Marshall: Yeah. Right.

Q4: Interesting set of stories there.

Q2: Did you keep a copy?

Marshall: I don't know, we had a copy for a while. I don't know where it is now.

The thing to comment about is the reactions of the Agency during this period. When I first came on board, Edward Proctor [Deputy Director for Intelligence] and others came down, pledging great cooperation, support, and everything, which quickly evaporated, I would say. Had a very, very poor response. It was kind of like my final disenchantment with these people, my feeling of how badly they compared, I think, to people that were there in the '50s and early '60s. Often the failures were just so obvious, but they never seemed to learn.

I remember one of the first things that happened when I was there, and I think it happened prior to Pickett's arrival, was that Kissinger had asked for a special study on the Chinese missile program, to make doubly sure that we really understood the Chinese missile program. There came back a report which said, "Everything is fine. We have multiple satellites up already, and when we get a new one up it's going to even be better. And we, with high confidence, can tell you that they don't have any missiles, yet." A few months later, it became clear they had missiles, that we had been missing them. In fact, one of the reasons given for why we were sure they didn't have them was precisely the reason we had missed them. The reason for absolute surety was that with both we and the Russians, the people who supposedly know the only way to do things, you have these things where you have versions of the way you're going to deploy them at the test sites, and then you see exactly what they look like, I mean the silos or this or whatever, the launching pads and so on, and then you are able to look and find them when they start deploying. Nonsense. That's not the way the Chinese did it. They obviously from the beginning had put a lot of emphasis on not being detected very easily. So some evidence appeared, as I say, a few months afterwards which showed they had them. Then when they went back, they were able to find them four or five years earlier. And yet, it never teaches them anything, as far as I can tell.

Q4: This is the group that when the Arab-Israeli war breaks out in October, it got off at about four o'clock on Saturday morning, at nine o'clock on Saturday morning in the Pentagon the strategic warning staff meets, this is the [indistinct] operation. And they have a debate that goes on from nine until 11 or 12 o'clock in the morning, and at that point they issued the DCI's official position that this is not a war, but simply a minor skirmish that is occurring and the Arabs and Israelis will go away.

But I think there is a theme that you want to sort of file away to raise. I think, maybe you disagree, there is this period between say 1967 and '69 on the one hand and, say, 1975, where lots of sort of major institutional traumas take place to the intelligence community, and I would guess probably the strongest period of sort of shock change to the organization. One of which is Schlesinger becomes DCI. He's the first outsider as the DCI. But then there's a whole bunch of other things. The failures in Vietnam intelligence and all the controversies...

Q2: The Church committees...

Q4: Two committees in the middle part of the '70s just absolutely blow apart the insulation of the intelligence community from the Congress and permanently set it aside.

Marshall: Yeah, yeah, right.

Q4: And there are a bunch of things we can talk about...

Marshall: Well, again, obviously we can come back to this later, but the CIA was a complete flop on the Soviet economic things, the defense burden, which was something I'll come back to a little later. Also indicative of the situation was that Kissinger was very interested in one of the products they produced, which were these psychological profiles that Jerry Post and some people out there did.[49] They were seen by the rest of the intelligence establishment as sort of very low products, I mean not our first-line estimates. And yet Kissinger very much wanted them, valued them, and you had to have a big fight with them to get them to really produce them, to do more than they were doing.

Another thing was that we wanted to get them to attach probabilities to their predictions, first to give alternative outcomes, not just one outcome as a forecast, and secondly, to attach probabilities. That was a great trauma for them and you never succeeded in dealing with it. During this period, RAND had been doing some work on a thing that's technically called "reproducible [reproducing] scoring systems."[50] It's a way of assigning a score for how close you are to the right answer that provides an incentive for someone supplying an estimate to provide, in fact, his best estimate, not to play games with the receiver of the thing. They had developed a really neat device of multiple-answer questions with three answers. They had developed it, as a demonstration, for a whole series of different fields, like Roman history or 17th-century music, or various kinds of things. The virtue of it with the three potential answers was that with a light pen you could register your relative probabilities among the three by the point on this triangle, the area of this triangle that you could get. If A, if you were 100 percent sure, you put it at…

Why don't we get back to that, I guess. Alright, maybe we just ought to take that up, the transition of Net Assessment from the NSC staff to the Defense Department. The main thing to say about the transition, I think, would be that Kissinger in the summer of '73 had been appointed the Secretary of State and had the two jobs, but he was beginning to spend more and more of his time over at the State Department. I guess this all happened at the same time that Schlesinger came to Defense. I guess he actually arrived on the 2nd of July. Fairly soon afterwards, he called me, and called me every week or two about coming over and setting up a net assessment office in Defense. I had been somewhat hesitant about doing that. In the first place, I wasn't sure I was going to stay in Washington—ha, ha, ha, ha, ha. Secondly, Mary for various reasons was against it and so on. But then what happened was really a negotiated deal with Kissinger to transfer the whole office, not just that I would come over and start over again, but that the whole office and responsibility for net assessment, national net assessments would be transferred with me to Defense.

Q2: Who negotiated that?

Marshall: Schlesinger and Kissinger between themselves. So that's what happened. In fact, someplace around here there probably is a National Security Study Memorandum that transfers the thing.

Q2: NSSM 186?

Marshall: No, no, there's several NSSMs. There's a NSSM that sets up the NSC net assessment process, which probably came out in the early part of '73, when we finally reached agreement about how we were going to proceed with net assessments from that point on, which was that they would be formulated through some process at the NSC level, with consultation and so on of the departments, and then they would be assigned to some department to carry out, depending on the character or focus that they would have. So there was a NSSM on that.[51] Then there was a NSSM, NSSM 186—that's where Project 186 got its number—which was the first net assessment that was to be done, which was on ground forces.[52] It wasn't completed until I came over to the Defense Department, when it was conducted or headed on the government side by Pat Parker, who went to work for, I forget the name of that fellow who was his superior, who…

Q4: Not in Command, Control, Communications and Intelligence? Assistant Secretary of Defense for Intelligence?

Marshall: Yeah.

Q2 or Q4: Parker.

Marshall: Yeah. Then there's another NSSM [NSDM—National Security Decision Memorandum] which transfers responsibility. That's later. To kind of sum up this NSC period, I would say— Well, let me go back before doing that.

One of the other things that happened, of course, in this period was that Schlesinger became the DCI. I sent him a letter, handwritten, so I don't have a copy of it, telling him what I thought the problems were that he was going to face. I remember the opening sentence or two says, "You're taking over an organization with very low morale that compares unfavorably with the people in the '50s." I made the point that then they were very activist, sought outside help, worked long hours; now it's become a routine eight-to-five job for these people. They seem uninterested in anybody else's information or knowledge, and so on. Schlesinger later told me, "You certainly were right about that." He said, "You could get trampled on the stairways at 4 o'clock. I don't know why they put lights in this building."

The other thing about Schlesinger, that was already mentioned, was that he was the first DCI from outside of the Agency. Well, that's not right. McCone had been from the outside.

Q4: McCone?

Marshall: Yes, he was a businessman.

Q4: But no intelligence experience in the war?

Marshall: No, I don't think so. At any rate, Schlesinger just scared them. In fact, there were a bunch of the middle managers, including Proctor, who just hid when he was out there. It's hard to know exactly why he was so frightened.

Q4: Well, his personality, as I recall, would lead one to feel that way. He did not tolerate fools, to put it mildly.

Marshall: Yeah. He did, of course, even before he went out there… Well, in the first place, let me just mention another part of the story. He and I had been talking for some time because, while he took over in, I think, February of '73, right after the election in '72, Nixon went up to Camp David, and then called the various heads of agencies up there. He called Helms and, at least second or third hand, told him he was not going to be continuing.[53] When Schlesinger came up there, to talk about the Atomic Energy Commission, which he was head of, he was told that he was going to take over. So he and I began talking right after that; this happened right after Thanksgiving. In December, he and I started talking about what do you do with this place, who are the good guys, and various things.[54] As I say, I wrote him this longish letter about the situation and things I thought that needed to be done for long-term goals. Then when he initially took over, I started spending weekends up there, meeting with him, helping him [indistinct].

Q2: Not literally the whole weekend?

Marshall: No, Saturdays. Then later when he took over as SecDef, I again wrote him something, and there's a three-, or four-, or five-page thing that's around, somehow that got typed up and there are copies around, about my perspective on…a longer-term look at the world and where things had [indistinct], issues that he ought to think about, and so on.[55]

To come back to this period from the NSC. The first thing I would say is my impression ended up that Kissinger, even going back to the initial studies he had me do and then this monitoring of the reorganization of the intelligence community and so on, that he fundamentally was not interested in all that. He was doing it, I think, because Nixon wanted something done. I think Kissinger thought the intelligence products were terrible, but he is not someone who is at all focused on improving organizations, changing organizations, making them better. His view, I think, of the NSC was, "We can get what we want out of the intelligence organizations, for the most part, so why do we care?" Schlesinger had a totally different view. Wherever he went he was very interested in the organization and changing or making it better.

In fact, that may be one of the big distinctions among officials, one way of thinking about the people who come into government. That is, there are some people who just want to do as well as they can during the couple of years they are there and are interested in performing well, or seeming to perform well. But they are totally uninterested in the organizations or the processes. Then there's a set of guys who are really interested in trying to change things or make them better.

The other thing is to look back at the performance of the intelligence community. Just really bad. Let me give you another example. I remember being called over and going into the Situation Room. Then Carl Duckett [CIA Deputy Director for Science and Technology] came down to break what was really bad news. In the negotiations with the Russians, we had put a lot of emphasis with respect to the SS-19, about what its size or volume could be and there were some specific provisions in the treaty.[56] Those provisions had been guided by past Agency estimates of what the volume or size of the SS-19 was. Duckett was there to tell us that the calculations had been wrong. There was an error in them that had not been detected for several years. It had been there for years. It was a simple thing of some number had been divided by another number rather than the square root of the number, or some damn thing of this sort. At any rate, the -19 was bigger, very likely bigger, enough bigger that it would cross this threshold that was set in the SALT I treaty. So if the Russians signed it as they probably did, or would, or had—I can't remember the date—they were already in violation of the thing, or they were intending to violate it. I remember Haig and I and, I forget, there were a couple of other people who were in there… I recall Haig being pretty upset about it all. But again, it was just symptomatic that here's this organization, probably the TRW guys had made this mistake, nobody in the Agency had ever checked the calculations or knew enough to check the calculations. Now, this in effect… not real serious…

Q2: Does this relate to the SS-13, and that the -19 could only be so much bigger than the -13 and there was an error on the size of the -13 or something like that?

Marshall: No, this, as far as I remember it, was that we had an estimate of what the -19 was, and the whole thing was an issue of trying to constrain...

Q2: You could only grow so much in volume.

Marshall: Yes, that's right. The point was that it was already bigger than that. And you had spent based on their estimates, had negotiated this set of constraints.

Q4: And we had it wrong.

Marshall: And we had it wrong.

Q4: Is this the -9 or the -19?

Marshall: No, it's the -19.

Q4: It was how much it could grow depending upon whether they became big missiles or small missiles.

Q2: This is a modernization issue and it had to do with the -19, which was not yet deployed, but was coming down the road, and I think was a modernization to the -13.

Marshall: Yeah, could be.

Q2: Then the issue was how much could it grow. And there was this error in calculations that I had known about, I mean I didn't know about this thing, there was this error in calculation, but I had thought it was based upon our estimated size of the -13 so that therefore the projected growth meant the -19 could be so big, but we were wrong about the size of the -13, which meant that the -19 was going to break the limit.

Marshall: It could be. My memory is not that good. It could be.

Q4: But the main point of it was that it was rotten arithmetic.

Marshall: Yeah, but the point was that it was one of these things where the arithmetic was wrong. And nobody had ever checked it; it had been around for several years.

Q2: Now, Henry had one of those, too, where his arithmetic was wrong, on equivalent megatonnage, as I recall, because it's the megatonnage to the two-thirds and he was doing the square root and I don't know where he got that from.

Marshall: Yeah, right. I guess one of the fun things about this period was the raising-of-the-Soviet submarine project.

Q4: Glomar Explorer.

Marshall: Yep. An interesting reflection: I talked with Schlesinger about it because there was a set of issues about whether to go forward with it.[57] And it goes back to an idea, which we'll maybe get into, of Schlesinger's, and mine, too, this business of thinking about things you do in terms of their impact on other people, on your opponents. His view was, yes, this is risky, but from time to time the U.S. should pull off some real exploit, just to shake up the other guy.

Q4: To keep them unbalanced.

Marshall: And this is sort of, therefore, the right kind of thing to do, not only for its purported immediate payoffs that you get, but its perceptions, psychological aspect of things.

Q4: Is this the period in which you were doing, they did the work on the anomalies, the 56, the 58 anomalies, whatever we called them. This is the things that the Soviets are doing which we can't find an adequate explanation for.

Marshall: Yes, I think that work was going on, and Herzfeld was much involved in that kind of thing. Yeah, I think so.

Anyway, the period, though, is kind of crazy, all this arms control nonsense going on. When you see it up close and see how irrational it all is, and how subject it is to these problems like this arithmetic about the size of the -19. And also, arms control has the problem of controlling some dimensions of things that don't have any impact on anything at all.

Q2: Or have negative impact.

Marshall: Yeah, right.

Q2: They just don't figure out what they're trying to fix.

Marshall: Right. No chance of [indistinct].

Q2: Like sausage. The more you see, the less…

Marshall: Right.

Let's talk a little bit about the Schlesinger period here and this transition. Schlesinger and I sat down and talked, once I got here, about what areas he's particularly interested in having some assessments done. There were three: the NATO-Warsaw Pact balance, the strategic forces balance, and the maritime balance. So we started off to do things in all those areas.

Q2: Who did you have working on those?

Marshall: I think that at that point, well, Chip Pickett started off on the NATO-Warsaw Pact.

Q4: When Marshall came over to the Pentagon he took Robin Pirie with him, actually moved everybody in the office except one, and there was just the same office replicated here. It was that way until…Lee Badgett?

Marshall: Yeah.

Q4: …joins in the middle of '74, something like that.

Marshall: Right.

Q2: And he was a Navy guy, right?

Marshall: Air Force.

Q2: He was here to do the strategic forces?

Marshall: Well, no. He came and because of his background, he was really more associated with the Soviet military budget and that sort of thing.

Q4: He came from the Air Force Academy faculty?

Marshall: Yes. And then James Roche replaced Pirie.

Q2: When was that?

Marshall: '75, the middle of '75, or spring of '75.

Q4: Robin retires and goes to the Congressional Budget Office or something.

Marshall: Right, yeah. Then we have some additional thing, I guess Peter Sharfman is the next one.

Q4: No, I think Bob Sorley comes in …

Q2: To replace Pickett?

Q4: …late. No, because he arrives, Sorley must arrive after Jim. Pickett leaves the summer of '76, Sorley must be there for six to nine months before then…

Marshall: Yes.

Q2: Did he pick up the NATO stuff?

Q4: I think, yeah, well, he was only there…

Marshall: But Sharfman came, because in the spring of '76 he was involved in these Israeli talks. Toward the fall, we started meeting twice a year with them.[58]

Q2: I thought Roche did those.

Marshall: He and Sharfman.

Q2: Oh, they both did.

Marshall: Anyway, Sharfman took over the strategic things. At some point, Fred Giessler shows up.

Q2: But when you first started out, Schlesinger wanted these three studies. You had Pickett working NATO-Warsaw Pact. Who was working strategic forces?

Marshall: Me, I guess.

Q2: And then Pirie did the maritime balance?

Marshall: Yeah, started that.

Q2: And then you did strategic forces. Who did you finally hand that off to?

Marshall: I handed it off to Sharfman.

Q4: But it doesn't get handed off for probably two years or more.

Marshall: Yeah.

Q4: Part of that is that nobody had…nobody in the office had his depth. So the others were kind of…the two of them, kind of going along after him, trying to keep him supported.

Q2: So Sharfman comes on board in '75.

Marshall: I think late '75.

Q2: After Roche.

Marshall: Yeah, probably.

Q2: But before Sorley.

Marshall: Yeah, right.

Q4: You have to check the dates. I'm just not sure.

Marshall: Yeah, right. You'd have to check those and so on.

OK, so we are doing these. We basically have three fundamental tasks. One was to do these assessments for the Secretary, proprietary assessments to him. Then the idea was that we were going to try to stimulate each of the Services to create a net assessment activity. The only one that succeeded was the Navy. The Army didn't do anything, they created a point of contact, which was Ed Vandiver. The Air Force didn't do anything except some special projects later, around '76, '77. Admiral Elmo Zumwalt, Chief of Naval Operations, created a net assessment activity.

Q2: Who headed that?

Marshall: Later it's Bill Manthorpe, but earlier it was this fellow who had a heart attack and died, didn't die but [indistinct]. Yeah, you'd have to go back, right.

At any rate, the third thing was to undertake a development program of tools, methods, databases, and so on.

In addition to these formal things, Schlesinger used me a lot as an intermediary with the intelligence community. Soon after he came here, he called Proctor down and said, "I just find it hard to believe that the Soviet military effort consists of six percent of their gross national product. Will you go take another look at that and tell me why any reasonable man should believe that?" When I came, he told me about that and then we had some further meetings. Then they finally decided they were going to write something and look at it, and they came back with a paper that was totally unconvincing, that said the most it could be is, like, about one percent off in either direction. Then in '76, of course, they… they'd known about it for about a year, but in '76 they had to change it and move it up to 12 to 13 percent. And I have the paper, the memos that I wrote for Schlesinger back to them.[59]

Q2: What forced them to raise it?

Q4: Oh, it was the inflation rate.

Marshall: No, what forced them to raise it was an emigre who came out, who had seen some books that were convincing to them.[60] Plus, when they went and did some work on the dollar-ruble exchange rates for some selected systems that they had some evidence on, they decided that they had been way off for modern systems.[61]

Q2: So that's in '76 that they raised it to 12 percent?

Marshall: Yeah, I think early '76.[62]

Q4: What I thought was interesting about then, and I was really on the periphery of it, was that there had to be some sort of seminal event, one, two, or three that would force this estimate, and these people to make a change. There would have to be an emigre who came and said, "I actually saw the books and it's wrong," or there was some very sensitive intelligence information in several cases that would enable them to get a window of insight, which was not a random sample window, but it was a window. And they have to go back and question themselves. But they couldn't bring themselves to go about doing it on their own.

Marshall: Yeah.

Q4: One of the problems with the community, it could not…it needed an external stimuli that was sort of uncontroversial in its implications.

Marshall: In some ways, there's not a lot to talk about, I think, about the Schlesinger period, in the sense that, the period of startup…Well, there is one other thing. As I mentioned, we had started this assessment on the ground forces. One of the first things that we did when we got here was to take over responsibility for that and having it finished. When the final report was being prepared, and we'd not had much to do with it up to that time…

[gap for tape change]

…it called for a listing or discussion of the strengths and weaknesses on both sides. When it came to the final thing, the intelligence people, the DIA guys, simply would not put down any weaknesses on the Soviet side. At one point they had made the preposterous explanation that, yes, they might have some weaknesses, but even if they did exist, they would notice them and fix them, so that it wasn't worth talking about them or taking much notice of them. I remember saying to them, "Would that that were so about our weaknesses." So it was resolved, finally, by Lt. Gen. Danny Graham, who had just taken over DIA. And talking with him… Danny and I had gotten to know one another, I guess, while I was still at the NSC. We had talked a lot about Soviet weaknesses, so he wasn't in any position to say that the Soviets didn't have any weaknesses. So he got it changed. But it's just symptomatic of these people. The constant weakness of the intelligence community, this unwillingness to really deal with, or to have an intelligent assessment of the opponent's weaknesses.

Q4: Let me give you an example. There were a whole set of four or five myths that they had about the Soviet Army. One of which was that the Russians are stupid because they design the T-62 tank and only somebody who was less than about six foot, five foot-eleven, or something like that could fit into it. And they made a big play out of it. And there was in a meeting with one of these intel guys and a couple of people from the Army made this comment, and Pickett said, "Pardon me, but do you have demographic data on the average height of the Soviet male?" "Well, yes we do." "What is it, what is the average height of the Soviet male?" "Probably about 5-foot-6." And Pickett said, "So are they designing the tank…" Pickett said, "What is the height of a male who can climb in the Army M-60?" And the answer was 6-foot-4. And Pickett said, "Where is that on the height distribution?" Guy on the Army staff said, "It's on the 99th percentile." Pickett said, "Shall we go back and try this all over again about who is stupid?" And it was just these… the guys who were doing [indistinct] analysis for the Defense Intelligence Agency, had just a very narrow focus on that. It was just absolutely phenomenal. And there are other stories.

Marshall: Yes. The other thing to say about this period is—I guess we talked about it before, the last time—that one of the things that Schlesinger was very interested in, and, in fact, both of us were because we had talked a lot about it back at RAND, was this whole aspect of the political-psychological impact or use of military forces, their role beyond fighting, or getting your way without fighting by looking impressive enough, and so on. I think we'd both been interested in that. I had been interested in that going way back because of Goldhamer and Leites and so on. Something that was very important for both Schlesinger and I, while we were both at RAND, was reading Robert Ardrey's *The Territorial Imperative*, which immediately raises the issue of the nonrational aspects of man and his behavior.[63] Our interest was based partly on an obvious fact of history, that coercion and display of forces and so on are used to gain your ends without fighting, and also on the possibility that this is not just a matter of, you show up and the other guy does a set of calculations and so on, but that there are other mechanisms at work and that you ought to understand that and so on.

So one of the other things I did for Schlesinger, after sort of getting on board, was to start this whole program of work on perceptions and so on. Now, actually, when I was working for Henry, I had already started down this road by making a request to the Agency—again, Henry was a naturally interested party in this kind of thing—and so we… Remember this Bob Levine?

Q4: Um-hum.

Marshall: So we assigned this one guy to do this work. The idea was to try to look at everything we could find from the Soviets to find out: "What is it that they pay attention to about us? What sorts of things seem to give us high scores in their books? What things don't matter or give us low scores?" In other words, how can we then start to do those things or show those characteristics which impress them and influence their assessments?

When I got here, Schlesinger was very, very interested in this. It was one of his absolutely fundamental views about things. So I started this work. I got some money from the Defense Advanced Research Projects Agency. Stephen Lukasik was the head of DARPA then, and I think, probably, Schlesinger had called him. So I went over, and one of the first projects we started was the stuff that started out at RAND with Herb Goldhamer in charge. There was a meeting at RAND.[64] That's when I first met Lionel Tiger, but I had read his book before that. I had read both *Men in Groups* and *The Imperial Animal. Men in Groups* probably came out in around '70 or '69, or something like that [1969].[65] *The Imperial Animal* came out in '73 [1971][66] We invited him and a bunch of others and started off this whole series of things.

Q2: Who was it— You said RAND was taking the lead on this?

Marshall: Yeah.

Q2: Who was doing it there?

Marshall: Herb Goldhamer. And he continued on with a whole bunch of things up to the time he died.[67]

The other thing to say about this period is, again, this kind of contrast between Schlesinger and Kissinger. Kissinger, I think, really had a very pessimistic view of America's future. He really did, and at that time, given what the hell was going on in this country, one had some rights to be pessimistic. But Schlesinger wasn't at all. [Indistinct] In fact, I think one of the most interesting things that Schlesinger ever said—I think it was in some *Time* magazine [11 Feb '74] article—which I think people never understood very well, was that he thought of himself as an "evangelist" [in the article, Schlesinger says "revivalist"].[68] A lot of the conflict between Kissinger and Schlesinger had to do with this very different view of what the long-term strengths and weaknesses were between ourselves and the Soviets. Once again, this was probably one of these areas where having some sense of economics and so on, even if it's the Agency's, which was pushing out all this bum dope about the Soviet economy, in the sense that [indistinct] the U.S. is not badly off, but a…

Q2: Were you involved in the move toward limited nuclear options at all at this point?[69]

Marshall: Yeah, right, although Donald Cotter, Assistant to the Secretary of Defense for Atomic Energy, was the main person involved in that.

Q2: But your strategic force assessment, I mean, did you have one out, or in some form that Schlesinger looked at before they started doing that, or was that preceded…

Marshall: No, no, no. That came about because of Schlesinger's own work when he was at RAND. One of the major things he did before leaving…well, it started back probably around '66, '67, '68.[70] One of the things he spent most of his time on was this issue of "the least miserable option," which was the way it was talked about. You know, if you really face up to the thing of what are you actually going to do after deterrence has failed, then this notion of simply unleashing the Single Integrated Operational Plan is—

Q2: —is silly.

Marshall: Absolutely, nonsense.

Q2: Cotter was at AE, or where was he?

Marshall: He was in AE. When Schlesinger was at the Atomic Energy Commission, he picked up a bunch of people who became very close to him and he brought along. One was Cotter. Cotter maybe had been down at Sandia National Laboratories, I forget exactly. Another was Hoffmann.

Q2: Fred?

Marshall: No, the fellow who became the Secretary of the Army later on.

Q2: Oh, Marty Hoffmann.

Marshall: Marty Hoffmann, who was a lawyer. He was the legal counsel, I think, at Atomic Energy and Schlesinger brought him over and]made him the legal counsel here. The secretary, Evelyn Irons, who he brought with… well, maybe she showed up first. She maybe was with OMB and then went with him to…yes, she was first with him in OMB. So there were a cluster of people that…

Q2: Were they your main allies in the Building [Eds: the Pentagon], or who were your main kind of two or three, besides the Secretary, who were you kind of—Zumwalt and you close at all?

Marshall: A little bit, and I was on that Chief of Naval Operations Executive Panel that he had formed. I joined it in '72, when I was still at the NSC, and continued while here. I would say that Cotter was a big ally. I liked him, although Lukasik was very helpful. Cotter and Pete Haas, Deputy Director for Science and Technology of the Defense Nuclear Agency…

See, the problem was that they wanted to do a whole bunch of things and I was in a very good bureaucratic position to help them, because, after all, Net Assessment can do anything, right? I mean look at almost any issue. So they wanted to do a lot of work on the Soviet doctrine, and exercises, and so on. It all started with a DNA-DARPA effort that began in probably '73, '72, and was called the New Alternatives thing. It was an effort to look at technologies and what could they do for you, and it was there that there was this first, not really first, but a major push toward precision-guidance weapons. By the time I moved here in '73, Cotter had, of course, come before me, and he and the DNA people very much wanted to look at how you would attack the Soviet forces. And there was this multiplicity of interests. On the one hand, our position in tactical nuclears meant that they had to remain dispersed until they tried to converge for attacks quickly, but then you could try to attack them so that [indistinct] disorganized [indistinct], but in order to do that you had to look a lot at how they would really come down, and find the details of their operational arrangements.

That's some of the things they wanted to do. It all verged on being intelligence and this and that, and they would be precluded from doing it. But if I tasked them to do it, then they could do that, or they could then get the intelligence people to go along with whatever. So they came seeking me. Also, Haas came offering lots of money, not only for the things that they wanted to do, but things I wanted to do. Additionally, I had no budget, so I got money for studies out of DNA and out of DARPA.

Q4: A thing about the bureaucratic aspects of this: somewhere in this same period there is, what's it called, Net Technical Assessment, NTA, run by Jim Wade at that point? '75? '74?

Marshall: No, it was that retired Army general, John something.

Q4: There was a Navy Admiral also. Maybe it was a retired Navy admiral at the time.

Marshall: Yes, oh that's right.

Q4: Yeah, it was a retired Navy admiral.

Marshall: Yes, right.

Q4: Then there's another guy named Don Marshall—

Marshall: Yes.

Q4: —who was in, I never did figure out where his position was [Special Assistant (Policy) to the Assistant to the Secretary of Defense for Atomic Energy], but he had assessment or net assessment with Schlesinger.

Marshall: Well, he had been out at DNA.

Q2: He was an Air Force colonel, I think.

Marshall: Yes, right.

Q4: This one was, he was an Army colonel, Don Marshall. And then there was the famous Tony Cordesman grab for Net Assessment which occurs in this period.

Marshall: That comes later, during the Rumsfeld period.

Q4: Oh, right.

Marshall: Which we haven't got to and before which we ought to break off.

Q4: You know, at some point in this story, I think you, hope [indistinct] would sort of do a, what did we learn from the tapestry of these anecdotes about net assessment and how it goes [?], and how you establish it, and so on and so forth. My footnote/comment at this point is that there is a story which has to do with how difficult it is to do this kind of analysis, and therefore how hard it is to convince others to engage in a particular [indistinct]. When we were trying to convince the Army that they ought to be doing things, we took on, sort of… For example, there was the tank crew training net assessment comparison…

Marshall: Yes, now that happened later… well, maybe not. I forget…

Q4: …well, the point was that there were people who were saying, "Why would we want to compare ourselves to the way the Soviets train their tank crews, because our training establishments don't fight. Our tanks fight. So who the hell cares if they train one way and we train another." It's obviously an extraordinarily, in my view, just simpleminded point for them to take, but you would say it was an example of how difficult it was to engage them.

There was another aspect of the net assessment one, sort of really subtext, was just simply getting people to compare things.

Marshall: Yes.

Q4: Because of what you might learn by asking, "Why are they different?" People were very reluctant to do that.

Marshall: Remember the other thing that Pickett was involved in we had this captured equipment.

Q2: Oh, from the '73 war.

Marshall: Yeah, right. And there were all kinds of stories about this stuff and how terrible it was, and how difficult it was, and so on. So…

Q4: In particular, one was this business about Russian tanks are so much better than American tanks. No, American tanks are so much better than Russian tanks.

Marshall: Yes, right.

Q4: So you basically forced a runoff between an [Eds: American] M-60 and a [Eds: Soviet] T-62 out at Aberdeen.

Marshall: Right.

Q4: And took an American crew and drove one, and took an American crew and drove the other and put them over various obstacle courses. And then they got back these observations like, "Well, yeah, you're right, the M-60 has an automatic transmission and the Russian one does not, and the M-60 has a steering wheel and the Russian has a sort of lever." But the driver said, "I mean it's no big deal. It doesn't require that much more force to shift gears," and so on and so forth. Then there was another one about, well, the crew compartment [Eds: in the T-62] is much too tight, and the comment from the loader or the tank commander was, "This is the only tank that I've ever been in where you can probably sleep while you are driving, while it was being driven across… because you are wedged in tight enough that you don't bang your head against every piece of the…" I mean it was just one thing after another one.

Marshall: Right, yeah, right. And Pickett went around and interviewed the people who were putting these things back in working order and had driven them a lot just to see… See, a lot of the claims were that these are so uncomfortable, so bad, that people, once they're in them for a while, just can't function. And one of the tests for the tank involved the fact that the scoring system was related to the accuracy and pointing and firing the gun; the notion was that they would get so tired that the scores would go way down. It didn't turn out to be true. And there were wild stories about, you know, people can't do it, their backs go out, this or that. Then there was the ZSU…

Q4: 23-4?

Marshall: Yeah.

Q4: Another sort of piece, there were a lot of stories, but one of the pieces that has to do with how the intelligence community does the analysis of these kinds of systems… For example, if you handed them a T-62 or you handed them a ZSU-23-4 and asked them for an appraisal of sort of the crew comfort and operational effectiveness, for example, what they gave you back were these very precise psychometric measures on vehicle, sort of, this will take a person who has a sitting height of 34.675 inches, to three-digit accuracies, which is the 60th percentile of the average height of the European male, and it has a leg room of 32.7215 inches, which will accommodate da-da, da-da. And after you got through reading this tremendous amount of data…

Q2: "What does that mean?"

Q4: And they were proud about the output, the people who produced this, and it was just absolutely just narrow, but they were very enthused about the output. But it had no relevance and it wasn't like it had relevance to anybody, you could determine. It was a repeat of what we had seen elsewhere, and people were producing things because that's what they had produced before, and the mere fact of the production was a reward in its own right. The fact that it wasn't very useful was irrelevant.

Marshall: I forget about timing of the tank crew thing, but it ended up in the hands of [Gen.] Paul Gorman, who was then at the Army Training and Doctrine Command. Best single study every done by any Service for me.

Q4: Gorman has this meeting with Andy. Door closes. And he looks at Andy, and he says, "Mr. Marshall, off the record, we have a national disaster here." That was his lead in, and it got better thereafter.

Marshall: Right.

Q4: Probably can't write that up.

Marshall: Right, right. "Give us time to fix it. Do you mind if the study is delayed?" [Indistinct] later fix this.

Q2: He was the National Intelligence Officer for General Purpose Forces...

Marshall: Yes, right.

Q2: ...in CIA when he was a two-star, so that must have been earlier...

Q4: He was a brigadier at this point.

Marshall: A one-star.

Q2: A one-star, so were you over here when he was at CIA?

Marshall: Yes.

Q2: And did you have any interactions with him?

Marshall: Yes I did.

Q2: He wrote, when he was the NIO, originally a classified *Studies in Intelligence* thing, which I think is still classified, on how you do net assessments.[71] A very good piece. Anything specific come to mind of your interaction with him when he was over there at CIA? Any projects you had undertaken with him?

Marshall: We didn't interact a lot. There were certain earlier relations, which were always very friendly. And there wasn't too much to interact about.

Q2: So you didn't provide specific support on the ground forces assessment or anything like that?

Marshall: No.

Q4: When does Project 186 and Phil Karber begin to appear on the horizon? After the Schlesinger period?

Marshall: No, it had started before. It had started before I even got here. Karber was put under contract by Pat Parker, Deputy Assistant Secretary of Defense for Intelligence Assessment, to start the thing. Then I just took it over when it... Right.

Q2: And you started Project Eager, when?

Marshall: Project Eager, Project Eager is...

Q2: ...is, was about '73, wasn't it?

Q4: What's Eager?

Q2: That's the thing that led into the SAIC Foreign Systems Research Center, wasn't it?

Marshall: Yeah. It's this effort to read all of the Soviet ops research military literature, and try to understand, to the extent that they are into analytical analyses, modeling and so on, how do they do that? And what does that tell you about how they think about, measure things?

Q2: So was that about this time or a little later?

Marshall: I think that was probably around '74, '75.

Q2: So Schlesinger was…

Marshall: I think that Schlesinger actually started down this road when he was at the CIA, and started it. It was originally a CIA project, and then I became joint with them, and finally, they dropped out and I was the supporter.

Q4: You know, Andy, one of the characteristics of this period [indistinct] I'm not sure on what the grounds [indistinct] it seems to be this constant pressure from you over literally years to force the agencies to stick the CIA to think about the Russians or about other opponents…

Marshall: Yes…

Q4: …with less attachment to the analytical constructs that they sort of were taught at college and applied to day-to-day living, and so on and so forth. I'd like to be able to say that that's probably the first time that the Agency's been put under constant pressure, and begin to get away from old-timer's analysis, but I find it hard to believe that they could be that insulated, and could be that insulated from pressures like that in the '50s and '60s.

Marshall: Well, I think in the '50s they were quite open. The reason I got to know a lot of the Agency guys was because they would come out to RAND. And they were interested in how we got this, or what analysis we used? It's true that once I got into government, and had customers who were the people that I was serving and I thought needed better analysis, then I was pressuring them and tasking them in various ways to try to do these things.

But to tell one last story: One of the things that I did for Schlesinger, that I think I mentioned, was this business of being an intermediary to the intelligence community. I mentioned my work with regard to the Soviet military burden and the cost of their military programs, but another case was this thing that flared up about Somalia. The Russians appeared to be taking over the use of this airfield and they had some ships tied up in the nearby port. The airfield was very close to the seashore, and they probably had missiles in there; they were constructing some buildings that looked like they might house the missiles.[72] So Schlesinger wanted an analysis by the people at the Agency about, "What do we know about the whole process by which the Russians go about acquiring access and the use of facilities? Are there some general patterns? Where are they in this thing?" and so on. So I went out there, and I think it may have even been Fritz Ermarth who was in charge of the group I contacted.

They started asking what I thought were kind of dumb questions, like, "We could do a better job if you could tell us what decision this is going to lead to," and so on. I said, "First, I'm here to tell you Schlesinger wants it. Why don't you just do it. Secondly, listening to you I come to the conclusion that you don't know anything about this government. There isn't any decision that's going to be made next Thursday or something like that. There's this controversy about what the significance of this is. Don't you know that on issues like this there's some kind of complex social-cultural-political process that takes place, and these assessments are part of an input to that process, and hopefully they make people reach conclusions that are wiser than if they don't have this input? In any case, Schlesinger already knows what he believes; you're not instructing him about what he believes. He wants you to produce something that's useful in this type of long-term effort to persuade others that maybe this could be something serious that they should pay attention to."

Q2: This goes back to the cooking-the-books argument. I mean, it's those people, "Tell me why you want to use the thing, ostensibly so I can shape my analytic effort, but really so that I can figure out what answer you are looking for so I can cook the books and give you the answer."

Marshall: Yeah. We better stop. When we take up Rumsfeld and Brown and some other...Look, thank you for coming.

Outline for 8 April 1994 Talk

[Note: Italicized items were not discussed during the interview.]

Introduction

1. Reviewed prior session
   — Got to Rumsfeld
   — Stories, some themes
2. Purpose
   — Emphasis on themes (intellectual)
   — Rumsfeld, Brown
   — People chip in—help me

Reflections

1. '70s period of big development for me (rich—Net Assessment created; needs lots of work to do it justice)
2. Value of responsibility
   — Long-term competition/RAND responsibility for strategic programs
   — Net Assessment
   — Time to think at NSC/Gen. Polk/Review state of models
   — What is real job, what can we responsibly do, honesty

Themes

1. Perceptions, behavioral, nonrational
   — Peacetime uses of military power (lack of analysis)
   — Manipulation of:
      - Deception
      - Demonstrations
      - Kissinger/Nixon views of crisis
   — What do Russians take note of, give high scores?
   (*747 story*/Space/Moon spin-offs)
2. Long-term competition/strategy
   — Rumsfeld/Roche
   — Economic aspects/Soviet military burden (Feshbach)
3. Judgment vs. models and calculations
   — History
   — Experience of military

Schlesinger Wrap-up (Nov '73–Nov '75)

1. Strategy
   — Develop tools
   — Change way balance looked at
2. Start with three areas
   — NATO/Warsaw Pact
   Strategic forces
   — Maritime
3. Three tasks
   — Assessments—>SecDef
   — Stimulate Services (net assessments)
   — Research program/tools and methods
4. Intermediary with Intelligence Community/CIA
   — Somalia (story)
   — Military burden
5. *Accuracy of ICBMs; U.S./Soviet—bias*
6. *Visit to Marines—role for Marines*

7. Intellectual aspects
   — Assessment focus
      • Net assessment role
      • Role of models
      • Data bases/trends/asymmetries
      • Perceptions
         –NATO story (meeting)
         –Nathan Leites/AWM—allies
         –Germany—will it fight
   — Importance of money for studies/independence
   — Soviet GNP/burden
   — Nathan Leites/[_____?] balance—allies; French ASW; Germany—unclear would fight

Rumsfeld: Nov '75–Jan '77
1. Lots of requests
   — Strategic/missiles
   — Deterrence
2. Short assessments
3. Rumsfeld questions on Navy
4. Initial effort at strategic planning July '76); Roche/Naval effort, July '76
5. Begin Israeli strategic dialogue

1973–1976[73]

Marshall: Let's start now. What the intention is, is to finish off the '70s. I read the record of the last session. We bogged down as we were just approaching the tenure of Secretary of Defense Donald Rumsfeld. But I also looked at it from the point of view that there are a lot of stories that are not as well organized as one might like about general themes. So what I thought I'd do this time is say a little bit more about the Secretary of Defense James Schlesinger period and also try to focus some on some major themes, then deal with Rumsfeld and Secretary of Defense Harold Brown.

Let me start by talking a little bit about the period as a whole, the '70s. It seems to me that a lot of things happened in the '70s. I'm not talking about events in the world, I'm talking about me and about the shape of this office which, in effect, was just set during that time period. At least net assessment as it currently exists was created intellectually in that period, to a large extent. So if you're looking at the book, the '70s really looks like a period that, when you get it all fleshed out, might end up being a couple of chapters or something like that, or more than just one chapter. I think it's hard to do it justice, and I haven't done enough digging back and so on to fully reconstruct the period.

The other thing about the '70s, as I thought about that period, and also the period really of '69, because I actually thought of this as starting in the summer or fall of '69… In thinking about it, what struck me about myself and what happened is how important assuming new responsibilities was in getting one to rethink how you really ought to think about some particular area. It certainly happened at RAND when I took over from Schlesinger to run the strategic forces program, because it made me focus on the question, "What in the hell should RAND really be doing? What should the program really look like? How should we be thinking about strategic forces?" And that led to this whole thing on the long-term competition and a kind of change in paradigm of, OK, the way you really want to look at this thing is very differently than we have been. In part, it's a recapturing of some style of thinking that existed at RAND at the very beginning but that had disappeared. And partly it reflected the influence of the business policy/strategy framework that had come up in my look at organizations and so on. Just the fact that I had this new responsibility required me to look at that.

A similar thing, I think, really happened with the whole net assessment business. In some ways, it was fortunate that during the early '70s, when I first went to the National Security Council staff, the main thing that—while nominally this was called the Net Assessment Group, and we were to start that activity or recreate something at the NSC level—the main tasks that we really spent most of our time on, particularly Chip Pickett and I, were on the implementation of the Nixon reorganization of the intelligence community and all that, which, in retrospect, wasn't intellectually very interesting. I learned something about how stubborn, stupid, organizations can be, the response of the Central Intelligence Agency to all this and their total failure to really fix anything, do anything. But that's an observation about the world in a certain sense. It didn't change the way I thought about things very much.

But there was this opportunity to really think a little bit at the NSC about, "Alright, we're supposed to do these net assessments: How should you really do these? If these things are to be serious and people are to base some decisions on them, how do you really do the best possible job?" So I had time, then. I think that it could have been that, given some responsibility and somebody say the first one has to be produced in six months, you would pull out...systems analysis and ... these models and say "Let's do something." And that's kind of intellectually respectable or acceptable to people, but may be very far from, if you're truly serious, what the thing ought to be like. So I think it was actually very fortunate that we didn't have to jump into it right away, which, as I mentioned the last time, came about because Secretary of Defense Melvin Laird threw up all kinds of bureaucratic obstacles. So I had time to conduct a leisurely review of the state of modeling, which I hadn't had much hope for anyway, but found to be dismal in offering a real basis for anything that anybody ought to trust.[74]

Q2: It's gotten so much better since then, hasn't it?

Marshall: No, not really. And then I also had time to think about what the whole strategy for doing these things might be.

I remember also how struck I was about how people thought this was very important. And people would come around and see me. I remember, in particular, General James Polk coming to see me. And again, right at the beginning I certainly was very interested in this notion of weaknesses as well as strengths on both sides, and he had a lot of interesting stories about how he had operated when he was the commander in Germany [Commander in Chief of the U.S. Army in Europe and Commander of the NATO Central Army Group], and had initially asked his intelligence officers to tell him about what the other people really were like and got absolutely nothing out of it, and so he chose what was a very effective strategy for him {...}. He sat down with the junior analysts and they told him stories about {...} and the screwups and so on. Out of that they got educated about the kind of things he was really interested in to base judgments on: How ready? How trained? What were the real capabilities of these other guys? And then they used to collect materials for him. He would come back every three months and sit down with them and hear about all this stuff.

Q3: Was this the time of the very focused look at one {...}?

Marshall: No, these were just general stories. Later we organized something like that, again during the '70s, as an adjunct to the [Gen. Paul] Gorman study of tank crew training {...} to look at things like that. At any rate, I just thought it's, in reflection, interesting that responsibility does make one more responsible and serious.

Q3: Sometimes.

Marshall: Sometimes.

Q3: For some people.

Marshall: For some people.

Q2: Would you also say that that period on the NSC shaped how you, how you went about conducting the business of this office over the subsequent decades...

Marshall: No. I don't think so. I thought a lot about how might you go about it, I mean intellectually. But as far as the office was concerned, that was created after we got here.

Q2: I mean how you went about doing the business of the office. One of the things that always struck me about this office was there was never any suspense date for any product. The office sort of worked the product until it was good. It wasn't like the office had to produce one by September or January. The office weren't driven by the other internals within the Department…

Marshall: Yeah, right.

Q2: …like, got to get an annual report out or things like that. When Jeff McKitrick replaced Barry Watts, you said, "Take six months to read everything you can in the office and then we'll start talking about…" Assignments where they said take six months and get smart as you can about these things were rare. Does that go back to your RAND days, or your own personal thing, or do you think it was shaped in some way by the NSC?

Marshall: No, mostly my own personal view of things. Also, we've had the luxury and been extraordinarily fortunate in just being allowed to do what I wanted to do, in a sense. Of course, there were times when people asked for something and then we did it more or less on a schedule. But my feeling was that, in the first place, you didn't have to update these balance assessments more than every couple of years. Things didn't happen, turn over, unless there was some major new source of information.

Q2: Or the collapse of the Soviet Union or something like that.

Marshall: The other thing was that you really wanted to produce a high-quality product, and so you took the time to do that.

Q3: Just following up on this business of how responsibility changes the way you think: Was it because you were given responsibility for certain intellectual tasks that you found yourself thinking about how to do that, or was it the fact that you were given responsibility for managing the activities of subordinates so you had to think about how you would break it up into tasks which people had to work on, or both?

Marshall: No, I think it was mainly the first.

Q3: So somebody gives you a subject and said understand…

Marshall: Well, in the one case, the RAND thing, I was to take over this program and looked at what people were doing. And I didn't think that amounted to anything. Then the question was, "What should we be doing?" I wanted to determine what a really adequate look at this would be. Then I came to this view that the way you needed to look at it was in terms of a long-term competition. Therefore, these studies of individual choices of, you know, this new bomber, really needed to be seen as moves in this competition, as part of the implementation of a strategy for achieving some set of goals that you really had, and that it just shouldn't be a matter of a more modern and up to-date way of killing the same targets that you had been "killing" for some while, or just a response to some change in the Soviet air defense system. My view was that you really need to think of this thing in a very different way, and therefore look at the decision about a bomber in terms of the whole force and the whole strategy that you were pursuing.

Of course, the fact was that, at a certain level, the U.S. never has strategies like that. Organizations can—The Air Force, say, was committed to the bombers and had some sense of the strategy, although it was never articulated very well. But my experience was that once you had articulated it that way, there were often people—here in the Pentagon it was particularly the guys in the Office of the Director of Defense Research and Engineering who came from the business world—who immediately grabbed onto that. Somebody like Dick DeLauer [Under Secretary of Defense for Research and Engineering]. You talked to him about how part of the thing that we're doing is driving up the cost of their staying in business, and he quickly had some of his guys do a rough calculation of the cost of the Soviet air defense system and what you might do, what's the next set of costs you can impose on them. In that sense, some of the civilians, particularly these guys from the business world, had far more the right framework for looking at this.

I think it really happened in the '80s, but I'll tell the story of General Lew Allen's [Air Force Chief of Staff, 1978–1982] reaction to the kind of competitive strategies sorts of things. He wasn't sure it was wise for two reasons. One was that we were too erratic, and therefore we might start something, they would react, and then we wouldn't continue on to the next stage, and we'd be worse off than if we'd not done anything, which is quite plausible. On the other hand, his view was that the costs are meaningless to them. It's as though they had infinite resources and there was no sense that they...It was as though...

Q3: There weren't opportunity costs.

Marshall: Yeah. His view suggested that in socialist countries there are no opportunity costs. They can always just go ahead and put more resources in, and there's no pain and nothing ever happened. Whereas other people have a sense of, "No, they have limited resources, too, and there are consequences," and so on.

OK. So let me then talk about the '70s and what are some of the interesting themes to me, from the intellectual side of things, apart from this whole invention of net assessment.

The thing that strikes me the most is the constant interest in, going back to this business of perceptions, concern for the calculations and perspectives of the other players. As I mentioned last time, that came out of interest that Schlesinger and I had at RAND, triggered by reading, at least initially, and I think for both of us, Robert Ardrey's *The Territorial Imperative*, and interest in, essentially, the peacetime aspects or uses of military power.[75] One of the things that struck me and struck him was that almost all the analysis about military forces goes into the war and how the fighting might be done. While people use words like "presence" and have some general sense that, yes, it's better to be strong than to be weak and things like that, there's very little analysis of this matter. In particular, Schlesinger was very interested in the question of, "alright, these things are going on, How do you do well at it? What actions, what ways of demonstration really are effective?"

And then it spills over into the whole area of deception and many other things, like Henry Kissinger's and Richard Nixon's views on crises as opportunities as well as problems.

Even before I got to the Pentagon, when I was working for Kissinger I tried to get the intelligence people to do some analyses, as best they could, of the Russians and what they pay attention to about us. What things get us good scores in their books as against things that get us bad or indifferent scores? And you know, you come up with some interesting things. Then later, there are still other developments, and you even get into issues of the prestige effects of reaching the moon and all of that.[76]

But clearly this theme of perceptions and the peacetime uses of military power is something that runs through a lot of what was done in the '70s.

The second thing is this long-term competition perspective, and in the beginning and in the middle '70s, in particular, this writing of various things on actual strategy or strategic planning sorts of things for Rumsfeld and for Brown, which, again, brought back in a lot of these economic dimensions of things, and led to the creation of the Investment Balance and some of these things. Those two things, plus all the things that get into the development of net assessment, are the main intellectual things, I think.

Let me just talk briefly about the Schlesinger period, to pick up some things that I think didn't come through clear enough during the last talk. This was a period, really, of the development of the state of the art of doing these things. We had the time to do them. Soon after I got here, Schlesinger and I talked, and there were these three areas that I mentioned last time. He was interested in assessments on the strategic balance, the NATO-Warsaw Pact balance, and the maritime balance. We started to work on that, and made, I think, the least progress on the strategic balance. I also did a bunch of special things for Schlesinger.

It also was after I got here that we changed the nature of the task, in some sense, for net assessment. That is, up till that time we were tasking the people in the Pentagon to do assessments from the NSC, because that's the way it was thought that it would be done. They were broad, general studies with many participants and task forces, which ended up—I came here before the first one was finished—with these brokered, bullshit analyses; you know, "the Soviets don't have any weaknesses" nonsense out of the Defense Intelligence Agency and so on. After we had finished off that first one, Schlesinger's attitude was, to put it in his way of speaking, "Enough of this shit. Let's get on and do something serious." So the whole thing then became that these were private assessments sent to him, and he read them, and he would then pass them on if he wanted to. The fact was that we were still learning how to do it, so we didn't have a lot of really great things to offer him, because of start-up times and the fact that once you started, you found that none of the right data existed, and all that. So we weren't so helpful.

Another thing I spent a lot of time on was serving as an intermediary for him with the intelligence community, trying to get special studies, special assessments done, trying to get them to look again and see whether they wouldn't find some different answers to things like the military burden of the Soviet Union and some things of that sort.[77]

But basically, the whole way of doing things began to take shape then. Certainly the avoidance of models, the reliance on judgment, the importance of data on long-term trends and asymmetries, the importance of really understanding the other side's doctrine and concepts, which would be a major aspect of the asymmetries, were all part of the net assessment approach. As I mentioned, there was a lot of attention to the psychological-political effects or uses of military forces. And it wasn't just the military forces. For example, when Schlesinger was Secretary, it became the U.S.'s turn to host one of these big, formal NATO meetings [Nuclear Planning Group meeting, 17 Jun 1975].[78] In discussion with him, we decided that… Again, one of the things your forces do is not just impress the enemy, but your allies. Well, it's not even just your forces. So we consciously chose to hold the meeting in California, so that the defense ministers and other allied officials could fly across the whole country, because most of these people never get beyond Washington. And so it's held at the Naval Postgraduate School, which I knew had appropriate facilities. And then they were taken down one day to the missile-launching sites and tours of the silos at Vandenberg Air Force Base and things of that sort. All, in part, based on bucking up the allies. So Schlesinger was very interested in all of these dimensions, as I was.

Q3: Did you get any feedback?

Marshall: No, it's hard to know. I don't know, it didn't, I think…I even found a Robinson Jeffers poem for Schlesinger to use in his speech at this meeting. At any rate, he was very, very interested in all that.

Another thing that we did along this line, I mean while we were thinking a lot about the Soviets, how they score us, how they see us—again to follow up on the business of the allies—I undertook, with Nathan Leites's help, to try to find out, If we're going to try to impress the allies as well as the Russians, what do they notice? What do they pay attention to? I used my connections into the French to talk with them and found some interesting things, which was that what appeared to weigh very highly with them, because of the importance of their own nuclear submarines to them, was essentially antisubmarine warfare, in the sense of who was ahead in that. That seemed to get very big weight. Nathan went off to interview the British, and the Germans, in particular.[79] Then subsequently, he did another thing on Germany, which we had to circumscribe all publication of, because, in essence, it said it is very doubtful that the Germans will actually fight. But Nathan had some very keen insights into the Germans.[80]

Q3: Is this the time when he came up with this image of "pygmies, midgets, and giants?"

Marshall: No, that was just verbally. That antedates this, I think. It goes back to… Right. The other thing that had a big impact as far as the office was concerned, I think, was that we got a reasonable start on getting funding for studies through the Defense Advanced Research Projects Agency and through the Defense Nuclear Agency. Again, one of the things that happened during this time was that Don

Cotter was here and wanted to do various kind of studies, and Pete Haas [Deputy Director for Science and Technology] of DNA came around and asked what they could do for me. That was, in part, what he would have done anyway, but it also, I'm sure, was seen as something they could use to then get me to authorize or request DNA to do things that it wanted to for Cotter, but that Cotter didn't necessarily have the authority or the right portfolio to get into. For example, Cotter was very interested in this whole business of analyzing Soviet exercises in Europe, and in the conventional aspects of that as well as the nuclear, but given his portfolio, it was hard for him to get into that and hard for DNA to get into it.

Q2: Where was he?

Marshall: He was the Assistant to the Secretary of Defense for Atomic Energy.

Q2: AE?

Marshall: AE, yeah. But Net Assessment had the virtue that we could use them for anything. We could say, "We want you to do those things." So a lot of that was done. But pretty much from the beginning, we got a fair amount of money, and that got us into this pattern of extensive contracting for outside work, which was very important, particularly in some of these cases where you didn't believe the CIA answers to a lot of questions. You'd go get somebody intelligent to look at those questions.

OK. Let's go on to Rumsfeld. Rumsfeld came in around late November or early December of '75 [20 Nov 1975]. He was here during '76, so he was here just for essentially a little over one year. I was very impressed with Rumsfeld. I think, in fact, in some ways, of all the people I've worked for, that he was potentially the best Secretary of Defense.

Q2: That's a lot of people. I mean, that's pretty…

Marshall: I think that because of his political skills and ability and interest in using good people. Schlesinger, of course, was a close friend of mine, but he scared too many people. People were frightened of him. I told the story last time that when he was the Director of Central Intelligence, a large part of the middle management hid from him out at CIA. He and I always got along terrifically, but lots of people were just frightened of him.

Q3: He was on a visiting committee for a Harvard thing and he had the same effect at Harvard. People were terrorized—"Oh no! We're going to have to give… Schlesinger is going to be there and he is going say some goddam thing and we're not going to know what to say."

Marshall: Yeah.

Q2: Of course that's true.

Q3: Yes. I got along fine with him.

Marshall: He and I got to be very close at RAND. And there was a set of people that he dealt with, Cotter and others. But kind of the ordinary bureaucrat was just terrified.

Q2: So you think that it was Rumsfeld's bureaucratic and organizational, people skills…

Marshall: Yeah, that's right. He seemed also, as I say, to use people extraordinarily well, all kinds of people. And then he also was politically astute, you know. But he didn't have a long stay.

He largely left us alone, except that he would make requests, I mean when there was some question or some issue that he was interested in or concerned with. He had some questions, I remember, about the strategic balance and what did I think, or the nature of deterrence, or something like that, and I would write him a paper. We also created some short, four- or five-page assessments, kind of summary stuff. But basically we were left alone and we went ahead and did that.

The other thing that impressed me about him was that he had this kind of strategic sense of, "Let's think about where we're going." And he tried to do something for the Navy, on the naval programs. I remember he asked for a paper overnight. Jim Roche and I prepared something that he liked a lot. I can remember the meeting that took place. It was on a Saturday morning. Rumsfeld had a bunch of people in. And an Office of Program Analysis and Evaluation guy got up to talk about things and was going on and obviously had no sense at all about what kind of perspective Rumsfeld was interested in.

So about half way through the thing, Rumsfeld says, "Why don't we stop looking at our shoelaces and raise our sights and see where we are going."

Q3: This is an example of not scaring people…

Marshall: And it was because of that, that and our own interests, that Roche and I wrote this July '76 thing, saying, "What would a strategy for defense really look like? How would you organize it, using business analogies of, What 'businesses' are you in? What are you trying to do in that 'business?' What are the major trends and other things that you ought to take account of in thinking about that part of the 'business?'" That's how we got into doing something beyond the more narrow assessment. That carried over later, which we will get to, to Brown, who was also interested in this kind of thing. In fact, while we wrote this thing for Rumsfeld, that is the July '76 thing, and he looked at it and responded favorably, Jimmy Carter came in as President and there wasn't much time to do anything about it. Then we sent it to Harold Brown, and he liked it, and then we did some strategic planning experiments for him.[81] And we'll get to that later.

The other interesting thing that happened during the Rumsfeld period was the start of this strategic dialogue with the Israelis, which ran from April of '76 through about 1980.[82] The Israelis had approached Schlesinger twice. He may have been approached through other things, but I knew of two approaches because they came through me.

Something else which is worth mentioning about the Schlesinger period was that shortly after I got here—I came here right after the Israeli-Arab War, Yom Kippur War—Schlesinger put me in charge of a quickie "lessons learned" effort. Each of the Services was going to undertake a much longer term, thorough effort, and the Joint Staff ran off and did something, but he wanted an effort to try to get any lessons learned that could be useful in any way. So we organized an effort, and I borrowed Tony Cordesman from ARPA to go over there and gather materials and so on. And we had tried to send S.L.A. Marshall there. He made a quickie trip and talked to the Israeli military that he knew best, came back, wrote something. Schlesinger liked him a lot and wanted to send him back in, but then the Israelis wouldn't let him back in the country.

Q2: Jeez. Why?

Marshall: The basic reason was, I think, that… Look, from the Israeli point of view, that war was very traumatic, because it showed up deficiencies in their planning, in their thinking. So there was a big internal quarrel in Israel afterwards. There was this big investigation. I forget the name of that commission that they set up.

Q3: Agranat?

Marshall: Yeah. So it was very controversial, and Marshall probably was most associated with one of the factions, and so they didn't want him back in the country. So we couldn't get him to do it, so we organized this other effort and, as I say, put Cordesman on it.

When Cordesman came back, he relayed this request by the Israelis that they wanted to set up a direct channel between the Ministry of Defense and the Defense Department. We found out later, I think, that part of it was related to the fact that they did not trust their ambassador here [Simcha Dinitz], who they felt was interfering with the appropriate discussions between the two military establishments. Schlesinger checked with Kissinger, who was at State, and the thing was kind of vetoed. I don't think Schlesinger was strongly tempted anyway, and it was a little unspecific as to what they wanted. There was a similar kind of request that happened later.

But after Schlesinger had left and Rumsfeld was already on board, a man named Saadia Amiel came to see me, and he saw other people, again requesting the establishment of this kind of dialogue. In the aftermath of the '73 war, the Israelis were trying to create more of an Office of the Secretary of Defense-like structure to both oversee the forces, but also they became aware—well, it goes back to the '67 war, when the French had cut them off after having been their main supplier—that they were beginning to do more of their own weaponry, and, I guess, after the '73 war, they decided they needed to do more thinking, planning in order to decide what they ought to invest in.

So Amiel came by, and again I kind of put him off, but he finally convinced Deputy Secretary of Defense Robert Ellsworth that some such interchange should take place. Rumsfeld agreed, and I was asked to be the point of contact for doing this. They wanted to keep these discussions out of any forum which the Israelis could use to argue for, you know, "What about the $200 million we asked for?" "Where's the shipment of so and so?" So it was explicitly understood that this was not to be a forum for any of that. It was to be a discussion with them, and also in some ways a channel for advice to them, as to how they ought to think about their own future, and do a better job of planning and so on. Also, the thing was to be kept very quiet. And in fact, no one knew about this, which was a problem, because the people who nominally do all the relations with the Israelis, got their nose got out of joint and so on. But initially, we had a kind of cover for doing it, which was,

"We're going there just to get information."

Q2: So State Department did not approve that?

Marshall: I don't know whether that was ever checked with State. I have no idea. But when Brown came in, he was told about this. The thing was done on a basis of twice a year, and by that time, they came here in April of '76 and we went there in September, October, something like that, of '76. In the Carter Administration, Brown then asked about it, and he, Secretary of State Cyrus Vance, and Assistant to the President for National Security Affairs Zbigniew Brzezinski knew about it and were the only... And we would send him reports of the discussions. And, I don't know, I presume he told them, at some breakfast meetings or something like that, about things.

Q2: Who within the Defense Department knew?

Marshall: Well, people then sort of knew it was happening, but they got none of the reports.

Q2: They didn't know what it was?

Marshall: They sort of knew what it was, but they didn't... It ended in the last year of the Carter Administration because Assistant Secretary of Defense for International Security Affairs David McGiffert and company couldn't stand it any longer, that is, to be excluded.

Q2: How's the intelligence community play on that?

Marshall: Not at all.

Q2: None of them involved, they got none of the reports, they never saw you over there talking to the Israelis and wondered, "What the hell?"

Marshall: Well, the attachés knew we were over there, and they would hear something about it. What they reported, I don't know. We didn't pay much attention to that. The ambassador there we would talk to; during part of the time it was Sam Lewis. My counterpart, initially—I forget the man's name, a guy who was a special assistant to Shimon Peres, who was the defense minister when the thing started, and had been the ambassador to France, and then later when the Likud came in, was the Deputy Minister of Defense, and his name was Mordechai...Zippori.

So at any rate, there isn't too much to say, as I say, about the Rumsfeld period, I think, as far as interests that continued to develop.

1969–1973[83]

Q1: Did you begin working for Henry Kissinger because he and President Nixon were dissatisfied with the National Intelligence Estimates they were receiving?

Marshall: [Name of person unclear] and Mitchell conveyed the same comments about the NIEs. It happened before I got involved. What I was told by Kissinger was that shortly after the Administration was in place, Nixon had read a few NIEs. He thought they were poor. He thought they were much worse than the ones he remembered from when he had been the vice president—I don't know whether that's really true. At any rate, he was greatly dissatisfied. So, he got Mitchell and Kissinger to kind of look at things. And I think the three of them had a meeting to compare notes on, "OK, what's wrong with these things. Why don't we like them?"

And that was one of the outstanding complaints—this lack of the separation of the fact and opinion.[84] Not just that they don't separate them, but the thing is written in a way that you don't know whether that's a fact, or a judgment, and that you needed clearly stated what things are facts, and what things are inferences, judgments, and so on.

Now Kissinger, when he first talked to me, had other complaints, in the sense that… He said to me, "Look, on the intelligence reports with respect to technology and so on, I'm not able to judge…but in the political stuff I feel I can judge and they're terrible. They're not better than, and indeed they're probably worse than the kind of stuff you just read in the newspapers every day. They have no real depth…. There's nothing behind them. The people writing them don't seem to have any real understanding of what's going on. They report what's going on but then they go off and they have their speculations and so on and they're just not really worth anything."

Q1: Was Kissinger really interested in intelligence reform?

Marshall: Now, of course, part of the problem was Henry was… He would talk with somebody like me or someone on the staff about how terrible, rotten it all was, but then he would never really say that to the intelligence people. I think it showed up in the whole way in which he conducted business while he was in government, certainly in the State Department, where he gathered a set of people around him and then shut off the rest of the State Department.

I don't know whether Nixon said he wanted someone to look into this, and that's why Henry called up and had me do these initial studies…

Q1: Did Fred Iklé recommended you to Kissinger for the work on intelligence?

Marshall: Well, I think so, I mean you might talk with Fred, I never really asked him.

Q1: When were you first asked to work for Kissinger?

Marshall: I must have got this call in September of '69 because I dropped by to see Kissinger. Actually, the call came from Al Haig, asking me whether I'd come by and see Kissinger. I was on my way to Europe for a lengthy vacation, probably around the middle or the third week in September. So I dropped by Washington on my way and had a meeting with Kissinger. And then, I think I was away about a month, and came back. Then had to disengage myself from the stuff I was doing at RAND. I ended up going there in the very early part of December.

Q1: What can you say about your study of intelligence flow to the White House?

Marshall: The first thing that I did was this study tracing the whole flow of intelligence into the White House. How it happened, how decisions were made about what to send. First I found out, "Where does this stuff actually come from?" Then I interviewed people from each of these places to look into how it was they went about deciding.

NSA had a very organized kind of interface and they had some people who would come down essentially once a month and sit down… Not just with the guy who ran the Situation Room, who was in fact a guy from NSA, but to sit around with some other people and try to say, "Well, OK, what are the things that are of most interest…"

Q1: But the NSA representatives would meet with people from the NSC staff? Who was talking to whom?

Marshall: Well, it really was with the people in… My impression was they met with the two people who ran the Situation Room and some of the other analysts who were in the Situation Room. And it was probably through them that they understood what people on the staff wanted, were requesting, had an interest in, etc., etc. Whereas the CIA did none of that.

No, somebody told me that that the CIA took its cues from *The New York Times* in deciding what intelligence to send to the White House; I don't know whether he was being dismissive. But their [Eds: the CIA's] whole idea was that they know what's interesting and important, I mean you just have to read the newspapers or something like that. It was, "We have this high-quality product…"

Now, a lot of that was particularly targeted on the issue of what went into the daily President's report, but just more generally, they just felt that through the attendance of Helms and others at various meetings and other things that they just didn't need any special apparatus or to spend much real effort at finding out. They just didn't spend much effort of any systematic sort.

Q1: It was a long-standing problem?

Marshall: I think so. I talked with members of the Kennedy/Johnson NSC staffs and they had similar complaints about intelligence. But again, one of the things I said is how little they seemed to care about the customers, even though they protest that, "We're here to serve," particularly these top level people would say this. One of the first things I did when I went to work for Kissinger permanently was to go through a number of books that Kissinger recently had written and extracted from them sections that I thought provided the most insight into how he looked at the world, what he thought, the kind of information that would be useful to him, and sent that around to them. But that's the kind of thing, in my view, if they were serious about their offers of help, they would do for themselves. But they didn't do any of that. Just really crazy.

So, to a large extent they thought that they had their way of doing things and it was the best way and they didn't—despite their talk of wanting to help, they were very unresponsive to what customers actually wanted, even somebody like Kissinger.

Q1: Did the CIA interface with groups at Harvard, Stanford, and elsewhere?

Marshall: The thing that struck me at the time was how at that point the CIA had kind of shut itself off from much connection with any other organizations doing stuff that could potentially have been interesting to it. And they kind of seemed to have, for the while, just adopted the notion of, "We're the real experts, we don't need to talk to other people." Which was very different than the situation in the '50s.

Q1: How did the CIA of the '50s differ from that of the late '60s?

Marshall: Well, I guess my own experience of it was really being at RAND… It may have been particular to RAND, but not exclusively. Because they had this special group up at MIT. I forget the guy's name now who was up there… But they had a number of connections out. Now, that was a kind of contracting arrangement to get certain kinds of studies… But it was a way of having an interface with people in academia, working in the economics and political science worlds. In the case of RAND, people came out, particularly Frank Walters and Wally Seidel, who were analysts working on the Soviet missile programs, aircraft programs. And in the middle '50s, '56 or so, they started to come out to try to understand what RAND was doing in the systems analysis-type of analysis techniques, how they might be used in what they were doing. Also in a certain sense the beginnings of their effort to cost the Soviet programs, to do a comprehensive effort, that a guy named Jerry Godaire, who was out at the Agency.[85]… some of the ideas came from people like Loftus and me at RAND. That you needed to do something like that to check on the overall resource demands implied by various kinds of programs that the Soviets were undertaking. Anyway, there was quite active interchange with them.

And then somehow, by the late '60s, it had pretty much dried up, I think. And I think it came about, in part, because of, maybe this sense that, "We're the real professionals and we don't really need to keep up these connections." Secondly, the people in the '50s— The whole place was made up of people who were brought in just a few years before. But by the late '60s, you were beginning to have people rise fairly high in the bureaucracy who had come in and their whole careers had been within the Agency.

And I remember, in fact, in the '60s, when I was brought on… Well, I was an Agency consultant from about '56, I guess, something. Anyway…I suppose it would have been the early '60s… Under McCone an organization was formed called National Intelligence Programs Evaluation (NIPE), headed by John Bross, and it was to look at the broadest management issues of the community as a whole. Bross for some reason knew something about me—I don't know exactly how that had come about—but he invited me to come in to see him, and then he made me a consultant of his. In fact, that was one of the most active relationships [indistinct] in the '60s. And he said to me once that it really worried him that this combination of their having moved out to McLean and the fact that people in the Agency had spent their whole professional lives in the Agency and didn't know anybody, let alone in the rest of the city, but in, you know, New York or wherever, really worried him. He used to shake his head about what that was going to lead to, because of his own experience and that it was very valuable to have contacts in other worlds, whether it was on Wall Street or wherever it is that you can get information from. And these people going up with these new careers didn't have any of that.

Q1: Did the situation improve?

Marshall: "Consumer research" by the intelligence community has gotten a little better over time. They now do some surveys of users; I don't know that it really influences what they do. It was striking how little interface or real knowledge they had of what consumers wanted.

One of the things I did was to conduct a survey of people on the NSC staff about what they wanted. Strengths, weaknesses, what things weren't they getting that they might want. Found they wanted, if they could get it, some of the products to be focused on longer term trends, basic things that were happening, not the day-to-day events, more speculative about the future, but really thoughtful.

Producing such things runs up against the ethos of people in the intelligence community, that they shouldn't be speculating, that they should be getting hard information and then at most interpreting that for the readers. But the actual consumers wanted something rather different, at least in part. But the consumers also wanted to know where there were controversies and differences of opinion within the intelligence community.

But again, at that point, the community was devoted to the notion that they should resolve their differences internally and kind of speak with one voice. But also, I guess the thing that drove some of this stuff about these new institutions, like the NSCIC [Eds: National Security Council Intelligence Committee] , was this dissatisfaction of Kissinger and Nixon with the products, that they just didn't think were very good and lacking real intellectual quality.

Q1: What was the background of your study of the quality of intelligence?

Marshall: When I was about to start work, I had another meeting with Kissinger in which we reviewed again what he wanted done. And after discussion, one of the things that I was asked to do as a second study was to make a judgment after really looking at the assessments, estimates of the missile programs, because that was the big thing that was happening… It had been left slightly open what particular area to pick to do this quality assessment. And it was then that essentially the Soviet strategic missiles programs were picked. SS-9 was one of the biggest of the issues. Questions about how many there were, what the characteristics of the new systems were. And it wasn't only just the technical things, but there were issues of whether they had reloads, what was the motivation for moving to the new systems…

Q1: I was just interested in why Kissinger was interested in the missiles assessment.

Marshall: Well, I think I proposed it, because I knew that it was an area of interest to him.

Q1: And you had a competence in the area.

Marshall: And he agreed with that.

Q1: What were your findings?

Marshall: I thought there were just a lot of limitations, I guess in the NIEs on Soviet strategic forces. I'm a little vague at this point on my findings. It wasn't a very favorable report. In the sense that... There were things that they did report on that represented the very skillful effort that was being made through various collection devices to analyze the programs, I mean in the sense of, you know, a new silo was going in here or there, or, you know, "We see this at the test facilities at Tyuratam," and so on. Or various other technical analyses about the performance of these missiles. But beyond that, there just wasn't anything.

And I remember later, when I was working for Kissinger, not when I was doing this study, there was a book called *The Soviet Space Bluff* [*The Russian Space Bluff*], written by a defector [Leonid Vladimirov], which was filled with fascinating stuff, giving you the names of the various designers. I mean the intelligence community was very late in providing any information about—I brought this book to the attention of people in the Agency, but they never did anything about it, to go interview this guy, or anything. I mean, he was living in England.

So, any sense of the personnel rivalries, the process by which these things were designed, or about the organizations and so on...they didn't provide anything. Or they...We didn't have any broad sweeping recommendations. We had some recommendations which came out of this look at missile things, which led to this proposal to do this prototype SS-9 study as an exemplar of the kind of the level of analysis you would like to see.

Q1: Did your findings have an impact?

Marshall: Mmmm, ah, maybe a slight bit of improvement in subsequent NIEs as a result of this exercise. I don't know. The real improvement came with Larry Gershwin going out there. I mean it took that to make a real difference. He didn't go out there until Harry Rowen went out there. So it was the very early '80s.

Q1: What other areas of intelligence analysis needed improvement, in your opinion?

Marshall: One of the things I tried to get them to look at, like when they did this special SS-9 study, was...try to get at things like the manpower costs associated with different systems—Was that a motive for switching to newer systems or not? They seemed to be content to just report, "Well, two more silos have started," with no interpretation of, "Why is that happening? What does that suggest?" Couldn't provide any insight about any of the internal processes that led to any of these things.

Because I had requested it, they at least made an attempt to answer some of these other kinds of questions. But again, it seemed to me that it was very important to understand the economics of this, I mean, what did these things really cost? So, anyway, they made a stab at these things. Also, more about the relation of the SS-9 to the SS-7 and things like that, so you get some sense of the evolution of their systems and so on.

Q1: How long did the SS-9 study take?

Marshall: It was more like a year study, from spring 1970 to fall 1971.

Q1: Among the other special products for Kissinger, the main thing was the analysis of key personalities?

Marshall: Biographic intelligence guidelines, if I remember correctly, it was kind of a two-page form to be filled in that had headings, specifying the kind of information we wanted about the person. But I don't think it ever had any impact at all.

In the first place, Henry had this great interest in these people that he might have to deal with in one fashion or the other. One of the first things that came up was this interest that he had in these psychological profiles that had been prepared by Jerry Post and other people out at the CIA, which he wanted more of. He even had the idea of some psychiatrist up in the Boston area, that was connected with a very famous...this was a professional psychiatric clinic for the treatment of mental diseases...I know about it from the work that I'd done with Herbert Goldhamer. It existed way back in the 1830s, 1840s and it was a private place where people with mental problems could go, as contrasted with the state system of asylums.[86] Anyway, so Kissinger even wanted to get more work and thought it would be interesting for this guy to somehow get involved in this. So he wanted more effort. People in the Agency were extremely reluctant, just didn't consider it one of their top-of-the-line products, a little disreputable...it's speculative, it's a little oddball, etc.

The thing that was striking to me was, OK, here's something that Henry liked, thought was useful, wanted more of, but essentially couldn't get it.

And then when this coup took place, he was again... "Well, who are these people..." And you have the background in the files, which is about, as I put it in a kind of cynical way, sarcastic way... I mean it's fine if you're going to entertain them, they like scotch or bourbon. It's stuff that's presumably acquired or written down based on social contacts with these people, some sense of them, whether they're womanizers or not, but there's nothing about what are they really like psychologically, as personalities, except sort of crude descriptors, you know, "A strong person." Nothing about ideological beliefs. So it's very limited, it doesn't show any insight into the person at all. Just puts him in a category. You know, "Catholic," "military officer," "likes bourbon" or "likes rock-and-roll music." No information about his thinking, or how he might behave in certain circumstances.

Q1: The coup was in Panama?

Marshall: Maybe a "coup" is too dramatic a word. Takeover of the government by a military officer, I think.

So Kissinger gets this stuff and they barely know much about these people. Maybe only for one or two of them there's even any data at all and what it is this piffling material. So I get asked, "What could be done about this?" So I looked into this, tried to understand a little more about how the thing worked, but only for a few days. Then said, "Well, one thing you could do at least is write down what it is you want about these people, what would be useful to you." And we did that and sent it over.

Q1: Would greater use of the Bayesian method for analysis and communication have improved intelligence analysis?

Marshall: Well again, that came up right at the beginning, because one of the things that also came out of this early technique [?] was the way that uncertainties were handled in this kind of backhanded way of, in paragraph 28 they say one thing and then they halfway take it back in paragraph 57. Uncertainty should be more squarely faced, they should look for ways of presenting the thing. And Kissinger was in favor of this even in saying, "Well, there are six alternatives and here are our best guesses about the probabilities of those." Well, they didn't like that at all.

Q1: Did they make some effort?

Marshall: Well, there were a couple of experiments to do that. But I think they were kind of halfhearted. They were being forced to do something they didn't like. At the level of the Board of National Estimates, it was just seen as too crude and not literary. That's not the way you would try to convey any of these things. The Bayesian method is partly how you do analysis and partly how do you present the stuff.

Q1: Was the use of Bayesian methods driven by the studies by Kelly and Peterson?

Marshall: So those guys were…I guess Kelly, when I first met him, was at DARPA, then he went off into this— There were a couple of small firms that were pushing this sort of analysis and how you would do that kind of thing. There was another guy that I knew from RAND that was at Michigan…Ward Edwards…a set of these people. There was both the presentational kind of thing…the Bayesian kind of thing… But then there was also a set of people out there that were working on situations where you get information over time…and supposing you have a set of estimates, the likelihood of various alternatives, and you get a new piece of information, how do you modify each of these… In fact one thing I did was I sponsored a conference out in Coronado in California. There people who had been working in this field were all brought together…

Q1: Now, was this while you were at the NSC or afterward?

Marshall: Probably was after, because I think I got DARPA to put up the money, yeah, probably soon after I got to the Pentagon. But Pat Crecine I remember was there, this guy from Michigan.

Q1: Now what did Kissinger think of some of these experiments or the results of these experiments?

Marshall: I forget now. I know that we got some and shipped them into him. But I don't think we got much of a reaction. I think we tried it out on some of the other people and they sort of liked it, but it wasn't as though they were entranced by it.

Q1: Were there other ways in which you tried to improve intelligence analysis?

Marshall: I made this big effort on organizational behavior, but I described some of that in the earlier interview or session.

Q1: While you were at the NSC?

Marshall: Yes, right.

Q1: Covering which topic?

Marshall: Well, the thing took two forms. One, there were a few people at the Agency who were quite interested in it, particularly some of the people over in the clandestine services side, that I had gotten to know while I was at RAND, because they had come to me to try to get some information about a country that I had a lot of contact with, and so I had talked to them somewhat in the course of that. So, after I was at the NSC, I said, "OK, look, here's a way of doing analysis that might be useful. I will try to tutor, in a way, although I don't know what the right word for it would be, some set of analysts, or some people, to try to make sure they understand what it is they ought to try look at when they look at these organizations. What information do you really want to have?" And so on. So, three or four of them came down several times and spent time talking about this. And then... That was one thing. There were these people who were very interested in it, who kind of picked it up, did whatever they did with it, I don't know, they kind of disappeared back into the clandestine services. Never had much effect on the people in the Directorate of Intelligence. There may be some papers around from that.

Q1: There was an informal group of CIA producers you convened in late 1972 to determine data needed to support intelligence analyses based on theories of organizational and human behavior. One member of the group looked at a case of using those methodologies to understand the French position in GATT [General Agreement on Tariffs and Trade] discussions in late 1973. Was that the same group?

Marshall: Well, it was kind of the second phase of that. Because after we'd done this other thing, just talking... What would you really want to know? I decided it would be good to have a case that would be done, where they could show what it was they could do. And it would be a kind of test of could they carry out something like this. Mainly the earlier discussions have been about, "Well, what do you want to know? What are the kind of things you want to find out?" And so on. You'd want to find out about the processes, who's involved, things about the internal procedures. But then it came to a point where we thought about, "Well, let's do a kind of test about what they can do."

And the idea was—Or my idea was, "Well, look, a country I know about is France, so let's pick France." It's also, in principle, a country that we should be able to find out a lot about. It's not like Russia or China... So I went to see Ken [Kenneth W.] Dam to discuss with him... He was a kind of special assistant to George Shultz, who was then the Secretary of the Treasury, I think. But Dam had an office in the White House...I don't remember whether Shultz did. ...I knew Dam quite well, so I went to him and said, "What is a topic? What is something we could get them to focus on?" And he said, "Well, look, something that would be interesting is French agricultural policy." And I know a lot about that... Because the idea was to pick something that we [?] could judge how good this thing was. So we picked that.

I then proposed that—and I forget who it was that exactly I proposed it to—It might have been Ed Proctor, who was still the head of the intelligence directorate.

And he picked a guy who was retired. …I can't remember the guy's name now, but he's somebody who I'd known for a long time, who had an economics background and had been in the Agency from way back in the '50s, had recently retired at this point. Who had been a close friend to Joe Loftus. So had been up to RAND several times. So I really knew him pretty well. So Proctor proposed that they get this guy back out of retirement, and give him this job. And the idea was for the person to draw on—not to have to do collection so much—what already was in the Agency. What did they really know? Pull together everything they knew that would fit this particular topic of their agricultural policy and their negotiations stances. And he did this paper, which I thought was not really very good, and neither did Ken Dam, because we both looked at it. Because it really didn't take over this organizational behavior thing, didn't get down to the level that you need. But, anyway, it wasn't pursued anymore. I took it to be, "Look, these guys may not be up to doing this."

Q1: How were intelligence products delivered to users?

Marshall: At the time, there was almost an exclusive reliance on the publications. I think there always was a certain trickle of people who came down and met face-to-face with people. I don't mean briefings, but coming to the office, talking with them and so on. I think Frank Walters at the time felt there was too little of this, and it's not just that the intelligence person making some sort of presentation can communicate more about what's behind the thing, or can answer questions, but also can come away with a better sense of what the people he's talking to really are interested in and care about.

Another thing I tried to have done—it kind of didn't get anywhere—I thought that there were some kind of topics that actually short videos would be better. For example, instead of trying to describe, through writing or even verbally, how a multiple-warhead system worked, that a short video that showed how it worked would communicate a lot more. Or, I felt—and again, neither of these things was done very well—thought that… Well, one of the things that seemed to be happening at the time was that… Because of the split with the Russians, the Chinese were moving their forces, and repositioning them more to the north and so on, etc. And that, for example, if one, again, had a kind of even like a cartoon version…or a map, and time-phasing showing, OK, here's where all of these forces were in the early '50s, here is… These units have moved from here, they're now over here… Or people have been taken from here and moved to there… Fortification programs begun and various other things. That you would be able to see if that kind of thing was just… The time-sequencing—You would see the adjustment of the forces. You would see also the big expansion of the Russian forces along the—That would tell you more… First, the analyst would probably see things he might not otherwise see. But they would be a wonderful way of presenting things to people at the top level.

I did have a primitive version of the MIRV thing done, that was never quite good enough to show Henry, and the other one I kept asking people to do and never succeeded in having it done. But I think there are just terrific possibilities in the presentation of stuff other than having people just read these things or look at static maps and so on. Everything we know about man's evolutionary history is just tremendously focused on the eyes. Lionel Tiger says one of the reasons television is so compelling to people, as against reading, and is in some sense ruining American education, as it is everyplace else, I guess, by competing for time with people, is because people so prefer to get information… And the problem on television is the information they get is so both inadequate and misguided that…

Q1: Any other presentation techniques?

Marshall: No, those were the main ones I think about.

Frank Walters was a CIA analyst, one of the good guys from the '50s that I knew very well. I can't remember exactly what his ideas were, but they're probably like this {…} briefings, face-to- face contact.

Q1: Could you describe the impact of the Schlesinger report on reform of the intelligence community and the role you played?

Marshall: I think it was in December of '70 that Nixon kind of instructed Shultz and Kissinger to undertake this review. And then I think initially there was a March reporting date, and then some decisions were supposed to be made in April-May or something. Then it got postponed till later. And so you had the March 1971 Schlesinger report, which was kind of an interim report. Then there was further work and a final report which made…[87] And as I look at this thing it looks kind of descriptive. And the thing that happened between this and the reports that came out later… The later ones are focused more on recommendations about what to do and kind of the rationale for the changes and so on. And while I was the NSC representative on the thing in an actual way, I guess nominally Wayne Smith was, but then he asked me to do it because I'd done these earlier intelligence studies for Henry. And the bulk of the effort was done by the OMB people. What happened at the end was, however, that Schlesinger had gone over to the AEC [Atomic Energy Commission]—I forget the dates of that—I guess probably in September, I wrote the final kind of summary that went up to Nixon. You know, Here's what is recommended, and also attached were the things he would sign, like a general memo… A memo to Helms as to his new duties and what he was directed to do. And I drafted all of that, because there wasn't a comparable person over at OMB because Schlesinger had left. …And there may have been the start of some kind of a draft…I think there was a partial thing that…May well have been that Bill Kaufmann had—Because he was acting as a consultant to Schlesinger at the time. But all of that stopped when Schlesinger left and I took over to finish it. Then Haig and Smith undertook to send it around to the OMB people, to whoever else they thought they needed to go to…Bronson Tweedy…I'm not sure all the people. And then a few changes and it was finalized… In fact, I prepared the final versions and it was sent up through Smith and Kissinger, Haig, to Nixon to sign off on.

Q1: What happened after the recommendations went to Nixon?

Marshall: After the reorganization was put in place, improving community management, we were involved in monitoring the implementation of the reorganization, and secondly, reviewing in selected areas the quality of the products.

I spent a lot of time talking with people up near Helms, who were involved in making the changes, particularly a man named Bronson Tweedy and just following what they did. They had to set up this new IC staff, there were personnel issues. We'd monitor that and periodically make reports to Kissinger about it. We were often consulted by those people. That was the main thing.

We tried to do some things in the area of the training and management of analysis. One of the things we found was there was really no investment in the training of analysts, or what there was was not very good.

On the management side, we looked at the issue of career lines. And only in a few areas did people stay in their particular area. That happened in things connected with the Soviet Union, but in other regional offices, people had very little background. But again, even in the Soviet Union they had a group of people dealing with the Soviet navy, but they turned over so rapidly that nobody there ever got to have real expertise in the Soviet navy. So there's this kind of incompatibility of the personnel system with getting really skilled people, knowledgeable people doing analysis. We recommended that they look at that and see what they wanted to do about that. We didn't have any particular recommendations, we just pointed out that there were really problems in terms of getting people to have the intellectual capital to do a good job.

I mean, what it comes back to in my mind— It shows also the virtue of lateral entry. I mean, again, one of the big problems with the career system out there is it's kind of based on the idea of getting younger guys—of course, it wasn't formed that way as an agency—and that's one of the reasons the Agency was a lot better in the '50s than I think it was later. They saddled themselves for security reasons, I guess, among other things, with a career system in which people were supposed to come in from universities. And a number of them just had undergraduate degrees.

And kind of progress up and kind of learn on the job, and so on. Well, you're not necessarily going to get great analysts out of them.

Q1: Particularly if you're not paying attention to training.

Marshall: That's right. So the big thing about Larry Gershwin was, first, he was technically better trained than most of the people they'd had looking at the missile programs, or the strategic forces programs and he'd shown real analytic ability, and was brought in by Harry Rowen at my suggestion. They have a fundamental problem of just the intellectual capacity and analytic capabilities of the people they have. That means you can organize, you can invest, you can do certain kinds of things well with a career system of that sort, but there are other things you can't do very well. So you can run collection systems and so forth...American intelligence is relatively strong on collection and weak on analysis.

Q1: To this day?

Marshall: I think so, but I don't pay very much attention to it anymore. But I look at it and it doesn't seem any better, but I don't pay that much attention.

And then the other thing we called their attention to was some possibilities of new training type techniques, this method developed at RAND, reproducing scoring systems, to judge how much people really knew about areas, how good they were at reporting, how certain or uncertain they were about the answers they gave. But all that just frightened them, they didn't want anything to do with anything. This was people down in the organization, not Tweedy.

Later when I was working for Henry, and Chip Pickett and I—in fact you might talk to Chip about it—we had all of these misgivings about the personnel system, about training, and we tried to communicate some of those in a variety of ways to the people out there. But, you know, they were not really that inclined to do anything. I mean, that meant changing something that was very basic in the organization and it would take a lot of effort, it would be a really big change.

When we did encourage some of this lateral entry—I encouraged Schlesinger in that direction and he did a little bit of it... And then later, particularly under Casey, still more of it was done. I don't know what's happening now.

Q1: Did you have any role in preparing the new National Security Intelligence Directives (NSCIDs) that followed the Schlesinger report?

Marshall: Not that I remember, I don't think so. It might well have been the special group over in OMB, but I'm not actually sure who did the drafting.

Q1: The NSCIC just met once or twice?

Marshall: Yeah, probably. It may have met some after Kissinger and I left. But that's right, I can only remember one meeting, actually, when I was there.

Q1: Did the infrequent meetings of the NSCIC result in a lack of direction given to the intelligence community by decisionmakers?

Marshall: When I tried to get from Kissinger, "Well, what do you really want" and so on, we didn't get very much. I mean we'd get these kind of gripes about how bad the stuff is and so on, but you couldn't get them to sit down and put together... You might talk with Chip Pickett about that.

He was around. I can't remember whether we tried to send out some kind of request for a statement of what they most wanted. The only thing that, as I say, worked was this survey I did of the NSC staff, and through them you get what they would have liked and what they thought Kissinger would find useful. I sent the results out to the people at CIA.

Q1: What did they do with it?

Marshall: At the moment, not very much. There was one thing that was requested by the people who followed the Soviet Union, which we put a special effort on. That was to create a new product, which was called something like "The Forum" or something like that. Which was to come out maybe once a month and contain an expression of views of people on some major controversy that existed. And they hated it, and they continued it for a while and dropped it at the first opportunity. But we did get to do it for about four, five, six months, something like that.

Q1: So there was just massive resistance across the board, or inertia?

Marshall: Right. Well, over time I think they've improved. They do some of their own surveys now. I don't think they're as—And they have sent people around to talk with people. And so they've picked up on some of it. I don't know that it really affects what they end up doing, that is the kind of products they produce.

But they go through the process anyway. They didn't do anything. They didn't do anything before.

Q1: You mentioned earlier that the Net Assessment Group was included in the intelligence reorganization document when it was brokered around by Smith and Haig.

Marshall: But it was not in my draft.

Q1: The March 1971 version of the Schlesinger report calls for a net assessment function.

Marshall: I had forgotten about that, but it wasn't, I know, in the other version that I drafted that included recommendations.

Q1: Lionel Olmer gave a talk at Harvard in 1980 in which he claimed that the President's Foreign Intelligence Advisory Board (PFIAB) wanted net assessment performed by the NSC, but that Kissinger was resistant.

Marshall: Well, as I told you, there were several sources of a push to do that. I think I told you of these people who were involved in the Blue Ribbon Panel... The three people that visited me from the Blue Ribbon Panel were on the staff. After the fact, I talked with Ruben Mettler because I had been told that for the Blue Ribbon Panel report that he had been the one that had pushed for this thing [net assessment] within the DoD structure. But these people came over...and it was before I finished. So it was early '70, probably March, maybe April. They had somehow heard the fact that Henry had me doing this, came over, and lobbied for me to put in my recommendations to him this reestablishment of a net assessment group at the NSC. But I didn't have any strong sense of— I didn't personally feel that that was the big thing. Maybe it would be good, maybe not, I don't know. But it didn't fit with the things I was actually coming to, issues I was then addressing.

Q1: Was the 6 December 1971 directive from Secretary of Defense Melvin Laird what you were referring to when you said that Laird issued a directive creating a Director of Net Assessment at the Pentagon the day after Nixon's intelligence memo came out?

Marshall: Maybe we didn't see this and maybe it took this time... But we knew right away that he had decided to do this. ...I forget how we knew.

Q1: Before that he hadn't acted at all on the recommendations of the Blue Ribbon Panel?

Marshall: No, he had not.

Q1: Why do you think he was so opposed to setting up a net assessment function?

Marshall: I don't know that he was strongly opposed. But he probably wasn't strongly for it. I don't know. Actually, the most interesting guy to talk to about it is dead—Fred Buzhardt. When I was over at the NSC trying to get the net assessment process started, he was one of the really helpful people at the Pentagon and he was very close to Laird.... But I don't know who you could talk to about that.

Q1: What's your general assessment of the result of the 1971 reorganization?

Marshall: Well, it's so long ago, and so many other things have intervened. I don't know. It's hard to—Certainly it changed eventually, but only eventually. I think the top-level management of the community—It's hard to say that it's—on issues that matter, or the thing I have the best sense about, the quality of analysis—It's hard to say that it's had that much effect. I mean in some areas—But it's more this matter of personnel, various people, like Gershwin, or some of the NIOs [National Intelligence Officers] they've had in the general-purpose forces, I mean, Gorman was there for a while—sensible thing—the current guy— But I'm not close enough to that set of organizations anymore to judge. But it certainly happened very slowly.

Q1: But I mean from your perch at the NSC, when you were responsible for monitoring what was happening?

Marshall: Oh, well, then, not very much happened at all. I mean it went very, very slowly. I mean because Helms was being asked to take on responsibilities that he didn't really want. Set up new organizations like the IC staff. But it all went very slowly, very slowly. And, as I say, two years later, which was essentially when I left, not a lot had happened, I would say.

Q1: This was very helpful.

Marshall: Well, we can have further sessions.

1969–1973[88]

Q1: So you joined the NSC full time in 1972 as head of the Net Assessment Group?

Marshall: 'Course, I mean, initially it was just me. And then Carol Doheny. And then I recruited Chip Pickett, who must have come in the summer [July '72], and then a few months later, Robin Pirie. And then we had this other–Then Carol had gotten Jo Hunerwadel assigned. But she was someone who came on as one of the regular NSC secretaries. Then when Carol left, then she moved up to be the senior secretary and we got this other secretary, Lexi Crow. And the five of us moved over here.

Let me say that one of the things that happened was that when I got the position as head of the NSC Net Assessment Group, a number of people came around to talk to me. Kind of give me advice. One of the most interesting was General Polk, and I mentioned some other things about his insight from his own time as the commander of the ground forces in Europe. His approach was driven by the fact that he thought it was very important to know how effective these other forces really were, something about their typical mistakes or failings, and so on, and etc.

And I guess the things I remember the most him talking about were these stories that he got, and were indicative of the difficulties they had in movement—When a unit was going to move from one place to the other, they would send out ahead of them people that were almost the equivalent of traffic directors to make sure they make the right turn at that crossroad, this crossroad. And then episodes of them still getting lost, or making the wrong turn. These were all kind of indicators to him of particular failings, problems that they were having. His point was that, here he was, he was the commander of the American ground forces, and when he asked his—down through his intelligence organizations, to tell him about how good the opposing forces were, I mean qualitatively and so on, he was very dissatisfied with what he got back. And that's why he undertook this personal effort to find out for himself. And I forget whether it was he or George Blanchard [a subsequent commander of the American ground forces] telling me later of things they thought about, planned to impress the Russian commander. Efforts were made to invite the counterpart commander to come over and then to either in a seemingly casual way or in a kind of staged way to kind of do things that would impress him, like the speed at which certain kinds of actions could be taken and executed and so on.

I don't remember specifically who [other than General] Polk came by to see me]. But I was struck by the fact that all of a sudden there were a bunch of people coming by to see me.

Q1: Could you talk about the review of combat models you did as part of the net assessment work at NSC?

Marshall: What I looked at was—To look at combat models—Well, I thought that given what the fundamental task was, I wanted to see whether the models of combat, and you know, at the, I suppose, division and above level, not very micro combat, what the state of affairs was. Were they adequate? Would they reflect all of the factors that really ought to go into such an assessment? And so, I guess I got briefed on the state of modeling and the existing models and fairly—First, I more or less decided on my own, but I consulted with a number of people as to their appraisal, and would the things really deal with the full range of factors that ought to be taken into account, and decided that they weren't going to be able to. That, in particular, they didn't take into account these kinds of levels of training, morale, leadership, any of the more qualitative factors. I remember talking with General Polk about the state of the models.

Q1: You also conducted several intelligence case studies?

Marshall: The idea was—You know, it was one of the general tasks of this new NSC Intelligence Committee to do these appraisals of the performance of the intelligence community. And we had started out with the performance in these crises. The studies of crises would have been done from...Beginning spring or summer of '72 and done into the early part of '73—I mean the ones we had planned.

The attempted overthrow of King Hussein and outside intervention in that crisis was one of the case studies. The 1971 Indo-Pak war was another one. The other case that I remember that we looked at was this...I forget the name of it. But there was this period when there was this canal zone area in which the Egyptians were supposedly not to—That the demilitarized strip, I guess, on the other side of the canal—And they were not supposed to move into this area. But they began surreptitiously to do so. And the issue was how well American intelligence was able to monitor that—as I remember that. You ought to talk with Chip about that, but I think...But I don't remember NSSM 69 as being something that we were that involved in. But there was a phrase for this encroachment back into this demilitarized zone.

I don't remember the specific insights from the case studies. But the things led to a certain number of proposals of things that ought to be done in order to do better in the future and so on. But as to the major insights, I guess that I felt that what came out of it was this thing you mentioned [that the analysts did not have the same perspective as Nixon and Kissinger]. But also how difficult it was to get the intelligence community people to really do objective assessments of their own performance. The first draft almost uniformly came back, Given all the circumstances, we did well, even if it was manifest that they hadn't done very well. They always had some out; you know, the circumstances prevented them from doing well. So we usually had to go through two or three drafts before you could finally get them to be a bit more objective and really do an appraisal from which there would be some admission that this or that didn't go so well and maybe something that could be done about it.

Q1: You commented in the past that the post-mortems found a lack of objectivity by the staff drawn from operating organizations; perishable data; lack of access to top officials; sensitivity of certain activities.

Marshall: Yeah, I think there was a memo around here, there was probably something written toward the very end of—You know, what did we learn from this whole thing and so on.

Q1: And the case studies were followed by the study of intelligence in the 1973 Arab-Israeli War?

Marshall: And then along came this other event in '73, that we then took advantage of. And we did the Arab-Israeli War study because one of the difficulties that we'd found in doing the earlier assessments was that you often couldn't really find out which intelligence document or bits of things had actually reached particular people, whether they had read them or not, and so on. So this was an attempt to have a better feel for what materials actually reached the relevant people. And then also to interview them while the thing's going on as to how useful and so on the stuff was. The study was done almost in real time. Talk more with Chip Pickett. And actually, you see, Chip stayed on after the rest of the group came over here, because our movement over here was not too much afterwards. And he stayed on to finish the report and then came over. So he was a lot more involved. He'd be the better person to talk to about that.

But at any rate, so we'd been looking at one aspect of the performance of the community, and I think at some meeting of the second level of this thing, the working group [of the NSCIC], discussed, Well, what other areas could we look at. And one area was the quality of political and economic intelligence. And then the question was, OK, we need to pick a country or some place to look at. And then several proposals were made and Yugoslavia was picked, because at the time, even though it didn't happen until much later, again there was some speculation that Tito—I forget whether it was his health or what it was—but he might leave soon. And so, it was a kind of case of, or a situation in which you would like to have good information about these things.

So, we then undertook this effort and, I guess as I remember it, we collected all this stuff. You know, recent intelligence or probably from the prior couple of years or so. And the materials, both various studies, or major papers, as well as current intelligence…We went and tried to understand, "Who were the people who prepared this?" particularly at CIA, and some of their characteristics, I mean, did they really know anything about Yugoslavia. I mean, what was the basis of their knowledge.

Q1: What yardsticks did you use?

Marshall: Well, whether they had previous experience, had ever lived in the country, knew any of the languages, you know. But we had this body of material. And then we had some people from the outside—and I forget who they all were, except for a couple of them that I—although they weren't the only ones—For the economics stuff, the one person I remember involving was Joe Kirchheimer, who was a friend of mine and secondly, a partner in Burnham & Company on Wall Street, had been in the CIA in the '50s, had been in military intelligence in the war.

Q1: In other words, he was very qualified.

Marshall: Yes, right. He was one of the people we had look at the—But he wasn't the only one. There were two or three other people. We had a little panel of two or three people for each, separately, for the economics and the political. And on the political side, I remember that one of the people was Nathan Leites. But I can't quite remember who the other people were.

And, in essence, we got pretty negative reviews from the panel after reading all this stuff. In both cases. In the case of the economic material...just thought that there was better material that could be available to people in the government from the banks on Wall Street, is the most particular thing I meant, but there were undoubtedly other sources. And then for the political analysis, it was just felt that there was very good reporting in Swiss and German newspapers, and just translating the articles from the better newspapers would be a lot better than the stuff there. Now, of course, there would be a tiny bit of material based on some special intelligence, say, that the newspapers wouldn't necessarily have. But it would be—The bulk of what these people tried to write about was done far better by people out in the real world.

Q1: And these outside sources had been ignored, neglected by the intelligence community?

Marshall: Well, probably. Yeah, I doubt that they read them.

Well, in the first place, again, Yugoslavia—Well, it turned out that the people who did this were part of the current intelligence people at CIA. Again, I think the reporting from the clandestine service people at the embassy and so on would have not had some of these difficulties. But these panels didn't look at that. They looked at the part that came out of the so-called intelligence analysis...But that was done by people in the current intelligence part of CIA, and there, particularly for countries like Yugoslavia, it was treated like almost a newspaper. There was the Yugoslavian desk, and there would be two or three people on it. But they moved from that to other jobs, or from other jobs to that. But they had no prior experience dealing with Yugoslavia. They only tended to—At the moment we wrote the report, the years of even being on that desk, the average for the three people was a year and a half. So they weren't there long enough to get great know-how. They probably didn't read German. So sending them to German and Swiss newspapers was not going to help unless there were some readily available translations. They didn't know the languages.

And it was again an insight to me that—And we talked to one of the people on the NSC, who actually was the recipient of this stuff, the intelligence stuff, who had twice served in Yugoslavia, in the embassy, and so he had about nine years or so of on-the-spot experience in Yugoslavia. So it was an example to me of how wrong—There was this book called *Organizational Intelligence* or something, I forget the name of the author now.

Q1: Wilensky?

Marshall: Yeah, some name like that. But the point is—Walinsky or Wilensky—[89] See the people...The kind of academic theorists sort of treat the world as being that there are these people, the real experts, and things are filtered up to other people who are not really in touch with the real world and also don't know very much about it. And that the problem in organizations is that the people who really know certain kinds of things, some part of it gets filtered out as it moves up to the top.

Here you had this case of the people down below just don't know very much, don't have much interpretive skill, sending stuff up to, in this case, someone who actually had a good deal of experience and knew more about the country than they did. And it's the product of the personnel systems in the intelligence community for these smaller countries. I mean in the case of—If you have other kinds of situations, where people can specialize for a long time, as in the case of Russia and so on, it's not as big a contrast.

Q1: Back on the Yugoslavia review. When was that done roughly?

Marshall: I think that was probably done in the spring of '73, would be my guess.

Q1: What impact do you think these different studies had or how were they used?

Marshall: Well, you know, again, they were sent out from this NSC committee. First, to the members on the committee itself, because they were done by the working group. So all the major organizations were represented there…Well, I guess not all, because I guess NSA wasn't on it or so on. But then it went out. But you know, again, it's just another report and I don't think very much happened as a result of that. But Chip would be good on this, I mean he was the main collaborator…

Q1: On all of these studies, rather than Robin Pirie?

Marshall: Oh yeah, Robin was off…We were trying to get the net assessment part of it going.

Q1: In your notes, you wrote: "History: — HAK • GPF [general-purpose forces] back-up plan, leads to CIA work."

Marshall: Well, I would guess—There had been this work that was done for Kissinger in the summer of '70, you know, the initial net assessment. And the whole motive that he had was that he wanted this study done in order to have some things in his hip pocket, so to speak, of programs that could be either augmented or so on, etc., that would be worthwhile, if things did not go well with the Soviets and new monies were available and…

So part of the background was that, OK, if you're going to have these—This idea that if things go badly you're going to start some new things, they might as well be things that would be impressive.

Q1: Now is this if things went badly in SALT or just a general deterioration in our relations with…?

Marshall: I forget now what was worrying Kissinger. I suppose it was SALT or just general behavior that kind of things…

Q1: There was a stalemate in the SALT talks around the time you started work on the net assessment study and also there were the crises discussed earlier.

Marshall: But in any case, I had had this—From the beginning on this idea of the net assessments—That is, that clearly one of the ways…One of our goals was deterring the Soviets from various kinds of actions, encroachments, and so on. And that it would be important to understand, what did they pay attention to, where did they give us good marks, where did they not give us good marks. And so, once I had joined Kissinger, I thought more about that. And then asked that this study be done to make an attempt to answer those kinds of questions.

Q1: And what were the findings? I know it [the study] was completed after you moved to the Pentagon.

Marshall: Well, as I remember it, I think there were a couple of reports that came out…Again, I remember that there were some things that came out…There certainly was some insights as to what it was they seemed particularly to pay attention to and so on. But I don't remember the details of the thing. But as I remember, I thought the things were OK, and yet they were seen as difficult to do, and the guy who did them I think moved on to something else, and they were never kind of continued.

Q1: But there was just one person working on them?

Marshall: Yeah, right.[90]

Q1: Do you remember why Kissinger wanted the study of the Chinese ballistic missile program done?

Marshall: No. In fact, it had been ordered before I had come on board. But it was one of the first things that came in, when I came on board in early January of '72. And it was one of the things that I was asked to evaluate. Well, also it was a special study that had been asked for by him. And I don't remember what triggered it. I remember reviewing it and sending up an appraisal of it to Kissinger. And then…The thing that was so ironic about it was that, here in this study they, in effect, say that they haven't deployed the missiles yet…At that time they believed that the Chinese hadn't deployed any missiles [MRBMs, IRBMs or ICBMs]. That they hadn't been deployed, and went into a long story about why they were so sure in saying, we would have seen this, we would have seen that. And all based on the assumption that they would proceed very much the way we and the Russians had. And then a relatively few months—within six months, I think—they had to come in and say, "Whoops, we now know they've been deployed more than that. They have been deployed since about 1965. Now that we see what they're doing" and so on, etc.

Q1: This is because they're road-mobile missiles and they're dispersed?

Marshall: Well, they're different in their—Right—and they're hidden in a variety of ways and so on, etc. And obviously the Chinese had taken great care to be far more deceptive about the whole business than either we or the Russians actually had been. And, very likely, in my judgment, because they were actually apprehensive about the Russian reaction had it been detected.

Q1: Possible attack.

Marshall: Yeah.

Q1: But the problem, was it that the CIA was saying, "Well, here's the setup of the missile at the test site, and that we don't see that duplicated in the field. And so, therefore, it hasn't been deployed?"

Marshall: Well, or that they've had certain kinds of tests, but, you know, that they probably would deploy—We would see at the test site the kind of silos, the kind of things that they would eventually deploy. And we don't see that. We don't see any deployed systems that match anything. We don't see anything that looks like what we see at the test sites, and so on, etc.

Q1: Do you remember if the misestimate of the size of the SS-19 ICBM occurred during the May '72 Moscow summit?

Marshall: No, I don't think it occurred during it. But it—The reason I remember it was there was a strong reaction on Haig's part, I think, because we were making, in the negotiations, a big point of some particular measure of size, and if what this new revelation meant, and everything, that this had been set—the size thing—had been set based on our notions of the earlier estimate of the size, which was smaller...And so, in effect, we either said, we were kind of complicating the negotiation by trying to force on them a size category which meant that the -19 would be, in fact, a big missile, or you're forcing them to lie and to... At any rate...

Q1: And Haig was upset about this?

Marshall: Yes, right.

Q1: But the main point was that these were just examples of where the intelligence community got things wrong?

Marshall: Yeah, right.

Q1: What were the general contents of the five-page handwritten letter you wrote to Schlesinger when he became Director of Central Intelligence?

Marshall: Well, the main thing I guess I recall about it was kind of saying to him—'Course Schlesinger already, because of his previous position at OMB, where he'd overseen a lot of—I mean the people who monitored intelligence were under him and so on. And I, of course, had talked with him from time to time anyway about it. But I guess the main thrust of the letter was to say to him, "Look, this is an organization which isn't as good as it used to be. It has a lot of problems," and try to outline a number of problems based on the experience that I had had of the inadequacies in the analysis area, its tendency to isolate itself from the outside world in terms of very little lateral entry, all of that. I just kind of laid all that out for him, as to the kind of place he was taking over and what seemed to be the kind of major problems, the obstacles to change, and so on.

Q1: Even at the NSC, you were criticizing the intelligence community for its estimates of the Soviet defense burden.

Marshall: Yes.

Q1: And what do you remember about your work on that subject while you were at the NSC?

Marshall: Well, it was...I mean I didn't do very much about it there, as compared with later, when I came here. But I had had doubts about it back at RAND. In fact, Schlesinger and I had talked some about that. And we also talked about it when we did that net assessment for Kissinger in the summer of '70, or probably early fall. And it just was unbelievable that they could produce the size of military effort, be modernizing at the kind of pace they seemed to be, all on six percent of a GNP which was smaller than ours. Again, it was based on the notion that it was about half ours, for the moment, accepting that the GNP was about the size that the CIA said it was, which would make it about half of the U.S. GNP. With only six percent of that being allocated, which would be roughly the equivalent of three percent of our GNP, that they could produce this very large set of military programs. So Schlesinger and I felt that there was something doubtful about that, going way back. And nothing I learned made me feel any better about it.

Q1: But this wasn't something that you studied, I mean a formal study or asking the intelligence community to relook, look at it again at that time.

Marshall: No. That was done largely here, initially when Schlesinger came here and we started pressing this as one of the issues with them. Which led to their relook at it, and coming back to us in, I suppose, '74, with this special memorandum, it wasn't a regular report, it came back initially as just a memorandum to Schlesinger, that we've looked at it, and we think the errors, we still believe the number and the errors would be very small on either side of it.

Q1: You had thought that the 1966 paper on *Problems of Estimating Military Power* might have been related to work for Fred Wikner.

Marshall: Yes.

Q1: But really the other work that you did—and I'll talk with him—

Marshall: Is later.

Q1: Much, much later.

Marshall: Well, so probably wasn't, yeah. So, I suppose it must have just come out of the, at one level, the work that I had been doing for a long time with Loftus, although that had pretty much stopped at that point. Loftus and I put out some final reports I think that we did together in '63. Then I went off to France.

Didn't return until February or March of '65. And one of the main things that I wanted to pick up at that point was trying to apply whatever was known about organizations, organizational behavior. Which interest, in a way, had started in the work with Loftus, had started back in the late '50s. And we did spend a lot of time, or I spent a lot of time, thinking about how to understand the Soviets from that perspective. But exactly why that thing [*Problems of Estimating Military Power*] looks the way it is, I don't know. Or why the topic is exactly what it is.

Q1: As Director of Strategic Studies at RAND, you were responsible for sort of integrating the efforts related to strategic forces that were done in the different departments at RAND?

Marshall: Yes.

Q1: So this was a new layer of management?

Marshall: Yes, it was a new thing that had come out of discussions that Schlesinger and I had had with Rowen, which led to Rowen initially asking Schlesinger to undertake this. And he did that for, maybe realistically, for six months. He hadn't gotten very far along in it when the 1968 election came along, and then he came back on some of the transition team things and then ended up going into the government here. So Rowen asked me to take it on. That was in early '69, I think, because initially, when Schlesinger came back here during the transition period, it wasn't known whether he'd come back. The expectation initially was that he would probably come back. So he [Schlesinger] probably knew about the offer to be assistant director in the Bureau of the Budget and accepted it in, I don't know, middle of January, late January, and shortly afterwards Rowen asked me to take on the thing.

And so I looked at—I'm not sure exactly all of what Schlesinger had done, or I don't remember exactly. But anyway, one of the things I did was to get a kind of a readout of, Well, who is doing work at RAND, and charging it off against, or claiming that he's working on strategic nuclear forces? And it was a very motley assortment, because there had been no overall real direction.

And Schlesinger, I think, hadn't gotten far enough along to really take hold of this. And so I then started to think about, Well, what should a program in this area really look at? What are the issues that you would want to have work being done on? And so on, etc. And then it all kind of came out kind of naturally. And then I had formed a small kind of discussion and planning group of not everyone who claimed he was working in this area, but—and I even forget exactly who was on it, but it was a small group of three or four people, probably. I had these ideas and tried them out there, and then said, more or less said, Well, here's the way I think you really ought to look at this thing. Here's a—In some sense we've been in this kind of competition in this area for twenty years or more already and it's going to go on for a long time and we maybe need to look at it a different way and so on. And so it kind of led to that set of ideas.

And internally…I don't know what the records show…My memory is that it…The whole thing, at least in an early form, was written down by the end of the summer of '69. And then I had this call from Kissinger's office asking me to stop by, and I was on my way to a vacation in France. And I stopped by in probably about the third week of September and had a session with Kissinger and then went on my way and was gone about a month and came back. And then when I came back I continued to try to sort out things on this program, but I was also being badgered by Kissinger to come back and get on with this study and so on, which I did in the early part of December of '69.

Q1: Do you remember what kind of reaction you got from the Air Force, or from the small group at RAND that you were meeting with?

Marshall: Well, they…I don't know…The group at RAND, I mean the people I talked to, thought that was interesting. But I think it was never particularly picked up or taken anywhere at RAND. I got several letters and notes from people in the Air Force who liked it, thought it was interesting. But that's about it, as I remember. But, as I say, my memory is—

I don't know whether anything shows up in that summer of '69. But essentially it seems in my own mind a lot of it was done then, or the essence of it had been worked out.[91] But then I went away. Did some more probably when I got back. And there was also a lot of practical stuff of trying to get the monies allocated, because I, in principle, now controlled the budget. "In principle," I say, because there was not a very effective apparatus for doing that. And then I went away. And then over '70, I mean I did the thing for Henry, the two reports, the last report was finished very early in May. It was essentially the same time as the Cambodian enterprise was started. I then came back to RAND. But then quite soon there was this...Two things: one, there was the SS-9 study that he accepted, and I ran that, coming back from time to time and doing it, and then this effort that started in the summer of '70 on these standby programs and the net assessment.

And then in December came this direction from Nixon to Henry and Shultz about the reform of the intelligence community. And that started another cycle of my commuting back and spending a lot of time there.

Q1: So how much time did you really have to devote to shaping the interdepartmental research program on strategic studies?

Marshall: Well, I mean it was hard, given all of these other things, and probably, you know—

Q1: How did you decide on the topic for the NSSM 186, the first net assessment?

Marshall: Well, we had had a long struggle to get the whole thing going, because Laird kind of dragged his feet on it, I talked a lot to Buzhardt, and also Laird had as a special assistant this guy who...I forget his name...Baroody...But I dealt mainly with Buzhardt, but I had some meetings with Baroody.

And both of them, I think, wanted to move ahead or be reasonable, but they—I never really talked to Laird about it, but I mean he clearly resisted the idea of the thing. So in the end, the way the things were going to be done was, while they would be run from the NSC, the actual work would be farmed out to the most relevant and interested party. We had a little working group...A guy named Al Hall, who was here in the Pentagon...

Q1: He was like the Assistant Secretary for Intelligence.

Marshall: Yeah. And he also had been involved in this working group, for this NSCIC.

So when we finally worked out a process that Defense would buy, then came the question of picking a topic. And I'm not sure exactly why we picked the ground forces. It may have been, Well, this is an area, let's just start with it, and some things probably favored it and so on. More work had been done looking at—not real net assessments—but looking at the nuclear forces... So, anyway, that ended up being picked as a reasonable one to start with, and see how this whole thing was going to go.

But it was started very late. It was started probably just...at most six months before I moved over here.

Q1: In fact, this NSSM 186 is dated 1 September 1973.

Marshall: Hmmm, yeah, right. So anyway, just a few months, right. But probably—It had been so slow to have the whole thing come together that the people were more or less in place, and a contract was put out to BDM to do a lot of stuff. That's how Phil Karber and all these guys... They had an office down in the basement someplace here and they got going.

Q1: Now did they do most of the work for the initial ground force net assessment?

Marshall: Well, a lot of it, yeah, right.

Q1: And then they continued after that?

Marshall: Yeah, right. That's right, that contract. And I took that over and it became a long-standing contract.

Q1: Do you know when you first started thinking about net assessment?

Marshall: I mainly wrote the 1970 net assessment, although Schlesinger spent a little time on it. And there were people around—I mean, again, Charlie Herzfeld was around, always pushing the idea of doing it. So I was thinking about it, and I'd done that earlier paper on measuring military power, and so on.

In a more systematic way, it would be for Wikner back in… [when Marshall wrote his report on comparisons and R&D strategy, part of a larger RAND effort.] And at the time I did not particularly aspire to do that and so on. I thought it was interesting, important, to try to have some way of measuring the comparative capabilities, and I clearly didn't think most of these simplified approaches did very much. And I guess I had—A lot of my interest in it grew out of this interest in the Soviet Union, but in this organizational assessment, I mean of how much of both what people did, how effective they were, depended on these organizational matters. But I hadn't any, did not envisage, or I wasn't particularly pushing the idea of setting up an office to do these things and so on. [indistinct] a good idea that somebody ought to do, but I didn't—At the time I was far more interested in the intelligence problems and in this…different ways of looking at these issues of "doing well," like the long-term competition set of ideas and in this application of these ideas coming out of an organizational, almost social-cultural kind of perspective on these issues.

Q1: Did you also write a study of possible U.S. responses to Soviet threats against China with Richard Kennedy while working for Kissinger?

Marshall: It was almost a drafting of a little paper on the thing, appraisal for—Well, and a recommendation about what we should do, I guess, which I didn't think I had any special competence in, but got dragged into by Kennedy.

Q1: In Kissinger's memoirs, and other sources, he mentions that at this time, early 1970, the NSC was doing a study of contingency planning for what the United States should do if there were a war between the Soviet Union and China.

Marshall: Yes.

Q1: Now was this small effort that you and Kennedy worked on related to that?

Marshall: Well, I presume it was right at the beginning, because what he and I did, as I remember, took place in December of '69. I came to kind of start this other stuff, and was just kind of getting organized to do all of these studies and so on, etc.

And this issue came up of the Russian—I don't know, I forget—approaches to us, but anyway, threats against the Chinese.

Q1: That was really in the late summer and fall of '69.

Marshall: That's right. But somehow or another it had become an issue just at that moment. And so, Kennedy indicated to me that Kissinger would like me to participate. So it was kind of like being dragged off to the side, from the main thing I was there to do, to just think through with him, you know, how should we react, what might we do and so on, and a little paper was prepared, which I participated in the drafting of. And it was sent into Kissinger, and after that I never knew what happened.

Q1: But what did you recommend in this short paper?

Marshall: I'm a little vague on that, you know. But my guess is that one of the things was, Well, you really ought to start some planning, some thinking what do you really do if…I think it was kind of to discourage the Russians, but also think about what you can do. But I don't have any clear memory of that.

Q1: This is just for completeness. Make sure that what you want to say is in the book.

Marshall: Yes, right.

Q1: Could you talk about your work on deception?

Marshall: I'd had this long-term interest in deception as a very important thing. And probably when I was at the NSC, I forget whether I had any money to do it, but I encouraged the people out at the Naval Postgraduate School, there were several people out there who were interested in it, ended up with a—you know there's a volume on deception and so on.

Q1: So that grew out of your support for—

Marshall: Well encouragement, and then later I think I gave them money, I forget— Right, yeah.

Q1: Anything else you want to say on that?

Marshall: No.

Q1: Alright.

Marshall: Except that it's a big, important topic in which the intelligence community has been very slow and resistant to really looking at. Not stuff that we would do, but the fact that other people may do it, and one of the problems in the intelligence world is how to deal with that. It was one of the good things about Casey, he understood what the real world was like.

Q1: Could you explain the line "Kissinger contrast of Chinese and Soviets" from the notes for one of your talks on this period?

Marshall: Well, I remember when Kissinger came back from the '72—I mean the first big official one that he went with Nixon… He had a meeting at which he briefed a number of the people on the NSC. Maybe it was only those who had somehow been involved in assisting the preparations. And I remember I did a few small things. And he made this point about how good it was to deal with the Chinese, who were, the way he put it, better than their word, as compared with the Soviets, who are violating the treaty at the same moment they're signing it. You know, I mean [Marshall laughs]. And also what a contrast…Not much…And also intellectually more interesting, better.

Q1: Who was the "AF gen" mentioned in the line "Tweedy/Helms reactions/role of DCI/Intelligence Community staff, AF gen" from the notes for one of your talks on this period?

Marshall: It may be a reference to Lew Allen, who was the first head of the Intelligence Community Staff and then later head of NSA. Both Schlesinger and I had come to know him because he had been involved in the NRO space operations and so on.[92]

Q1: How much contact did you have with Kissinger before 1969?

Marshall: Well, one that I—Well, very slim. I met him…He'd come to RAND…I forget when it would be, probably the late '50s. I remember that he had been invited, and there was a party for him up at the Wohlstetter's and I remember being there, shaking his hand and so on. But he'd also been at RAND, I mean that particular visit, so I presume that would have been in… probably after this book of his in '57 [*Nuclear Weapons and Foreign Policy*] so it probably was the '58–'59 time period. But that's the only two things I can remember. I mean—Right. But I mean I had no real personal connection with him.

Q1: Can you think anything else about this period you'd like to…?

Marshall: [Laughs] No, I don't think so.

Q1: I've asked enough.

Endnotes

1 Eds: This chapter includes four sections. The first two draw on consecutive interviews Guthe conducted with Marshall in 1993 and 1994—the first covers 1969–1975, and is an interview from 14 December 1993, the second covers 1973–1976, and is excerpted from an interview from 8 April 1994. The first outline was prepared for the first interview, while the second outline is an excerpt of the relevant sections from the outline prepared for the second interview. The third section is an interview conducted on 13 July 1999, and the fourth section is an interview conducted on 20 July 1999. Both cover 1969–1973. Because the third and fourth sections draw on follow-up interviews and lack the narrative flow of the original interviews, the editors have chosen to place them at the end of the chapter even though they cover an earlier time period than the second section. See also transcripts of interviews of Andrew Marshall by Alfred Goldberg and Maurice Matloff (Historical Office, Office of the Secretary of Defense), 1 Jun 1992, 13–23; 15 Jun 1992, 1–10.

2 Eds: U.S. Department of Defense, Deployment of Military Personnel by Country (1969), accessed at www.dmdc.osd.mil/appj/dwp/dwp_reports.jsp.

3 Eds: Andrew F. Krepinevich, Jr., *The Army in Vietnam* (Baltimore: Johns Hopkins University Press, 1986), 237–251.

4 Eds: National Security Council, "Vietnamizing the War," National Security Study Memorandum 36 (10 Apr 1969).

5 Eds: Public Papers of the Presidents of the United Sates: Richard Nixon, 1969 (Washington: GPO, 1971), 544–556. See also Richard M. Nixon, "Address to the Nation on the War in Vietnam," (3 Nov 1969) P-691101, Nixon Presidential Library, available at www.nixonlibrary.gov/sites/default/files/2018-08/silentmajority_transcript.pdf.

6 Eds: Robert S. Norris and Hans M. Kristensen, "Global Nuclear Weapons Inventories, 1945-2010," *Bulletin of the Atomic Scientists* (July/Aug 2010): 81.

7 Eds: Richard A. Hunt, *Melvin Laird and Nixon's Quest for a Post-Vietnam Foreign Policy, 1969-1973,* Cold War Foreign Policy Series, Special Study 6 (Washington: Historical Office, Office of the Secretary of Defense, 2014), 16–18.

8 Eds: Lawrence Freedman, *The Evolution of Nuclear Strategy*, 2nd ed. (New York: St. Martin's Press, 1989), 354–357; Richard A. Hunt, *Melvin Laird and the Foundation of the Post-Vietnam Military, 1969-1973,* Secretaries of Defense Historical Series, Vol. 8 (Washington: Historical Office, Office of the Secretary of Defense, 2015), 428–429.

9 Eds: National Defense Budget Estimates for FY 2020 (Washington: Office of the Undersecretary of Defense (Comptroller), 2019), 134–35, 260–61; United States Air Force Statistical Digest, FY 1969 (Washington: Comptroller of the Air Force, 1970), 3;United States Air Force Statistical Digest, FY 1976 (Washington: Comptroller of the Air Force, 1977), 4; Naval History and Heritage Command, "U.S. Navy Active Ship Force Levels," www.history.navy.mil/research/histories/ship-histories/us-ship-force-levels.html; John Lewis Gaddis, Strategies of Containment: A Critical Reappraisal of Postwar American National Security Policy (Oxford: Oxford University Press, 1982), 323.

10 Eds: Walter S. Poole, *The Decline of Détente: Elliot Richardson, James Schlesinger, and Donald Rumsfeld, 1973-1977,* Cold War Foreign Policy Series, Special Study 7 (Washington: Historical Office, Office of the Secretary of Defense, 2015), 9–11.

11 Eds: Norris and Kristensen, "Global Nuclear Weapons Inventories, 1945-2010," 81.

12 Eds: Hunt, *Melvin Laird and the Foundation of the Post-Vietnam Military…*, 330, 355–59.

13 Eds: J.R. Transue et al., *Assessment of the Weapons and Tactics Used in the October 1973 Middle East War,* WSEG Report 249 (Arlington: Weapons Systems Evaluation Group, Oct. 1974), declassified, 3–5.

14 Eds: Lloyd R. Leavitt, *Following the Flag: An Air Force Officer Provides an Eyewitness View of Major Events and Policies during the Cold War* (Maxwell Air Force Base: Air University Press, 2010), 474–489.

15 Eds: Interview conducted on 14 December 1993.

16 A.W. Marshall and J.F. Digby, *Indications of Imminent Hostilities (U)*, 0-1477 (Santa Monica: RAND Corp., 29 Dec 1952), CONFIDENTIAL; A.W. Marshall and J.F. Digby, *The Military Value of Advanced Warning of Hostilities and Its Implication for Intelligence Indicators (U)*, SM-14 (Santa Monica: RAND Corp., 20 Jul 1953), TOP SECRET. Digby discusses this work briefly in "Contributions of RAND to Strategy in the 1950s," in Andrew W. Marshall, J.J. Martin, and Henry S. Rowen, eds., *On Not Confusing Ourselves: Essays on National Strategy in Honor of Albert and Roberta Wohlstetter* (Boulder: Westview Press, 1991), 22. A short summary of SM-14 appears in Merton E. Davies and William R. Harris, *RAND's Role in the Evolution of Balloon and Satellite Observation Systems and Related Space Technology*, R-3692- RC (Santa Monica: RAND Corp., Sep 1988), 49.

17 J.E. Loftus and A.W. Marshall, *The Limitations and Potentialities of the U.S. Strategic Warning System (U)*, D(L)-4944, TS-1710, Classified. This paper is referred to in Joseph E. Loftus and Andrew W. Marshall, *RAND Research on the Soviet Military*, D-4943-PR (Santa Monica: RAND Corp., 21 Feb 1958); J.E. Loftus and A.W. Marshall, *Strategic Warning: A Proposed Collection Concept (U)*, D-6048 (Santa Monica: RAND Corp., 20 Feb 1959), Classified; J.E. Loftus and A.W. Marshall, *The Structure of the U.S. Warning System (U)*, D-6064 (Santa Monica: RAND Corp., 26 Feb 1959), Classified; J.E. Loftus and A.W. Marshall, *Soviet Passive Detection Capability (U)*, 0-6300 (Santa Monica: RAND Corp., 15 May 1959), Classified; Russell T. Nichols, *Economics Division Progress Report*, 0-3740 (Santa Monica: RAND Corp., 15 Jul 1956), 7; T.E. Greene and J. Vogel, *Engineering Division Activity Report, First Half of 1956*, D-3750 (Santa Monica: RAND Corp., 15 Jul 1956), 46; Russell T. Nichols, *Economic Analysis Department Progress Report, January 15, 1957*, D-4052 (Santa Monica: RAND Corp., 7 Jan 1957); Russell T. Nichols, *Economic Analysis Department Progress Report*, 0-4398 (Santa Monica: RAND Corp., 15 Jul 1957), 16–17, 19.

18 J.E. Loftus and AW. Marshall, *Research on Soviet Military Programs (U)*, D(L)-6500 (Santa Monica: RAND Corp., 26 Jun 1959), SECRET; J. E. Loftus and A.W. Marshall, *On Improving Midterm Estimates of Soviet Military Capability for Use in U.S. Military Planning (U)*, RM-2892-PR (Santa Monica: RAND Corp., Aug 1962), SECRET; J.E. Loftus and A.W. Marshall, *Forecasting Soviet Force Structure: The Importance and Budgetary Constraints (U)*, RM-3612-PR (Santa Monica: RAND Corp., Jul 1963), SECRET; Andrew Marshall, "Propositions from Case Study on Forecasting Soviet Force Structure," 20 Jul 1970, in Hans Heymann, Jr., rapporteur , *Bureaucracy and National Security Policy: A Conference Report* (Draft), R-644-PR/ARPA (Santa Monica: RAND Corp., Feb 1971), 46–47; Andrew W. Marshall, "Arms Competitions: The Status of Analysis," in Uwe Nerlich, ed., *The Western Panacea: Constraining Soviet Power through Negotiation, Volume II of Soviet Power and Western Negotiating Policies* (Cambridge: Ballinger Publishing Co., 1983), 6.

19 Bross's position was Deputy to the DCI for National Intelligence Programs Evaluation. On the NIPE staff, see Senate Select Committee to Study Governmental Operations With Respect to Intelligence Activities, *Book I: Foreign and Military Intelligence*, 94th Cong., 2nd sess. (Washington: GPO, 1976), 117, *Book IV: Supplementary Detailed Staff Reports on Foreign and Military Intelligence*, 76, 100; and Ray S. Cline, *Secrets, Spies and Scholars: Blueprint of the Essential CIA* (Washington: Acropolis Books, Ltd., 1976), 207–209.

20 A.W. Marshall, *Comparisons, R&D Strategy, and Policy Issues (U)*, WN-7630-DDRE (Santa Monica: RAND Corp., Oct 1971), SECRET.

21 Report to the President by the Security Resources Panel of the Office of Defense Mobilization Science Advisory Committee, "Deterrence and Survival in the Nuclear Age," 7 Nov 1957 (TOP SECRET; Declassified 10 Jan 1973), in Department of State, *Foreign Relations of the U11ited States, 1955-1957, Vol. 19: National Security Policy* (Washington: GPO, 1990), 639–661. While on the staff of the Gaither Committee, Marshall was a member of a "small group [of] people with systems analysis-analytical kinds of skills [that] worked as consultants and experts to people in [the] other panels [and did] various kinds of tasks." Transcript of interview of Marshall by Goldberg and Matloff, 1 Jun 1992, 8. An organization chart for the committee shows an "Evaluation Quantitative Assessments" group headed by Robert C. Prim (Bell Telephone Labs) and Stanley J. Lawwill (Strategic Air Command).

22 A.W. Marshall and J.W. Tukey, *A Stringent Inspection Scheme Directed Against the Danger of Surprise Attack (U),* SAW/180, 2 Nov 1958, SECRET; Memo, Andrew Marshall, Vincent V. McRae, and Albert Wohlstetter to Mr. Foster, subj: The "Accident" Problem, 19 Nov 1958, with attachments, "Part I: The Problem of Nuclear Weapon Accidents and Other Causes of Accidental War"; "Part II: Some Comments on Specific Statements by Eastern Experts" [SAC/00/34, File 9T.2 US Experts b. Inter-office Memos SAW/00/1-50, Meetings-Talks-Conferences, Box 99, Lot No. 58D133, Records Relating to Disarmament, 1942-1962, 1953-62 Period, General Records of the Department of State, Record Group 59, National Archives]; Summary Minutes, Five Power Working Croup Meeting on Systems of Control, 22 Oct 1958 (CONFIDENTIAL; Declassified 4 Oct 1994) [SAW/FPC/4, File 9T.1k. Five Power Conversations SAW/FPC/1-11, Box 98, Lot 58D133, RG 59, National Archives]. Marshall was a member of the Instruments of Detection and Control Systems Groups of the U.S. delegation. For discussions of the Surprise Attack Conference, see Bernhard G. Bechhoefer, *Postwar Negotiations for Arms Control* (Washington: Brookings Institution, 1961), 464–487; Johan J. Holst, "Strategic Arms Control and Stability: A Retrospective Look," in Holst and William Schneider, Jr., *Why ABM? Policy Issues in the Missile Defense Controversy* (New York: Pergamon Press, 1969), 245–284; and Memo, Rear Adm. Paul L. Dudley, USN to the Joint Chiefs of Staff, subj: Report on the Conference on the Problem of Surprise Attack, Geneva, Switzerland, 30 Dec 1958 (SECRET//NOFORN; Declassified 4 Oct 1994), with attachment, "A Summary of the Preparatory and Conference Activities of the U.S. Delegation to the Conference on the Problem of Surprise Attack" [File 9T.4r. Proposed Basis for Examination of Possible Systems for Observation and Inspection of Naval Force, Box 109, Lot No. 58D133, RG 59, National Archives].

23 This group was called the Consultative Group on War by Accident, Miscalculation or Surprise Attack. The McCloy arms control study, of which this group was a part, may have been run by the U.S. Disarmament Administration in the Department of State. An initial search of State Department records at the National Archives failed to turn up material related to the study. In addition, staff members at the John F. Kennedy Library were unable to find any relevant material in their files.

24 The group was chaired by Thomas Schelling (Center for International Affairs, Harvard University). The other members were: Helmut Sonnenfeldt (Policy Staff, U.S. Disarmament Administration), Lt. Col. R.A. Boyd, USAF (Plans and Policy, Organization of the Joint Chiefs of Staff), Marvin Stern (Office of the Director of Research and Engineering, Department of Defense), Brig. Gen . Ariel W. Nielsen, USAF (Plans and Policy, Organization of the Joint Chiefs of Staff), Henry D. Owen (Policy Planning Council, Department of State), and J.J. Hitchcock (National Indications Center, Central Intelligence Agency).

25 Marshall worked on this task with Thomas W. Wolfe, then an Air Force officer at the RAND Corp. Much of their work involved reviewing and commenting on various material. Rowen then was Deputy Assistant Secretary of Defense for International Security Affairs. In Nov 1962, OSD/ISA produced a draft paper titled, "Some Lessons from Cuba" (SECRET; Declassified). The paper is reproduced in Laurence Chang and Peter Kornbluth, eds., *The Cuban Missile Crisis: A National Security Archive Reader* (New York: The New Press, 1992), 308–318. Paul Nitze, ASD/ISA at the time, in his memoirs briefly discusses "a postmortem [on the crisis] prepared in February 1963 by Walt W. Rostow's Policy Planning Staff in the State Department and some of [Nitze's] staff in ISA." Nitze, with Ann M. Smith and Steven L. Rearden, *From Hiroshima to Glasnost: At the Center of Decision—A Memoir* (New York: Grove Weidenfeld, 1989), 237. The Nov 1962 draft was part of the Feb 1963 postmortem.

26 B. Klein, C. Schubert, O. Hoeffding, F. Hoffman, A. Marshall, *The NATO Force Planning Exercise: Scope, Aims, and Organization (U),* D(L)-12548-ISA (Santa Monica: RAND Corp., 15 Jun 1964), Classified; Section on NATO Force Planning Exercise in "Some RAND Products," paper in "RAND Research Results II" notebook, 16 Sep 1969 [RAND Archives/Robert Specht Files]; Ltr, James F. Digby to Henry S. Rowen, Dep. Asst. Secretary of Defense for International Security Affairs (Planning & NSC Office), 6 Dec 1963, with encl, B.H. Klein, "Progress Report on RAND Participation in the Stikker Exercise (U)," 14 Nov 1963 (Confidential; Declassified 23 Dec 1975).

27 Ltr, F.R. Collbohm to Henry S. Rowen, Deputy Assistant Secretary of Defense for International Security Affairs (Planning and North Atlantic Affairs), 2 Dec 1964 [RAND Archives/J. Digby Files/Box: Chronological Files; Memos: Outgoing (Defense), 1949–1951 & 1959–1960]

28 Rowen was Assistant Director of the Bureau of the Budget from 1965 to 1966.

29 A.W. Marshall, *Cost/Benefit Analysis in Health*, P-3274 (Santa Monica: RAND Corp., Dec 1965). The paper was delivered to a "special session of the Defense Management Systems Course sponsored by the Bureau of the Budget. Attending…were personnel from many United States Government Departments and Agencies attempting to introduce program budgeting and cost-effectiveness methods" 1.

30 C. Wolf, Jr., "Working Group I—Improving Policy Analysis: Introductory Paper," 17 Jul 1970, A.W. Marshall, "Notes on Improving Policy Analysis: Attachment to Introductory Paper (Working Group I)," 17 Jul 1970, and "Working Group I: Improving Policy Analysis: Summary of Findings," 23 Jul 190, in Hans Heymann, Jr., rapporteur, *Bureaucracy and National Security Policy: A Conference Report* (Draft), R-644-PR/ARPA (Santa Monica: RAND Corp., Feb 1971), 57–58, 60–63.

31 Charles J. Hitch and Roland N. McKean, *The Economics of Defense in the Nuclear Age* (Cambridge: Harvard University Press, 1960), 125–128.

32 Alain C. Enthoven and K. Wayne Smith, *How Much Is Enough? Shaping the Defense Program, 1961-1969* (New York: Harper & Row, Publishers, 1971).

33 Thomas Powers, *The Man Who Kept The Secrets: Richard Helms and the CIA* (New York: Alfred A. Knopf, 1979), 200–206.

34 In early 1970, the NSC staff developed contingency plans for a Sino-Soviet war. Henry Kissinger, *The White House Years* (Boston: Little, Brown and Co., 1979), 693. See also Philip Odeen, in Nuclear History Program, *The Evolution of Selective Options Strategy*, NHP Oral History Transcript #4 (College Park: Center for International Security Studies, University of Maryland, 1993), 5.

35 See comments by Russell Jack Smith (then Deputy Director for Intelligence) in his *The Unknown CIA: My Three Decades with the Agency* (Washington: Pergamon-Brassey's, 1989), 205.

36 For a discussion of the Office of National Estimates and its decline, see Harold P. Ford, *Estimative Intelligence: The Purposes and Problems of National Intelligence Estimating*, revised edition (Lanham: University Press of America, 1993), 81–105. See also William Colby and Peter Forbath, *Honorable Men: My Life in the CIA* (New York: Simon and Schuster, 1978), 351.

37 The SS-9 study was begun in the spring of 1970 and completed in the fall of 1971.

38 Department of Defense, *Report to the President and the Secretary of Defense on the Department of Defense by the Blue Ribbon Defense Panel* (Washington: GPO, 1 Jul 1970), 7, 31.

39 From 1953 to 1964, the Net Evaluation Subcommittee of the NSC prepared assessments of the nuclear capabilities and vulnerabilities of the United States and the Soviet Union. Gen. Leon Johnson (USAF, Ret.), *The History of the NESC, 1961-1964* (Arlington: Stanford Research Institute, 9 Dec 1968); *Secrets, Spies and Scholars*, 140–143; Net Evaluation Subcommittee Folders, 1955–1964, Box 66, Records of the National Security Council, Record Group 273, National Archives].

40 Ruben F. Mettler, the President of TRW, was a member of the Blue Ribbon Panel. *Report to the President and the Secretary of Defense on the Department of Defense by the Blue Ribbon Defense Panel*, iii.

41 Office of Management and Budget, *A Review of the Intelligence Community*, 10 Mar 1971 (Classified; Sanitized) ("Schlesinger Report") [National Security Archive, The U.S. Intelligence Community, 1947–1989: Printed Guided and Index, Microfiche #00867].

42 "Announcement Outlining Management Steps for Improving the Effectiveness of the Intelligence Community," 5 Nov 1971, in *Weekly Compilation of Presidential Documents*, 7, no. 45 (8 Nov 1971), 1482. See also *Book IV: Supplementary Detailed Staff Reports on Foreign and military Intelligence*, 84; and Robert M. Macy, "Issues on Intelligence Resource Management," in *Report of the Commission on the Organization of the Government for the Conduct of Foreign Policy, Volume VII, Appendix U: Intelligence Functions Analysis* (Washington: CPO, Jun 1975), 52–58.

43 Tweedy replaced Bross in Jan 1971.

44 *Honorable Men: My Life in the CIA*, 293
45 Graham T. Allison, Joseph L. Bower, C. Roland Christensen, James G. March, A.W. Marshall, and Andris Trapans, *Organizational Behavior of the Soviet Military Bureaucracy (U)*, D(L)-17608-ISA (Santa Monica: RAND Corp., 14 Aug 1968), SECRET//NOFORN; A.W. Marshall, *The Improvement in Intelligence Estimates Through Study of Organizational Behavior*, D-16858-PR (Santa Monica: RAND Corp., 15 Mar 1968) (CONFIDENTIAL; Declassified 14 Jan 1977), 7–10.
46 Graham T. Allison, *Essence of Decision: Explaining the Cuban Missile* Crisis (Boston: Little, Brown and Co., 1971).
47 Andrew Marshall, "Intelligence and Crisis Management," in Gen. Jack N. Merritt, Gen. Robert Reed, and Roger Weissinger-Baylon, eds., *Crisis Decision Making in the Atlantic Alliance: Perspectives on Deterrence* (Menlo Park: Strategic Decisions Press, 1988), 8-1–8-2.
48 Robert F. Ellsworth and Kenneth L. Adelman, "Foolish Intelligence," *Foreign Policy*, no. 36 (Fall 1979): 153.
49 On the work of Post and the CIA's Center for the Analysis of Personality and Political Behavior, see Thomas Omestad, "Psychology and the CIA: Leaders on the Couch," *Foreign Policy*, no. 95 (Summer 1994): 105–122; and Jerrold M. Post, letter to the editor, *Foreign Policy*, no. 96 (Fall 1994): 194–197.
50 Thomas A. Brown, *Probabilistic Forecasts and Reproducing Scoring Systems*, RM-6299-ARPA (Santa Monica: RAND Corp., Jun 1970); Thomas A. Brown and Emir H. Shuford, *Quantifying Uncertainty Into Numerical Probabilities for the Reporting of Intelligence*, R-1185-ARPA (Santa Monica: RAND Corp., Jul 1973); Thomas A. Brown, *An Experiment in Probabilistic Forecasting*, R-944-ARPA.
51 National Security Study Memorandum 178, Henry A. Kissinger to Secretary of State, Secretary of Defense, Director of Central Intelligence, subj: Program for National Net Assessment, 29 Mar 1973 (SECRET; Declassified 26 Jan 1994) [NSSM 178, National Security Study Memorandum (NSSM), Box 3, Records of the National Security Council, Record Group 273, National Archives].
52 National Security Study Memorandum 186, Henry A. Kissinger to Secretary of State, Secretary of Defense, Director of Central Intelligence, subj: National Net Assessment of the Comparative Costs and Capabilities of U.S. and Soviet Military Establishments, 1 Sep 1973 (SECRET; Declassified 10 Mar 1994) NSSM 186, National Security Study Memorandum (NSSM), Box 3, Records of the National Security Council, Record Group 273, National Archives].
53 *The Man Who Kept the Secrets: Richard Helms and the CIA*, 242–243; *Honorable Men: My Life in the CIA*, 328.
54 Schlesinger was appointed DCI on 21 Dec 1972. History Staff, Central Intelligence Agency, *Directors and Deputy Directors of Central Intelligence* (Washington: CIA, 1989), 21.
55 A declassified copy of this memo can be found in James Digby's papers at RAND [RAND Archives/J. Digby Files/Box: Unclassified, 1951–1955, 1958, 1972–1976.
56 In the SALT I negotiations, the United States hoped to prevent the replacement of the Soviet SS-11 "light" (70 cubic meters in volume) ICBM with the SS-19 "heavy" (100 cubic meters in volume) ICBM, a missile with substantially greater throw-weight. Toward this end, the U.S. delegation made a statement (Unilateral Statement D) declaring that although "the Soviet Delegation [had] not been willing to agree on a common definition of a heavy missile…[t]he United States would consider any ICBM having a volume significantly greater than the largest light ICBM [then] operational on either side [the SS-11] to be a heavy ICBM." The SALT I Interim Agreement (Agreed Statement C and Common Understanding A) required that "in the process of modernization and replacement the dimensions of land-based ICBM silo launchers [would] not be significantly increased," with "significantly increased" defined as "an increase…greater than 10–15 percent of the present dimensions of land-based ICBM silo launchers." Arms Control and Disarmament Agency, *Arms Control and Disarmament Agreements: Texts and Histories of the Negotiations* (Washington: ACDA, 1990), 173–174, 176. These limits were ineffective in preventing the replacement of SS-11s by SS-19s. For a discussion of

this issue, see Raymond L. Garthoff, *Détente and Confrontation: American-Soviet Relations from Nixon to Reagan*, revised edition (Washington: Brookings Institution, 1994), 191–197.

57 On 10 Aug 1974, President Ford met with Schlesinger, Secretary of State Henry Kissinger, Assistant for National Security Affairs Brent Scowcroft, and Director of Central Intelligence William Colby. The purpose of the meeting was to decide whether to go ahead with the Glomar Explorer operation. The ship was in position to retrieve a sunken Golf-class ballistic missile submarine, but was being trailed by a Soviet trawler. U.S. officials were concerned that the trawler might attempt to stop the unprotected ship from raising the submarine. Ford decided the operation should proceed. Sections of the submarine were recovered and a "significant amount of information was retrieved." Gerald R. Ford, *A Time to Heal: The Autobiography of Gerald* R. *Ford* (New York: Harper & Row Publishers, 1979), 135–136.

58 For a brief discussion of these talks based on interviews with Israeli officials (Mordechai Zippori, Menachem Meron, and Elyakim Rubenstein), see Karen L. Puschel, *US-Israeli Strategic Cooperation in the Post-Cold War Era: An American Perspective* (Boulder: Westview Press, 1992), 28–29. See also Howard Teicher and Gayle Radley Teicher, *Twin Pillars to Desert Storm: America's Flawed Vision in the Middle East from Nixon to Bush* (New York: William Morrow and Co., 1993), 90–92, 141.

59 Ltr, Marshall to Richard Kaufman, 20 Oct 1975, with attachment, "Comparisons of US and SU Defense Expenditures," 16 Sep 1975, in Joint Economic Committee, *Allocation of Resources in the Soviet Union and China—1975*, Part 1, 94th Cong., 1st sess. (Washington: GPO, 1975), 153–177.

60 Prepared statement of Lt. Gen. Samuel V. Wilson, Director, DIA, 15 Jun 76, in Joint Economic Committee, *Allocation of Resources in the Soviet Union and China—1976*, Part 2, 94th Cong, 2nd sess. (Washington: GPO, 1976), 82; Lt. Gen. Daniel O. Graham, USA (Ret.), "The Soviet Military Budget Controversy," *Air Force Magazine*, 59, no. 5 (May 1976): 36; Henry S. Bradsher, "CIA Uncovers Soviet Arms Shift," *Washington Star*, 15 Feb 1976, A11.

61 Prepared statement of George Bush, DCI, 24 May 1976, in *Allocation of Resources in the Soviet Union and China—1976*, 20, 22.

62 "When measured according to a definition of defense activities roughly comparable to that used in the U.S., the Soviet defense effort absorbs some 11–12 percent of Soviet GNP. When the calculation is based on the broader definition of defense, the share increases to about 12–13 percent.... The new estimate of the share of defense in the Soviet GNP is almost twice as high as the 6-8 percent previously estimated." Ibid., 24. See also John W. Finney, "Soviet Arms Outlay May Take a Bigger Share Than CIA Had Estimated," *New York Times*, 23 Feb 1976, 13 (mentions Marshall and Schlesinger criticisms of CIA estimates of Soviet defense burden).

63 Robert Ardrey, *The Territorial Imperative: A Personal Inquiry into the Animal Origins of Property and Nations* (New York: Atheneum, 1966).

64 Paul Y. Hammond, Charles Wolf, Jr., and Peter de Leon, *Perceptions of the Military Balance: Report of a Workshop, August 26-27, 1974*, WN-8876-ARPA (Santa Monica: RAND Corp., Oct 1974), FOUO.

65 Lionel Tiger, *Men in Groups* (New York: Random House, 1969).

66 Lionel Tiger and Robin Fox, *The Imperial Animal* (New York: Holt, Rinehart and Winston, 1971).

67 Herbert Goldhamer, *The Economist's Perception of the U.S.-Soviet Strategic Balance*, WN-8991-ARPA (Santa Monica: RAND Corp., Feb 1975), FOUO; Herbert Goldhamer, *Le Monde's Perception of the U.S.-Soviet Military Balance, 1948–1973*, WN-9318-ARPA (Santa Monica: RAND Corp., Nov 1975), FOUO; Herbert Goldhamer, "The US-Soviet Strategic Balance as seen from London and Paris," *Survival*, 19, no. 5 (Sep/Oct 1977): 202–207 (summarizes WN-8991-ARPA and WN-9318-ARPA); Herbert Goldhamer (edited by Joan Goldhamer), *Reality and Belief in Military Affairs: A First Draft (June 1977)*, R-2448-NA (Santa Monica: RAND Corp., Feb 1979).

68 "As Secretary of Defense, [Schlesinger] recognizes that his function is to provide both intellectual and administrative leadership. 'You cannot control a department of this size; you have to guide it,' he said last week. 'I'm really a revivalist.'" "Watching Birds and Budgets," *Time*, 103, no. 6 (11 Feb 1974): 16.

69 Remarks by Secretary of Defense James R. Schlesinger at the Overseas Writers Association Luncheon, International Club, Washington, 10 Jan 1974, Defense Department Office of Public Affairs; Senate Foreign Relations Committee, *U.S.-U.S.S.R. Strategic Policies,* 93rd Cong., 2nd sess. (Washington: GPO, 1974); Senate Foreign Relations Committee, *Briefing on Counterforce Attacks,* 93rd Cong., 2nd sess. (Washington: GPO, 1974). Donald Cotter, Gen. Russell Dougherty (USAF, Ret.), Philip Odeen, Leon Sloss, and Seymour Weiss in *The Evolution of Selective Options Strategy.*

70 James R. Schlesinger, *Rationale for NU-OPTS,* R-1608-PR (Santa Monica: RAND Corp., Dec 1968; reissued Sep 1974) (SECRET; Declassified).

71 Maj. Gen. Paul F. Gorman, USA, "Measuring the Military Balance in Central Europe (U)" *Studies in Intelligence,* (Winter 1979), (SECRET//NOFORN; Declassified), 1–37.

72 Schlesinger gave a briefing on the Soviet facilities, using U-2 photographs, to the Senate Armed Services Committee on 10 Jun 1975. See Senate Armed Services Committee, *Disapprove Construction Projects on the Island of Diego Garcia,* 94th Cong., 1st sess. (Washington: GPO, 1975), 12–21. Schlesinger recounts the episode in Ben Rich and Leo Janos, *Skunk Works* (Boston: Little Brown and Co., 1994), 188–190.

73 Eds: Excerpted from an interview conducted 8 April 1994. The remainder of the interview covered the Jimmy Carter Administration and can be found in Chapter 4.

74 See remarks by Marshall in SRI International, *Theater-Level Gaming and Analysis Workshop for Force Planning*, Vol. I, Proceedings, September 22–27, 1977, held at Xerox International Center for Training and Management Development, Leesburg, Virginia (Menlo Park: SRI International, 1977) [Available from Defense Technical Information Center, ADA 068177].

75 Robert Ardrey, *The Territorial Imperative: A Personal Inquiry into the Animal Origins of Property and Nations* (New York: Atheneum, 1966).

76 Herbert Goldhamer, *Le Monde's Perspective of the U.S.-Soviet Strategic Balance, 1948-1973*, WN- 9318-ARPA (Santa Monica: RAND Corp., Nov 1975), 31–32, FOUO.

77 Ltr, Marshall to Richard Kaufman, 20 Oct 1975, with attachment, "Comparisons of US and SU Defense Expenditures," 16 Sep 1975, in Joint Economic Committee, *Allocation of Resources in the Soviet Union and China-1975*, Part 1, 94th Cong., 1st sess. (Washington: GPO, 1975), 153–177.

78 *Texts of Final Communiques Issued by Ministerial Sessions of the North Atlantic Council, the Defence Planning Committee, and the Nuclear Planning Group, Volume II, 1975-1980* (Brussels: NATO Information Service, n.d.), 32–33.

79 Nathan Leites, *Western European Perceptions of the Strategic Balance (U)*, R-1772-ARPA (Santa Monica: RAND Corp., Aug 1975), SECRET.

80 Nathan Leites, *The Rejected War: The 'Case of Defense' Comes Upon the Federal Republic*, WN-9303-1-ARPA (Santa Monica: RAND Corp., Nov 1975), FOUO.

81 *The Maritime Balance Study: The Navy Strategic Planning Experiment, Executive Summary and Appendices* (Washington: Office of the Chief of Naval Operations, 15 Apr 1979); *A Report of the Defense Science Board Task Force on Strategic Planning and the Maritime Balance: An Experiment* (Washington: Office of the Under Secretary of Defense for Research and Engineering, Nov 1979).

82 For a brief discussion of these talks based on interviews with Israeli officials (Mordechai Zippori, Menachem Meron, and Elyakim Rubenstein), see Karen L Puschel, *US-Israeli Strategic Cooperation in the Post-Cold War era: An American Perspective* (Boulder: Westview Press, 1992), 28–29 , See also Howard Teicher and Gayle Radley Teicher, *Twin Pillars to Desert Storm: America's Flawed Vision in the Middle East from Nixon to Bush* (New York: William Morrow and Co., 1993), 90–92, 141.

83 Eds: Interview conducted 13 July 1999.

84 Russell Jack Smith, *The Unknown CIA: My Three Decades with the Agency* (Washington: Potomac Books, 1989), 205.

85 John G. "Jerry" Godaire, the first chief of the Military-Economics Branch, according to Noel Firth and James Noren, *Soviet Defense Spending: A History of CIA Estimates, 1950-1990* (College Station: Texas A&M University Press, 1998), 30.

86 "The only major private hospital operating throughout our period was the McLean Hospital in Boston; this was the psychiatric branch of the Massachusetts General Hospital." Herbert Goldhamer and Andrew W. Marshall, *Psychosis and Civilization: Two Studies in the Frequency of Mental Disease* (New York: Free Press, 1953), 27.

87 Marshall says there was a final version of the Schlesinger report done in August or September 1971.

88 Interview conducted 20 July 1999.

89 Harold L. Wilensky, *Organizational Intelligence: Knowledge and Policy in Government and Industry* (New York: Basic Books, 1967).

90 Marshall confirmed that the study of Soviet perceptions was done by Robert Levine at CIA. He also confirmed that it was a study of Soviet perceptions of U.S. general-purpose forces.

91 Marshall agreed that the fact that the 1972 long-term competition report came out so late was due to the slowness of RAND's publication process.

92 Marshall's interactions with Allen probably occurred while Allen was serving as the first head of the IC Staff.

4 The Carter Administration: 1977–1980[1]

Introduction

Jimmy Carter entered office determined to implement a new approach to national strategy. He wanted "to help create a stable, just, and peaceful world order" that was "more responsive to human aspirations."[2] Carter recognized that competition with the Soviet Union and the attendant risk of conflict were likely to persist but also envisioned the possibility of increased cooperation and the stabilization of U.S.-Soviet relations. Carter committed to counterbalancing Soviet military power and influence in key areas, such as Europe, the Middle East, and East Asia, by maintaining the overall balance of military power and competing politically with the Soviet Union. In the political competition, however, the United States would rely on a basic commitment to human rights and national independence. It would also seek Soviet cooperation to resolve regional conflicts and would encourage constructive Soviet engagement in global trade and development.[3]

As part of his new approach to foreign policy, Carter had pledged to pursue complete nuclear disarmament in his inaugural address.[4] The SALT II talks inherited from the Ford Administration became the vehicle the Carter Administration used to further reduce nuclear weapons. At the Vladivostok Summit in November 1974, Ford and Soviet leader Leonid Brezhnev had agreed on a basic framework for a SALT II agreement capping each side's strategic delivery vehicles at 2,400 and multiple independent reentry vehicles (MIRVs) at 1,320; however, questions, including how cruise missiles fit into the agreement and whether the Soviet Tu-22M Backfire bomber was a strategic weapon, remained and the negotiations languished. Carter sought to reinvigorate the talks by proposing drastic new limits on strategic nuclear delivery systems and MIRVs for both sides (1,800 to 2,000 strategic delivery vehicles and 1,100 to 1.200 MIRVs), but retaining the U.S.-preferred range definitions for cruise missiles in exchange for an agreement that the Backfire was not a strategic weapon. The United States also wanted to freeze deployment, limit testing, and ban modification of existing ICBMs and to halt development of new ones.[5] When Secretary of State Cyrus Vance presented the new proposal to Moscow in March 1977, the Soviets rejected it out of hand. Negotiations soon resumed, however, and Carter and Brezhnev signed the SALT II Treaty, which closely resembled the Vladivostok agreement, on 18 June 1979. Carter submitted the treaty to the Senate for ratification four days later, but withdrew it from consideration in January 1980 after the Soviets invaded Afghanistan.

Both the United States and the Soviet Union nevertheless agreed to abide by the treaty provisions.[6]

Any effort to reduce reliance on nuclear weapons would require improving conventional capabilities to defend Western Europe from Soviet attack. Heavy U.S. spending on military operations in Vietnam had led Congress to cut significantly defense spending for conventional force modernization and readiness. When Carter entered office, the U.S. forces assigned to the defense of Europe were using outdated equipment and their readiness was low. The Europeans were reluctant to divert money from social welfare spending to pay for defense when they considered a war unlikely. Meanwhile, the Soviet Union had significantly improved its conventional forces and enjoyed numerical superiority in everything except helicopters and tactical nuclear weapons. The Allies still retained a technological edge but it appeared to be fading. The consensus was that NATO was not prepared to counter a Soviet conventional attack in Central Europe.[7] The Chairman of the Joint Chiefs of Staff reported, "In Central Europe, the chance of NATO stopping a Warsaw Pact attack with minimal loss of territory and then achieving its full objective of recovering that land which had been lost appears remote at the present time." It was, perhaps, possible to stabilize a defensive line in Central Europe and prevent the Soviets from reaching the French border and the North Sea coast, but not without losing territory.[8] American policymakers were concerned that NATO's weakness would force it to resort to a nuclear defense, which risked escalation.

To improve NATO's collective capabilities, the Carter Administration pursued several options. It pushed for member nations to increase military spending by three percent (real growth) every year. The administration also pressed NATO to improve cooperation in research, development, production, and procurement of armaments with an emphasis on rationalization, standardization, and interoperability, and to formulate a long-term defense program to strengthen the alliance over time.[9] The Europeans were concerned, however, that improving their defensive capabilities while the Americans reduced their strategic nuclear weapons might allow the United States to withdraw its nuclear guarantee, leaving Western Europe susceptible to Soviet diplomatic and military pressure, particularly in light of the new, longer-range Soviet ballistic missile, the SS-20. Consequently, the United States agreed to discuss strengthening NATO's theater nuclear posture. After extensive negotiations, the NATO defense and foreign ministers announced in December 1979 that the Alliance would modernize its nuclear systems by deploying ground-launched cruise missiles and Pershing II ballistic missiles in Europe.[10]

In Asia, the United States sought to reduce its military presence and to explore expanding relations with China. During the 1976 presidential campaign, Carter promised to withdraw U.S. combat troops from South Korea. Soon after entering office, he ordered the withdrawal of two army brigades from South Korea before the end of his term in 1980.[11] The U.S. military and the intelligence community opposed the plan as did the South Koreans. They feared that the South Koreans would be unable to stop a North Korean attack without U.S. combat troops. Congress refused to fund either the withdrawal or the provision of security assistance to ease South Korea's concerns. In the end, the Administration withdrew just one surface-to-air missile battalion and 1,500 support troops. The assassination of South

Korean President Park Chun Hee on 26 October 1979, and the rise to power of military strongman Chun Do Hwan further complicated U.S.-South Korean relations.[12]

While the Carter Administration's policies regarding South Korea foundered, its attempts to improve relations with China were more successful. Carter's National Security Advisor, Zbigniew Brzezinski, wanted to normalize relations with China in order to pressure the Soviet Union. Secretary of Defense Harold Brown and Chairman of the Joint Chiefs of Staff General George Brown agreed. General Brown noted, "The PRC can play an important role in a U.S. worldwide strategy for dealing with the Soviet Union."[13] Washington and Beijing established full diplomatic relations in January 1979. After the Soviet invasion of Afghanistan in December 1979, U.S. representatives discussed the sale of nonlethal military equipment and advanced technology to China as well as how to respond to Soviet actions in Afghanistan and to support Pakistan. Normalizing relations with China also meant the end of the close U.S. military relationship with Taiwan. U.S. facilities were shuttered and all U.S. troops left the island soon after normalization; a one-year embargo was placed on U.S. arms sales to Taiwan, after which, the United States provided only older, defensive weapons to Taiwan.[14]

Events in the Middle East during the Carter Administration altered the strategic context dramatically. Carter met with Egyptian President Anwar Sadat and Israeli Prime Minister Menachem Begin soon after taking office and urged them to continue the reconciliation talks initiated during the Ford Administration. After Congress approved the sale of military aircraft to both Egypt and Israel in May 1978, Carter invited the leaders of the two countries to the presidential retreat at Camp David in the hope of advancing the peace process. The Camp David Accords signed in September 1978, built a framework for peace in Middle East and led to the Egypt-Israel Peace Treaty signed on 26 March 1979.

In Iran, the U.S.-backed Shah, Mohammed Reza Pahlavi, fell from power in January 1979. Pahlavi had introduced social and economic reforms in the 1960s, but refused to allow political freedom. His failure to maintain economic progress while spending vast sums on modern weaponry weakened his popular support and led to a revolution that forced him to leave the country. On 4 November 1979, soon after he came to the United States for cancer treatment, radical Iranian students seized the U.S. embassy in Tehran and took 66 hostages. Although the Iranians freed 14 hostages, negotiations to secure the release of the remaining 52 proved fruitless and an American rescue operation in April 1980 was unsuccessful. The hostages were not released until the day Carter left office; they had been in captivity for 444 days.

In Afghanistan, Marxist/Leninist politicians had overthrown the government in April 1978. When enthusiasm for the new government began to wane, the Soviet Union invaded to aid and sustain communist revolution. The invasion shocked Carter and prompted his administration not only to support Afghan rebels fighting the Soviets but also to adopt a much firmer stance against the Soviet Union around the world.[15] Coming on the heels of the seizure of the U.S. embassy in Iran, the Soviet invasion of Afghanistan dampened any optimism within the Carter Administration regarding the U.S. position in the Middle East or the relationship with the Soviet Union that the Egypt-Israel Peace and SALT II Treaties might have engendered.

Outline for 8 April 1994 Talk

[Note: Italicized items were not discussed during the interview.]

Brown period

1. Change in modus operandi/relation with Brown
   — Special character of Harold Brown; Brown as reader
   — Best, most productive period; final shape of net assessment
2. Israeli talks continue
   — Experience
   — *Nuclear case*
   — Failure in end/McGiffert
3. Special successes
   — ASW assessment (grade: A)/story—Northern Flank
   — *C3I—>change in doctrine* (B+)
   — Naval (A)
   — Power projection (B+)
   — Investment balance (A)
   — Also B+
      • '80s strategic forces
      • '80s NATO/Warsaw Pact
      • *Soviet assessments ('80s)*
   — Failures
      • Training
      • Maintenance/logistics/story of
4. Soviet economy
   — Igor Birman
   — Vladimir Shlapentokh
   — Feshbach (RAND work on demographics; Soviet problems—>health/ecology—long-term burden) [Arrow in handwritten outline indicates this point ties in with "Investment balance (A)" above]

Format developed [Common structure of assessments]

1. How to think about balance
2. Description of trends and asymmetries
3. Assessment
   — Deterrence/Soviet assessment
   — If war comes, how do we do
4. Issues

1977–1980

Marshall: Then we come to the Brown period, which in some respects is the best period this office has had in the sense of its impact, I think. We had some interesting products to send, and Brown was a fantastic reader. And he also wanted to act on, use these things. So if we would send papers up, he would look at them. By that time we had, as I say, assessments where we had interesting products. We would send up a package, which would be a cover memo of maybe two, three, or four pages, which would give the main points, but also maybe contained some things that were very sensitive; then there would be the main assessment, which would have an executive summary, which might be 10 pages or something like that; and then a longer thing, which might be 150 pages. You would send them up to Brown and three or four days later they'd come back with all kinds of annotations through the whole, nearly 200 pages. At least every two or three pages there was something—he had read the whole thing. Now, we didn't send them up every day; four or five a year, something like that. And then he would put a note on it about who else should go and see it and what action requests should be made based on it. And then we would prepare these memos for him to sign.

By that time, I would say we were doing the following: We did a strategic forces assessment, which I was never quite satisfied with. The only time I think we ever did a reasonably good job on that was in the early '80s. There was the strategic, naval or maritime, {...}. Oh, one of the things that happened back in the Rumsfeld period was a request that came from the Ford White House for a {...}. Then we did a second one for Brown, which we'll come back to, which was probably the most influential assessment we ever wrote, I think.

Q2: Who in the Ford White House asked for it?

Marshall: I don't remember.

Q2: Who got it?

Marshall: It would have gone up to the NSC.

Q2: Assistant to the President for National Security Affairs Brent Scowcroft?

Marshall: Probably, yeah.

OK. So in the Brown period, the balance assessments we were doing were the strategic, the maritime, {...}, power projection, NATO-Warsaw Pact—a kind of Central Region assessment—and we were doing a Northern Flank assessment, which we'll say a bit more about.

Beginning in '76, while Rumsfeld was still here, we got a request from William Clements, who was the Deputy Secretary of Defense then, relayed from the Congress, who were raising some problems about continuing to fund future programs like anti-jam or something. The Congress requested that there be a net assessment in the command, control, communications and intelligence area. So we did one that started in '76 and was finished in '77 or maybe early '78. Then we did another one right after it that further developed it.[16]

We started doing this Investment Balance. That might have started in the Rumsfeld period, but the first ones were really produced then. And we did, toward the end of the Brown period, a Korean assessment and had the beginnings of a broader Asian assessment.

Q2: Is that the set?

Marshall: I think so.

Q2: Strategic, naval, {…}, power projection, NATO-Warsaw Pact/Central Region, Northern Flank, C3I, investment, and Korean assessments. You did nine assessments in four years?

Marshall: Well, sometimes they were done twice.

Q2: I mean that's not only maybe the most, I mean that's not… You said the Brown period was the best period for Net Assessment in terms of the impact. It looks like also in terms of productivity. I don't think you ever had any four-year period where you produced nine assessments.

Marshall: Yeah, probably not. And so, maybe I should talk a little bit about some of those {…} We had planned, at one point, that there'd be a Central Front, a Northern Flank, and a Southern Flank, but we never— The Northern Flank thing not only was a natural thing to do, but Roche and I became convinced fairly early that people were looking at that region in the wrong way, in that primarily it needed to be looked as a maritime theater. And that, whereas people in NATO and the traditional things all looked at it in terms of how much of northern Norway one was going to lose, probably the Soviets didn't care very much about northern Norway, except for either putting it out of existence or moving in and using the air bases along the coast, and that the real focus of attention for them was this maritime region on their perimeter.

There was some interesting data that we knew of that suggested that if you looked back at their exercises, that you could see that there was some suggestion they were trying progressively over time to control a bigger and bigger area. Then along came some very interesting information, probably around '75, maybe '76, that came to us as information from analysis being done at BDM by Admiral Robert Welander, looking at Soviet exercises there, which were the first reason we had to believe that the whole notion of how the Russians were going to deploy and use their naval assets, as believed by the American Navy, was incorrect.[17] That they had at sometime—I don't know when analysts went back and really reconstructed exactly when it was—but by the middle '70s, they already had adopted this "bastion strategy," and that therefore all of the assets, surface and underwater, were going to be used, initially, primarily to defend this area and to protect their nuclear-powered ballistic missile submarines. And a little after that, we came to know that there was some actual intelligence that suggested that that was really right. And that all fitted with our notion about how they, in fact, looked at the Northern Flank.

Therefore {…} we tried to make this case that… By the way—I don't know the exact date of the conversion of the U.S. Navy—the U.S. Navy initially resisted this.

Q2: Strange.

Marshall: Well, they…

Q2: No, not strange at all.

Marshall: What Jim Roche and I saw right away was that if this is true, it is such a strategic advantage to us. Since the Navy was all primed to try to erect screens to keep the Soviet submarines from going into the Atlantic and sinking all of the convoys on the way over, if they're not even going to try that, for a variety of reasons, this is just terrific. And we need to do whatever it takes to keep them doing this. So we wrote this assessment, which I think in some ways is not only the most effective thing we did, but it also had only two readers, I mean outside of the Navy, which were Brown and Charles Duncan. In fact, the Navy turned down a request from Bill Perry to see it later.

Q2: Perry wanted to see it and the Navy wouldn't let him?

Q3: Forgive me, Duncan was at this time?

Marshall: The Deputy Secretary of Defense, Brown's deputy.

Q2: Perry was the Under Secretary of Defense for Research and Engineering.

Marshall: Yeah. And the Navy still is quite sensitive about this particular thing, because I asked recently, and they looked at it and they decided, "No… Out!" Anyway, it was very effective, because Brown read it and was convinced by what we said. And it was such a big shift to see that there was this strategic gain to be made by keeping them doing this, that we had this gain, unrecognized, but now you could exploit this. {…}

Q3: My inclination is to let you go on, but I remember you saying that the Investment Balance also had a big impact.

Marshall: Yeah, but that's a longer term and more diffuse effect.

Q3: Well, it had a direct effect in the sense of stopping the decline in U.S. budgets in the short term?

Marshall: Yeah, right.

The other thing that I know that Jim feels, or has some stories or feedback about via the Navy, is that the maritime assessments we did in the middle '70s also had a good effect, because we made the case that… Well, let me go back a bit.

One of things that had bothered me about some things that had happened in the very early '70s was that here as a country we had had a strategy relative to the Russians which relied to some extent on nuclear, strategic superiority. Then we get into these SALT negotiations with them and settle for parity. And then there's no reaction. I mean we didn't say, "Well, here we have our strategy; we had superiority here and that allowed us to maybe be inferior in certain areas," and so on. So we settled for parity there and make no correction in the rest of the thing. In fact, I asked William Hyland, who was helping Henry Kissinger write his book about his years at the White House and at State, and I said, "Look, I don't remember there ever being any kind of consideration of, 'Well, having given this up, what other adjustments do we make?'" And he said, "Yes, I've been back through all the papers and I don't see any, either." Which strikes me as just not the way to run a country, or to have a strategy where you do this.

So one of the arguments that Jim and I pushed is, "Here are these various balances. We're ahead in the maritime area. That's a good thing. We ought to be ahead in something. It's hard to say that we're ahead in the balance in Europe, or here and here. This is something we are ahead in, we ought to welcome that, and exploit that, and do something with it." And somehow that argument got to Brown, and he stopped hacking away at the Navy; I mean it's alleged that that happened. I don't know enough about the story to know whether that's right, but Jim thinks that that's so.

Q2: You know, you might ask Brown whether or not…

Marshall: Yeah.

Q2: …he thought, saw…I mean, because of the three Secretaries that you lay out as being sort of the best ones—Schlesinger, Rumsfeld, Brown—it's really on their watch that we went to nuclear parity and didn't change our military strategy. So you might go back and ask those guys, "Did this going to nuclear parity bother you?"

Marshall: Umm-hmm.

Q2: Did you do any thinking? I mean, maybe this is one of the things that caused Brown to…

Marshall: Yeah, I don't know. Well, it bothered Schlesinger. He wasn't in favor of it. People in the White House and so on wanted that. It wasn't anything he was in favor of.

Q2: Well, but the question really is, if the State Department and White House are going to do stupid things like that, what's the Defense Department…

Marshall: Yeah, I understand that. Yeah, right.

Q2: …going to try to compensate for it. I mean, limited nuclear options in one way, shape, or form.

Marshall: Again, Schlesinger was in favor of being ahead anyplace we could be ahead. So this didn't come up. The reason this came up was that in the Brown period there were people here—some guys in PA&E—who were for just kind of whacking away at the Navy, too. "We're spending too much! Why are we ahead anyplace?" was sort of their attitude. "Aren't we overinvesting? We're ahead! My God!"

So I would say the things that had some effect were certainly this{…} The naval, maritime thing had some effect. The power projection thing did, too, in that it had some role in the setting up of the Central Command—we became exercised enough about that. The Investment Balance, I think, defines the whole level, size, nature of the Soviet program.

Q5: Would you say a little… I mean, I'm just curious how you feel about that from, looking back now. I mean, it's the whole issue of, you know, going almost back to the beginning of this thing, you're starting to beat on, say, CIA—

Marshall: Yeah, right.

Q5: —about, you know, What's the real defense burden? As the Berlin Wall was falling, the last sort of set of testimony … I think maybe we got them up to 17 or 18 percent of gross national product or something. I mean that was DIA.[18]

Marshall: Right, yeah.

Q5: And I think your last view was...

Q2: ...33 percent?[19]

Marshall: Yeah, right.

Q5: Which is a, you know, really nontrivial difference, I mean...

Marshall: Yeah, right.

Q5: We did World War II in the '40s with that.

Marshall: Right.

Q5: So the difference between 11 and 12 percent, or maybe, occasionally, 16 or 17 percent and 30 is, or 33 is...

Marshall: I agree. I agree with that.

Q5: But, I mean, in a sense, it's sort of like the earlier comment about giving up nuclear parity. I mean, didn't in the aggregate the office sort of— While all the instincts were right, the office tried to get the right views and collect the data and everything, I mean in a way this was a failure. The office never convinced anybody, and they're an intelligence organization.

Marshall: Well, especially them, I guess.

Q2: Lastly them.

Marshall: But I think what was possible, in terms of convincing people, was to show that the size of their program relative to ours was substantial. The burden thing—I was probably never successful in quite getting them into that, to adopt the view that I would have had.

I guess the sequence for myself went back to the late '60s and early '70s. Both Schlesinger and I were quite sure that the intelligence estimates of the percentage of GNP going to the Soviet military were just crazy. They started out around six percent, in their view, and that just was ridiculous. Initially, the focus of criticism was that they were just underestimating the size of the Soviets' program and the cost of it. Then they changed and for a variety of reasons they moved it up to the 12-, 13-percent level.[20] So, much of the '70s was used up in beating on them about why they were underestimating all the things they were neglecting to include and so on.

You got more and more ammunition as you got the emigres coming out that had all kinds of very interesting anecdotes about things. One of the more striking ones I'm sure I've talked to all of you about was this naval architect who told this story about the tanker fleet to support the fishing fleet. You find out that the whole size and character is really determined by the Soviet Navy because they wanted to use them.[21] And the fact was that the whole Navy did use the merchant marine and other maritime assets, whose specifications they influenced, but they weren't counted as being in the Navy or part of the naval assets. Or this story about this roll-on/roll-off ship that had this extensive set of exhaust fans and so on that could only be required if you're going to have a whole bunch of things with their motors running at the same time in order to rush off the landing or something like that.

Q2: Well, and gun mounts on merchant, commercial ships…

Marshall: Yeah, or electromagnetic pulse hardening and…

Q3: I recall when you ran this sort of undercover intelligence operation. You had this guy pose as a guy looking to buy one of these ships. He went and said, "Gee, all your electrical cables are running in this armored conduit down the spine of the boat. Why do you do it that way?"

Q2: "Ah… to protect them from careless sailors."

Marshall: So, much of the '70s was spent beating on them about all the things they were ignoring and ought to be costing.

Q3: But just on the Investment Balance, in response to the earlier statement suggesting it did not change anyone's mind…I thought the big victory with the Investment Balance was in [indistinct] which was that people were, just by habit, "Well, here's our budget for 1976, here's their budget for 1976…" And the Investment Balance says these things are cumulative.

Marshall: Yes, right.

Q3: These things are cumulative and look at what it looks like when you put it all together. And people went…like that, as I recall.

Marshall: Yeah, right.

Q2: The other thing is the burden issue, though. In the '80s when the office was working on competitive strategies, one of the reasons the office was able to get Secretary of Defense Caspar Weinberger to do anything on that was Graham Allison. And Graham, you know, was very taken with this notion of, we were going to try—I mean it wasn't his idea, it was Andy's… And there was a meeting here one day with Graham when he said, "Andy you want to impose costs on those guys and take away every last ruble they've got." And Andy said, "Yes. That's right." But, I mean, if there's an intellectual one, in the sense that people… If it was only six percent of GNP, people would say, "This is hopeless"…low number, you know. But if it's 17 percent or 30 percent, people start to say, "Well, wait a minute. That's a pretty big burden, and if we impose a couple more costs on them, we can get them to allocate the resources differently or stop the competition or collapse or something." It became much more feasible and possible. So when you had people like Allison taken on board that can convince Weinberger that we should do something like competitive strategies because it can make a difference, then the Investment Balance has this indirect, you know, intellectual, psychological effect—although it still begs the question of, "What did you get out of competitive strategies?" and I won't get into that, but you know…good idea.

Q3: My memory, Andy—and you're much more intimately involved in this—is that one of the few marginalia that Weinberger ever passed back was on the Investment Balance, sort of to the effect, "Gee, are you one of these liberals who thinks that the Soviet military is full of obsolete equipment?" when your point was we should work to make their big investments obsolete.

Marshall: Right.

Q3: And he was saying, "Oh gee, you guys think they're no good."

Marshall: No, Weinberger never got it.

Q2: Well, he sent you marginalia on that *Washington Post* article, too, didn't he? About you and what you were doing in Net Assessment. That you had a great campaign under way to seize the intellectual leadership of the Department away from him.

Marshall: Right.

Q2: But, I mean, direct effects? Weinberger, and I'm trying [indistinct] through Graham and other people.

Marshall: Right. To go back to this business of the burden: My first thing was just to get them to get a more adequate accounting of the size of the Soviet effort and to count some of these things that were not included in their normal definition of what's in the budget. It was only really toward the very end of, like, '79, '80, and in the early '80s, that I became convinced—I hadn't really been looking at their estimate of the GNP so much—that that was also wrong. And you had this kind of crazy situation where they were underestimating the size of the military efforts and overestimating the size of the GNP. So you had kind of both effects, when you looked at the level, looked at the burden. And the first things there that were convincing to me was really some of the stuff that Igor Birman and some of the other emigres started talking about.

Q2: Vladimir Shlapentokh do some stuff for you, too?

Marshall: Shlapentokh, although we didn't get in touch with him until probably '81 or something like that.

Q3: Fyodor Kushnirsky and some of those people like that?

Marshall: I don't know. Birman is the one I remember the most. What was the name of… I'm trying to think of the name of one of the other emigres…Vladimir Bukovsky.

Q3: Bukovsky, yeah, the other guy. Kushnirsky did the stuff on inflation.

Marshall: Yeah, right, right. Of course, there are other things. For example, I give a lot of credit to Jim Billington, who I first met through Ellsworth; Schlesinger knew him, but I didn't. He came over and gave a talk on his view of the religious revival that was beginning in Russia, in the Soviet Union, probably in '76.

The other thing to mention about the Brown period in the '70s—I'd have to go back and check—was that we did start to look at Russian demographics. Initially, we got into that because of the impact of some stuff that Murray Feshbach did, when he was still in the government, about the changing racial and ethnic composition of the conscription cohort. It was rapidly going to be the situation that half or more of the men presenting themselves would be coming from these Asian ethnic groups, who had generally somewhat lower skills, but also whose Russian wasn't very good.

So we had these very good studies by Enders Wimbush and the people at RAND to try to understand, If that happened, what would be the impact?[22] What were these people used for? We were able to pin down the fact that a lot of them went off to serve in the railroad troops—

Q3: Transport companies.

Marshall: —transport companies. That they tended for a long time not to be used in combat elements, or not very much.[23] And again, we had a resounding flop with the people in the intelligence community, who pooh-poohed that that was any problem.

But then, because of this sense of the Soviet defense burden, when Feshbach left the Department of Commerce and the demographics group there, I started supporting his research on the Soviet health situation, because it seemed to me that, first, that was interesting because of its direct effects on the military, but also, from this point of view of the GNP, that if this was an area which had been slighted and underinvested in—and there was some evidence already that life expectancy was either stagnant or… I think it was only later when it started to come down, beginning in the late '60s or so…

Resources needed for health care would amount to a competing claim on investment. In the first place, if the health of the society went down, that in itself undermined their ability to do well in those areas. But also, human beings don't like it that an extraordinary proportion of their children are born with birth defects and so on. In fact, I tried to get the United States Information Agency people, in the beginning years of the Reagan Administration, to actually start broadcasting some of this material on health problems in the Soviet Union. I thought that one of the things we ought to be doing is stirring up these competing demands. But you could never get them to do that sort of thing. Part of the cost-imposing thing is to impose costs, drive up their costs, but also try to encourage demands that would compete for investment. Health care, and later, environmental pollution, etc.—which were related items—were things that they were underinvesting in, and which eventually, somehow or another, they would have to correct and would have to become, long term, up to something like Western levels of expenditure. Therefore, if you're looking at the future allocation of their resources, that was an interesting thing to take a look at. So I sponsored this stuff by Feshbach. Initially, of course, it was mainly in the health area, and then toward the end, I got him focused also on these environmental issues.

Q5: Seas going away and other minor details.

Marshall: Yeah. So that's about…

Q2: When was the Gorman armor…

Marshall: That was done in the middle '70s, around '75 [1974–1977).

Q2: Under Rumsfeld, then. Was he…or '74?

Marshall: The Gorman study may well have started under Schlesinger.

One of the other things that we felt needed a lot more work was the information on these qualitative performance aspects of military forces. So we tried to get each of the Services to undertake a study in the training area. Only the Army one succeeded. It was a tribute to Gorman. It's the best study any Service ever did for me. Really good. And as I mentioned, there was a complementary study {…} And there was an Army organization out there that then put all that stuff together in a big study that was very, very, very interesting. That was all part of that effort. The Air Force effort on tactical pilot training was poor. The Navy study was abysmal. I just never got much.

We made another effort under Brown and got him interested in this. We tried to get an effort looking at maintenance and repair and how that was done. Some of it came from the fact that Roche and I felt that one of the more interesting things about the Soviet Navy was the contrast to ours, where when ships leave port, their level of things that are available and working actually go up. In the Soviet Navy, the minute they leave port, things start shutting down, because they don't have the skilled people. They have a system where most of the maintenance is done by people ashore, where they can have the skilled people. The people who are actually on the ships are not trained well enough. Also, their ships are built, or were built then, in such a way that they didn't have the same space devoted to workshops. Also, the internal designs were such that there wasn't— Ours are designed so people can easily get around and maintain equipment and that's not true for them; they have much more close quarters. So their ships are fundamentally designed to be kept up in port and then they go. We wanted to understand more about that, get that down more. So we tried. In fact, we got Brown, I think, to send off a tasker to each of the Services. Again we got very little.

I was very impressed by one of the things that came out of the '73 war. The Israelis and the Egyptians started with essentially the same number of tanks. The Israeli tanks, on the average, got back into the battle three times. The Egyptians, once. And, therefore, if you're looking at the balance, thinking of actual combat, it was a terrific advantage for the Israelis, who had crews go out and collect the stuff, repair it, put new crews in, and so on, for the things that were reparable. So the Israelis' tanks rotated three times and the other guys' tanks rotated once. It was very important to know that, to understand that, to get that into your judgment and calculations—which nobody does—and, therefore, it's very important to understand how the Soviets would operate, etc.

We always got into a lot of that sort of thing. For example, these various practices on the Soviet side of the tanks that were going to fight being kept in garages and started maybe once a week and kept lubricated and so on, while there's another set of tanks that they actually use at training ranges and so on, in order to have tanks that are ready to go, that have very low mileage on them and therefore have a certain number of expected hours of performance. It's totally different from us. So we need to know all these things in order to understand, and think about, and prepare. And the whole Soviet system was an adaptation to the fact that it was a conscript army, [indistinct] population, relatively low-skill levels. This had some impact on the proportion of officers, which is much higher for them, and on everything—how the equipment is designed, the practices, everything.

Then there's another level, which has to do with social-cultural factors, because [indistinct] are designed to fight. Jim and I used to make something of the fact that on the Soviet naval ships you'd find delivery systems, weapons, that don't have matching sensors. We don't build ships like that because of our thinking that a ship can fight individually and the captain of a ship is a responsible, serious, decision-making person. On the Soviet side that's not the case. The thing is commanded by someone above him. The sensors may be on one ship, the strike weapon on another. The airplanes—similar things. You could see [that] where responsibility and decision-making lies was just totally different in comparison to our military, and so was reflected very much in the kind of weapons that are designed.

And in our tank platoons, everybody talks to everybody, and we share out targets, and so on. But in the Soviet small unit, only the platoon commander speaks and they all shoot at one target and then he shifts to another one.

At any rate, the one last thing to say is—it happened before the Brown period, or at the beginning of the Brown period—there came to be a general structure to all of the assessments. Each had a first section, which was on how to think about this balance. This was very important, I think, because the most important thing in some sense that we had to offer was a suggestion about, "Here's the way you ought to think about this." Then there was a section that's really information on trends and asymmetries; straight, descriptive material about what are these [indistinct] really like. And then there was an assessment section, which would focus, depending on what was said in how you ought to look at this balance and particularly how you ought to measure things… But there's no measures of merit. But very often, given our general strategy, it was deterrence. So a great deal of the assessment would be, "Are we deterring those guys? How do they look at it? How do they size things up?" And then there's some exploration of, "What if deterrence fails? How do we do against our measures?" Maybe two, three, or four contingencies would be examined. It was probably the weakest part of the assessment because we don't have models that are worth a goddam. If you had war games, if you had some way of…

Q2: Well, it's not just models. We really didn't understand, really, in many cases, how the other side was going to fight, you know. And so…

Marshall: Right.

Q2: …your confidence in… [indistinct] best models in the world if you're not reflecting how the other guy's really going to fight.

Marshall: Yeah, that's right. It's certainly very dependent on initial conditions, circumstances, precisely how they're going to fight and so on.

And then the assessment would address the issues. What are real, strategic, SecDef-level management issues? Because of the way we got started under Schlesinger, the whole notion was that we were doing something for the Secretary or other very high-level officials. Therefore, that meant that the kind of perspective you had, the kind of issues you wanted to illuminate, were strategic management issues. We weren't doing this assessment for the theater commander and his immediate problems.

Q3: I have one overall response or comment. A lot of the developments that you describe occurred as a result of happy accidents: that Schlesinger was brought in, then was succeeded by Rumsfeld, and was succeeded by Brown, and you were there at the time, and someone like Jim Roche was around, and other talented people.

Marshall: Right, yeah.

Q3: At another level, you could say that in a way there was a logic of history. This was the period during which, kind of thinking back to what the period was, people had a real sense of urgency and crisis. We were in big trouble at home, come out of a disastrous war, the Soviets were capitalizing on it by modernizing their forces while we were not...

Marshall: Right. They were running around the world knocking off this place or the other, right.

Q3: The Power Projection Balance emerges out of your sense that they have a real long-term way of thinking about this and that...

Marshall: Well, it's not only that, the power projection one. It's a certain amount of statements by some of the people that suggested they'd gone into a new business.

Q3: That's right. So listening to it... Again, I'm thinking about how a reader who is not intimately involved with this whole thing would— You look at the list of the balances that you put out and the kind of attention that they got, and it's really an extraordinarily productive and effective period for the office. To what extent was that the result of necessity, that there was this sense of there were real problems to be solved? Now, not everybody had this perception. I remember Harry Rowen going off to RAND saying, "Well, we've solved, intellectually, all these problems and now—we should shift RAND over to doing domestic policy analysis."

Marshall: Ah, yes. Right.

Q3: So, you know, that's kind of a different perspective. Again, from the standpoint of the book and the reader, it might be interesting to contrast the interaction of the perceived—from a strategic level—problems that the United States was facing, and your activity. Especially in contrast with the '60s, before the '70s, and the '80s, actually, because in the '80s, [indistinct] very roughly, people kind of said, "Yes, there is a problem but we know the answer. We'll just spend a lot of money. And therefore we don't have to be too sophisticated about it. We'll just spend a lot of money." And in the '60s, there was a sense, "Well, we'd solved the problem." Is that perception at all correct? That the impact of the office and the utility of it, was a result, in part, of the perceived, you know, pure crisis that people thought we were going through, or is that just reading too much logic back into this thing?

Marshall: No, I think there's something in that. I think that there certainly were fortunate accidents. I think that's true. The whole thing could have aborted in the Carter period had it not been for somebody like Brown as Secretary. And there certainly was a certain sense of crisis and concern among some parts of the establishment because of what seemed like evident Soviet successes. So I think there's something...

Q3: Paul Nitze cranks up the Committee on the Present Danger...

Q2: On the other hand, you had détente, you had large arms control—SALT, MBFR [Eds: Mutual and Balanced Force Reductions]; you name it, we're going to control it—

Q3: Right.

Q2: —you had Jimmy Carter going in and saying we'd had an "inordinate fear of communism" and now we're going to work with these guys, stuff like that.

Q3: I didn't say this perception was universally shared.

Q2: Well, not even at some of the highest levels of government. I mean, you've got Cy Vance in there, and then you've got a number of Presidents and Secretaries in other places that are kind of going, "Things aren't bad. We can work with these guys. We can cut deals. There's not a problem."

Marshall: I guess there's one other thing about the Brown period which is kind of interesting, and that was Sam Huntington at the NSC. The other things that were favorable were Brzezinski and Sam. They did this PRM-10 [Presidential Review Memorandum 10] effort that Sam led and William Odom was there helping him.[24] Because of the character of that thing, Brown asked us to be the point of contact for this net assessment thing they were doing. So we worked very closely with them, and that was another part of the context.

Q2: PD-59 Presidential Directive 59—you work on that, too?

Q3: Oh yeah, the Leon Sloss thing, countervailing strategy.

Marshall: Yes, right. In fact, Walter Slocombe [Principal Deputy Assistant Secretary of Defense for International Security Affairs] and I ran that, really, for Brown.[25] And it was another part of the general sorts of things that were...

Q2: I think we're back to luck.

Q3: No, no, no—I mean your point's well taken. Like for every Sam there was a Lynn Davis [Deputy Assistant Secretary of Defense for Policy Planning] and there is a Paul Warnke [Director of the Arms Control and Disarmament Agency] for every Zbig Brzezinski. But there was a significant community of people, you know, all through the '70s that said, "The United States really is in a very difficult situation, much worse than we've been since the end of World War II, in a way." And not everybody responded to this. But there was enough of a community to stand on procedure this way that there was support for and a market for the kinds of things that Andy was turning out. I suspect luck plays a role, but the fact there's a sense of urgency means that smart people, talented people tend to gravitate toward solving the problem and thinking about the problem, which, you know, may have played a role.

Q2: You had once said you thought that General David Jones [Chairman of the Joint Chiefs of Staff) was a very good...

Marshall: What I felt was...getting on all these guys...and the Committee on the Present Danger and all these things. I guess in response to some of the criticisms, I said that I thought that in the history books, people like Brown and Jones were going to look a lot better than these guys who were raising this level of alarm and so on.

Q2: Could you say a few things about what Jones did as Chairman that you found to be particularly, you know, hopeful or interesting.

Marshall: Well, I don't have a lot of specifics. He was certainly very open and helpful to me. Well, there were things that began, and, again, I don't know what we can do here... Because of some of these concerns we were speaking of, there was this sense—somewhat stimulated by the Investment Balance that we did—of, "These bastards are outspending us!" And that was certainly true for a while, I think. They may have done it very wastefully, but the size of their effort, when you think of the actual size of their economy, was just fantastic. Part of the countering of that involved ... There were new kinds of exercises on our part.

There was more of this conscious effort at trying to impact on their perceptions and General Jones was part of that. {...} And then there were all of these programs that Brown and Perry launched.

Q2: Like stealth.

Marshall: Stealth and precision weapons.

Q2: Were you involved in the stealth deliberations at all?

Marshall: No.

Q2: That has been, at least from the '80s, one of the problems that the office has had. That is, that the office just never really got access to a number of the special access programs and things that were going on.

Marshall: Right, right.

Q2: Now, in some ways that's a benefit, because you don't get captured by the "techno-weenies" that want to claim these magic widgets will solve all our problems. But, in other cases when there really is something that's a serious effort like stealth going on and, the office doesn't know about it, it's tough to make a really good judgment...

Marshall: Well, I personally knew something serious about the stealth area, I guess, in the early '80s, something like that.

Q5: It was made clear fairly early on that the office wasn't doing certain equipment/technical/hardware kinds of assessments. And yet, thinking about something like stealth, or thinking about the history of things like precision weapons, and how long it takes to get those things done...it seems like that's an aspect of the competition that it doesn't appear the office was ever real focused on. Maybe it kind of gets back to the point about, well, you got [indistinct] and you got an annual budget cycle and you do everything in weapons systems, and it's just hard to look beyond those sort of "cookie cutter" concepts, the way they divide the thing out.

Marshall: Yeah.

Q2: Well, for example, the F-117 achieved initial operational capability in '83 or something like that and the office in '85, '86, '87 is trying to do a NATO-Warsaw Pact Balance, part of which is looking at the air-to-air balance, but also air-to-ground and all that other stuff, and the office is not folding in the F-117s at all.

Q5: Well, the same thing is going on right now with the B-2.

Q2: Yeah, but people at least now know about the B-2. And I may not know all the characteristics and capabilities stuff, but you can...

Q5: But there's this perverse, sort of like bad aftertaste from all the various security guys, which is, Well, I could...for something like, say, seminar war-gaming you can play it. But in the more detailed models that go into things like we used and this stuff, if you continue to hold the real information that you need to actually be able to play it realistically SAR [special access required], and you never cough that up, then the modelers are kind of…

Q2: Just scratching their heads.

Q5: So there's another way in which things cannot be played or just can't reflect all the balances.

Marshall: Yeah. These things can have an effect. To move ahead a little bit, when we did the strategic balance thing in '82, '83, which I think on the whole was probably the best thing which we've done on the strategic balance thing, there were three big things that I knew about that couldn't be in the assessment. And so, I arranged to meet with Weinberger and to remind him—I mean he knew about these programs—that there were three other things that he knew about that made the situation much more favorable than this assessment indicated, which itself was fairly favorable. One was the B-2 and stealth, which I knew about at that time. {…} stuff we can't talk about. I indicated to him that these are also part of the picture, and you ought to understand that and read the thing in that light. So sometimes it's possible to do a little bit on that. But basically there just is this problem with special access programs, and sometimes they can have very great effects.

Q3: A question on a different subject. As you've said, this is the period in which the mode of operations of Net Assessment basically jelled.

Marshall: Yes, right.

Q3: Internally, what was most striking was the way in which you utilized the people in the office, the analysts, from the standpoint of, how to put it…

Q3: You would give absolutely minimal guidance.

Marshall: Yes.

Q3: And—it appeared at the time and afterwards—you didn't want people to kind of do what happened in other analytical offices, which is, figure out what the boss wants to hear, and how he wants it done, and then just do it that way.

Marshall: Yeah.

Q3: And what you wanted, you actually wanted, to the extent possible, you wanted people to actually think about these things.

Marshall: Yes, yes.

Q3: So that was deliberate?

Marshall: Oh yes.

Q3: OK.

Marshall: Well, I'd say two things. One is that some of the "crypticness" may just be personal, idiosyncratic. But the other part is that that was very conscious. I really wanted people to think about their particular balance. I'm not kidding when I say I think that this, "How to think about it," how to really structure the thing, is very important and a very productive thing for people to work on. Now, some people succeed at doing that, but not everybody does. I really looked at these things, in some way, as analogous to having people do PhD theses on some particular things. Of course, in lots of ways I had my own views, and I tried to nudge people in that direction. Also, once we had a relatively good assessment in an area, I always tried to encourage people to read the previous assessment and sometimes I was a little discouraged by the fact that they didn't seem to learn a lot by doing that. They didn't really take on board the...

Q3: Painful. Alright, yes...

Marshall: No, I mean that they didn't...

Q3: It's very striking because it's very different from the ways in which other analytical shops are set up, which they say, "Here's the methodology."

Q2: Especially in government.

Q3: And now, by the way, PhD dissertations are now done that way—"Here's the way you do it. You do it this way"—which has a very stifling effect. I guess I became self-conscious about this when I was talking to Ed Luttwak about something else—remember he was working with Martin van Creveld on some of those things—and he said, "Andy Marshall. You walk into his office and he kind of mumbles at you for a couple minutes and then you go away and do the best research you've ever done in your life. I don't understand how it works." [Indistinct]

Marshall: Yeah...

Q3: But again, it's useful because it just kind of helps illuminate how the office worked, but it's also helpful for other people trying to think about how to set up productive research organizations.

Marshall: Yes.

Q3: Because everybody sort of says, "What is net assessment? What's the 'net'? What is it?" ...don't know exactly. And as I said, my hunch had been it was partly deliberate strategy. It's a way of getting people to think outside of certain boxes.

Marshall: Yes. I also should have mentioned earlier that certainly part of my deliberate strategy, feeling, was to keep people away from proposing answers. That the real function of Net Assessment is diagnosis and analysis, assessment of the situation, and not to provide the answers. That it's dangerous if we go into the assessment already having made up our minds about what the answers are.

Q2: I always thought that was one of the key things about this office, that it didn't try to do the strategic planning or policy. It tried to do the assessment and then you'd give that— I mean, in part, you used to say that was because you didn't want people cooking the books, you know, you didn't want the guys sort of saying, "Well, I'm going to ignore that part of the information because it doesn't fit with my recommendation, you know, that I've already preconceived." But I think also there...very useful to approach that methodologically within the Department, because all these guys doing policy—and even if you had a strategic planning shop—very busy with sort of day-to-day, you know, even strategic planning guys, real...I don't want to say "real world," but things that Andy has kept out of Net Assessment, so to speak. So that people can be more reflective and think about these diagnostic kinds of things. And I've always felt that was one of the key factors in the success of the office.

Q3: There are several phrases engraved in my head as a result of this experience. One of them is, "You keep giving me solutions. Stop giving me solutions. Why don't you tell me what the problem is?" The other phrase, which is indelibly burned in my mind is, "Why don't you go out and do some research for a change?"

Q2: Well, Marshall told me that up front, "Take six months. Read everything."

Q3: No, no, no. This was after I'd taken six months. I thought I was doing analysis. Did you observe other people using this way of running analysts or is this something that you arrived at on your own by hit or miss? Or anything special? Personalities?

Marshall: I don't know. It's some combination of the last two, I guess. No, I haven't observed other people particularly doing this kind of thing. Although to the extent that there would be some formative things for me in that respect, it really would be RAND in the early days and people like Charlie Hitch and so on.

Q2: The other part of the office that always struck me as being really a significant departure from the way the rest of the Building works is in dealing with programs and projects.

Marshall: Right.

Q2: You know, 99.9 percent of the people in the Building deal with programs and budgets. This office doesn't. And one thing you get out of that is people, other organizations don't feel threatened, because you're not out there trying to take away their budget money or their program. So they're more willing to cooperate with you, especially at the lower levels, where you might find smart analysts and things like that. You know, they don't feel like they're betraying their Service or whatever, it's not going to cost them money, etc. Now, how did you settle on that kind of approach, I mean, because clearly you could see how— I mean, you still get from some people in Net Assessment today, "How's this going to affect the Program Objective Memorandum, and what does this mean for what we should be acquiring?" and things like that, you know. So I think they're all the wrong questions, but how come you decided to separate all that out?

Marshall: I think it's just part of focusing on the diagnosis: What is the problem? What are the issues? What are the opportunities? What is it that's really there or what are the things that are shaping the future? Partly, it was that I think I was very aware that it was tactically an advantage bureaucratically, but I was also very concerned about people having the answers, pushing programs, pushing things, and corrupting the assessment and not really focusing in on that. I had some concern about that.

Q2: Part of that kind of approach argues against getting access to special access programs and, you know, SAR kinds of things, because you don't want people doing assessments to think, "Aha! There really is a silver bullet out there that's going to solve the problem four years down the..."

Marshall: Well, I think it would be better if we had had a bit more access to the special programs, just the information.

Over time I've been involved in or had access to a lot of things that other people in the office did not. Some of it has come as a by-product of my sitting on the Chief of Naval Operations Executive Panel, where we got a bunch of stuff on that. Also, when I was at the NSC, I got involved in various kinds of things, largely in the intelligence world.

Q2: [Indistinct] people would like to draw lessons on how to do this right in the future and maybe one of the things is you get the access, but your analysts don't, but then you can say, "No, that's probably not a good...Go think about that some more." How and when did you get involved in the CNO Executive Panel?

Marshall: When I was on the NSC. It was formed in '72, in the spring or something. I was asked to join it in the summer, after the first meeting, actually. Exactly why that happened, I don't know. It started, of course, with a bunch of people who I knew very well that were on it, like Albert Wohlstetter and Rowen, and so on. So I was asked to join.

Q2: Who was the Chief of Naval Operations?

Marshall: Admiral Elmo Zumwalt. It was a much smaller group then, about 10, 12 people.

Q2: But there's really been no other Service—I mean, you played around with the Defense Science Board stuff and everything—but there's really been no other Service, other than the Navy and the CNO Executive Panel, that you really had that kind of relationship...

Marshall: Right.

Q3: From the standpoint of the book that this will turn into, you mentioned that this is a lot of material here and maybe there should be two chapters. I think that's right. And I think the logical division on the basis of this talk would be kind of the organizational and larger context which gave rise to the birth of Net Assessment—the discussion of Schlesinger and your work at NSC. And then the second chapter would be focusing on the balances themselves, their intellectual content, their impact, and so forth. That's very rough, but seems to me good enough. And since there's 10 balances, nine balances to go through, that's...

Marshall: Yeah. I think at one point, when you look at all that we ever did, there's probably something like 12 or 13.

Q2: What was the worst balance you ever did?

Marshall: The worst? I don't know. Well, the Southern Flank never got launched. We never really succeeded at all.

Q5: As I said earlier, we eventually collected...

Q3: Yeah, we never figured what that one was about.

Marshall: Yeah.

Q2: Was there something about that?

Marshall: And we never quite did anything in the area of space. We kept trying, you know, "Well, we ought to do that..."

Q2: Well, George Kraus, remember, had that one good piece that then the National Reconnaissance Office stole and said, "You're not going anywhere with this piece, thank you very much." It wasn't really an assessment…

Marshall: Yeah, right. Too much bureaucratic resistance. I guess I've never been entirely satisfied with the Strategic Balance.

Q2: That's probably a personal standard that goes back many decades.

Q5: Yeah, I think that's right.

Q2: And you'll never be satisfied with that one.

Q3: Are there things that in retrospect during this period you would have done differently, or you thought it was going to work out or turn out…on, or…I mean, it was a very productive period.

Marshall: No, I don't feel that way. I think things went extraordinarily well.

Q5: Yeah, I think all things considered, it turned out to be very…I don't know why you'd be depressed, or even not relatively pleased. Because there were certainly lots of opportunities for this whole thing to collapse, implode, be squashed out of existence.

Q2: We haven't got to the Weinberger years yet. We'll talk about that a little next time.

Q5: But you know, in a way, given what's happened since that, I mean, that too, since one of the sort of right hands for Weinberger who made my life unpleasant at a certain point later became Chairman of the Joint Chiefs of Staff, I mean Colin Powell, remember…

Marshall: Yes.

Q5: That's a more interesting personality than he seemed at the time…

Q3: Again, I'm listening to this, but also kind of keeping in mind about the broader audience. There really is this extraordinary sense of detachment. I mean, most people would look at the 1970s and say, well, the Vietnam War came to an end, and there's a lot of arms control going on, and the forces were downsizing. And the perspective for you yourself is you really pulled way back from that.

Marshall: Right.

Q3: And [indistinct] book suggests that Zumwalt had that approach also. Because he makes this in reference, in passing, that Zumwalt decided when he became CNO, that the Vietnam War was going to be lost anyway. No point spending his time worrying about it. Kind of assigned—that's for a junior officer. So let me think about the long term health of my organization and what are the issues relevant to that.

Marshall: Yes, right.

Q3: And that's why he spent so much time on race relations, because that was the key to getting recruitment of the kind of people he needed. But it will strike people who have lived through that period as, "Gee, this is a very different cut on the period, a very different perspective."

Q5: Well, if you think about it, we got through this whole period today without once mentioning Vietnam. At a certain point I thought about, well, should I raise…

mention something about that? But, yeah, you're right, I mean the way you put it is exactly right on. There was a much longer-term, broader perspective.

Marshall: Right. Well you know, when Schlesinger was here I used to talk with him about some aspects of Vietnam, but he really didn't want us to be particularly involved with that, we were doing other things. And also, I personally had not much interest in Vietnam and same was true when I was at RAND. A lot of RAND people got very lost in Vietnam. I didn't. I continued looking at the Soviets. I did get involved a little bit toward the end, when some people were starting up some "lessons learned" activities. I got a little bit involved in some of that. But personally, I didn't get very engaged with respect to the Vietnam War.

Q3: And my reading of that was that you thought it was peripheral to the central competition.

Marshall: Yeah. My feeling on Vietnam was that probably this was a strategic mistake to get involved, but if we were involved, we ought to flatten the place and stop screwing around. I couldn't understand why we didn't mine the harbors, sink the ships right from the beginning.

Q2: Right.

Q3: The office was productive in this period, partly because Andy was smart, he was thinking about [indistinct], partly obviously because he recruited a bunch of very talented people. And I'm trying to figure out... Again—pure luck? Or other factors which tend to make it more comprehensible that he got you, he got Jim Roche, he got...Greg Kuzniewski was here, I think in that period.

Marshall: Yeah, right. Jerry Dunne.

The other thing to mention was that in the Brown period, the office was the biggest it has ever been.

Q2: How many people?

Marshall: Seventeen.

Q2: Wow. Not in here? Seventeen people?

Marshall: Right.

Q2: You didn't have room for some of them?

Marshall: Yes there was room.

Q2: Where did you put them?

Marshall: I don't know where they all were.

Q3: Seventeen analysts?

Marshall: No, no, no. There were more secretaries and so on. All I know is the number. I don't know, I'd have to look at it. See, the office when we came over here to the Pentagon was initially five people, then six, and it kept creeping up to probably around 12, 13. Brown gave us an authorization of 17 people. And then it sort of went down like the defense budget.

Q5: I still feel very strongly that the larger theme there is that it never became an empire, never got to 50 or 100 or something like that, which means you just can't do anything like what went on.

Marshall: Yeah, right.

Q2: In fact, the interesting thing is that it got smaller. It's actually not a surprise that organizations get bigger. What's interesting is it got smaller, and stayed around [indistinct].

Q5: Those are real anomalies compared to almost all institutions or organizations.

Q2: They get big and worthless, you know. But it does help to explain how you're able to do nine assessments in four years, I mean, having 17 people helps you do that.

Q5: Well, I'm willing to at least throw on the table as a conjecture that in another 50 years if somebody goes back into the history of this, a lot of the military-technical revolution stuff that's gone on since the Cold War, there will be a very important chapter on all this. Now, we may not get it as a superpower and follow in the steps of Great Britain into obscurity. But at least a few people sat around and said, "There are other choices. Use your heads." But we may screw it up anyway.

Marshall: Yeah, right.

Q5: And it's very hard to figure out where that's all going to lead or not lead.

Marshall: Right, right. OK.

Q3: It was good.

Q2: And the next time is the '80s?

Marshall: Yeah

Endnotes

1 Eds: This chapter and the outline are excerpted from an interview conducted on 8 April 1994. The portions of the interview relevant to the Nixon and Ford Administrations can be found in Chapter 3. See also transcripts of interviews of Marshall by Alfred Goldberg and Maurice Matloff (Historical Office, Office of the Secretary of Defense), 1 June 1992, 18–23; 15 June 1992; 29 June 1992, 1–3.

2 Eds: Jimmy Carter, "United States Foreign Policy Remarks to People of Other Nations on Assuming Office" (20 January 1977), American Presidency Project, www.presidency.ucsb.edu/documents/united-states-foreign-policy-remarks-people-other-nations-assuming-office.

3 Eds: Presidential Directive 18, "U.S. National Strategy," (24 August 1977), Jimmy Carter Presidential Library, www.jimmycarterlibrary.gov/assets/documents/directives/pd18.pdf

4 Eds: Jimmy Carter, "Inaugural Address" (20 January 1977) American Presidency Project, www.presidency.ucsb.edu/documents/inaugural-address-0

5 Eds: Presidential Directive 7, "SALT Negotiations," (23 March 1977), Jimmy Carter Presidential Library, www.jimmycarterlibrary.gov/assets/documents/directives/pd07.pdf

6 Eds: Gordon S. Barrass, *The Great Cold War: A Journey Through the Hall of Mirrors* (Stanford: Stanford University Press, 2009), 206–11.

7 Eds: Edward C. Keefer, *Harold Brown and the Imperatives of Foreign Policy, 1977-1981,* Cold War Foreign Policy Series, Special Study 8 (Washington: Historical Office, Office of the Secretary of Defense, 2017), 8–10.

8 Eds: Chairman of the Joint Chiefs of Staff, "Military Strategy and Force Posture Review Final Report," Attachment to Presidential Review Memorandum 10, "Comprehensive Net Assessment and Military Force Posture Review" (18 February 1977) Jimmy Carter Presidential Library, www.jimmycarterlibrary.gov/assets/documents/memorandums/prm10.pdf

9 Eds: Frank Leith Jones, *Blowtorch: Robert Komer, Vietnam, and American Cold War Strategy* (Annapolis: Naval Institute Press, 2013), 223–33.

10 Eds: Edward C. Keefer, *Harold Brown: Offsetting the Soviet Military Challenge, 1977-1981,* Secretaries of Defense Historical Series, Vol. 9 (Washington: Historical Office, Office of the Secretary of Defense, 2017), 466–72.

11 Eds: Presidential Directive 12, "U.S. Policy in Korea," (5 May 1977), Jimmy Carter Presidential Library, www.jimmycarterlibrary.gov/assets/documents/directives/pd12.pdf

12 Eds: Keefer, *Harold Brown and the Imperatives of Foreign Policy,* 14–17.

13 Eds: Chairman of the Joint Chiefs of Staff, "Military Strategy and Force Posture Review Final Report."

14 Eds: Keefer, *Harold Brown: Offsetting the Soviet Military Challenge,* 395–405.

15 Eds: Keefer, *Harold Brown: Offsetting the Soviet Military Challenge,* 322–327.

16 *Report to the Congress on U.S. and Soviet Command, Control, and Communications, 1977; Report to the Congress on U.S. and Soviet Command, Control, and Communications,* 15 January 1978.

17 Welander reports listed in the collection of the Defense Technical Information Center: *Soviet Navy Declaratory Doctrine for Theater Nuclear Warfare,* DNA 4434T (McLean: BDM Corp., 30 Sep 1977) [ADA 058363]; *Assessment of Soviet Navy Doctrine for Theater Nuclear Warfare—Phase I* (U), DNA 4122F (McLean: BDM Corp., 7 Oct 1977), SECRET//FORMERLY RESTRICTED DATA [ADC 012703]; *Assessment of Soviet Navy Doctrine for Theater Nuclear Warfare—Phase I* (Sanitized), DNA 4122FSAN (McLean: BDM Corp., 7 Oct 1977) [ADA 955364]; *Guidelines for the Red Theater Command* (U), DNA 5229F (McLean: BDM Corp., 1 Mar 1980), SECRET//Controlled [ADC 956006].

18 In April 1989, CIA and DIA estimated the Soviet defense budget to be about 15 to 17 percent of GNP. In July 1990, the CIA gave the same estimate. Joint Economic Committee, *Allocation of Resources in the Soviet Union and China,* Part 14, 101st Cong., 1st sess. (Washington: GPO, 1990), 60 ; Joint Economic Committee, *Allocation of Resources in the Soviet Union and China,* Part 15, 101st Cong., 2nd sess. (Washington: GPO, 1991), 170.

19 Andrew W. Marshall, "Commentary," in Joint Economic Committee, *Gorbachev's Economic*

Plans, Vol. 1, 100th Cong., 1st sess. (Washington: GPO, 1987), 481–484 ; Andrew W. Marshall, remarks at 11 June 1987 seminar of the International Security Studies Program of the Wilson Center, in Security Digest #20, Jul/Aug 1987; David F. Epstein, "The Economic Cost of Soviet Security and Empire," in Henry S. Rowen and Charles Wolf, Jr., eds., *The Impoverished Superpower: Perestroika and the Soviet Military Burden* (San Francisco: JCS Press, 1990), 127–154.

20 "When measured according to a definition of defense activities roughly comparable to that used in the U.S., the Soviet defense effort absorbs some 11–12 percent of Soviet GNP. When the calculation is based on the broader definition of defense, the share increases to about 12–13 percent. The new estimate of the share of defense in the Soviet GNP is almost twice as high as the 6–8 percent previously estimated." Prepared statement of George Bush, DCI, 24 May 1976, in Joint Economic Committee, *Allocation of Resources in the Soviet Union and China-1976*, Part 2, 94th Cong., 2nd sess. (Washington: GPO, 1976), 24.

21 Marshall, "Commentary," 483.

22 Alexander R. Alexiev and S. Enders Wimbush, "The Ethnic Factor in the Soviet Armed Forces," and Edmund Brunner, Jr., "Soviet Demographic Trends and the Ethnic Composition of Draft-Age Males, 1980-1995," in Alexiev and Wimbush, eds., *Ethnic Minorities in the Red Army: Asset or Liability?* (Boulder: Westview Press, 1988).

23 Alexiev and Wimbush, "The Ethnic Factor in the Soviet Armed Forces," 143–150.

24 Presidential Review Memorandum/NSC-10, Jimmy Carter to Vice President; Secretary of State; Secretary of Defense; Director, Office of Management and Budget; Director, Arms Control and Disarmament Agency; Chairman, Joint Chiefs of Staff; Director of Central Intelligence; U.S. Representative to the United Nations, subj: Comprehensive Net Assessment and Military Force Posture Review, 18 Feb 1977 (SECRET; Declassified 6 Jan 1992) [PRM/NSC-10, Presidential Review Memorandum (PRM), Box 1, Records of the National Security Council, Record Group 273, National Archives]. The NSC study, "Comprehensive Net Assessment," Jul 1977 has not been declassified (Telecon with James Yancey, Jimmy Carter Library). Unclassified discussions of the study can be found in Memo, Zbigniew Brzezinski to Secretary of State; Secretary of Defense; Director, Office of Management and Budget; Director, Arms Control and Disarmament Agency; Director of Central Intelligence, subj: SCC Meeting on PRM-10, July 7, 1977, 25 Jul 1977 (SECRET; Declassified 29 Apr 1994) [Declassified Documents Collection, 1995/0598]; Zbigniew Brzezinski, *Power and Principle: Memoirs of the National Security Adviser, 1977-1981* (New York: Farrar, Straus, and Giroux, 1983), 177; Samuel P. Huntington, *American Military Strategy*, Policy Paper in International Affairs No. 28 (Berkeley: Institute of International Studies, University of California, Berkeley, 1986), 28–29; Hedrick Smith, "Carter Study Takes More Hopeful View of Strategy of U.S.," *New York Times*, 8 Jul 1977, Al. A sanitized version of the DoD study has been released. See *PRM/NSC-10 Military Strategy and Force Posture Review Final Report*, Jun 1977 (TOP SECRET; Sanitized 13 Aug 1993) [Jimmy Carter Library].

25 Remarks by Secretary of Defense Harold Brown at the Naval War College, 20 Aug 1980, Defense Department Office of Public Affairs; Harold Brown in Senate Foreign Relations Committee, Nuclear War Strategy, 96th Cong., 2nd sess. (Washington: GPO, 1981); Walter Slocombe, "The Countervailing Strategy," *International Security*, 5, no. 4 (Spring 1981): 18–27; Brzezinski, *Power and Principle: Memoirs of the National Security Advisor, 1977-1981*, 454–459; Leon Sloss and Marc Dean Millot, "U.S." Strategy in Evolution," *Strategic Review* 12, no. 1 (Winter 1984): 19–20.

5 The First Reagan Administration: 1981–1984[1]

Introduction

The election of Ronald Reagan exemplified a shift toward a more muscular American national security policy toward the Soviet Union. The shift began in the last year of the Carter Administration with the issuance of Presidential Directive 59 calling for a new "Countervailing Strategy": if the U.S. couldn't win a war with the Soviets, then at least the Soviets would not be able to win either. As the directive stated, "(I)f deterrence fails initially, we must be capable of fighting successfully so that the adversary would not achieve his war aims and would suffer costs that are unacceptable..."[2]

President Reagan took that concept a step further in National Security Decision Directive 13. "Should nuclear attack nonetheless occur, the United States and its Allies must prevail," he believed, not "countervail."[3] Reagan's theory of the Cold War was simple: "we win, they lose."[4] His administration's plans for modernizing strategic nuclear and conventional forces were intended to accomplish that goal. Reagan's 1983 Strategic Defense Initiative added a new dimension to the competition with the Soviet Union. While brandishing the stick of a military buildup, however, Reagan also proposed dramatically reducing strategic and intermediate range nuclear forces, both of which later resulted in arms control treaties with the Soviet Union.

The NATO allies had agreed to a major long-term military modernization program in 1979 that complemented the U.S. buildup. In 1981, the U.S. Army rolled out the AirLand Battle concept for fighting and winning a war in Central Europe, and by 1984 it became NATO doctrine as well. For its part, the U.S. Navy signed out a new, aggressive Maritime Strategy in 1984 intended to persuade the Soviets that, in the event of war, they would not be able to disrupt easily U.S. convoys reinforcing Europe nor be confident of the survivability of their ballistic missile submarine force.

The Soviets, meanwhile, were bogged down in Afghanistan after their 1979 invasion, and U.S-supplied Stinger missiles to the mujahideen contributed to the eventual failure and withdrawal of the Red Army ten years later. The aging Soviet leadership was in turmoil when Leonid Brezhnev died in 1982, Yuri Andropov died in 1984, and Konstantin Chernenko died in 1985. The frequent changes contributed to the weakening Soviet hold on Eastern Europe. Soviet leaders were unable to deal effectively with rising national expectations for increased freedoms, as exemplified by the 1981 formation of the social labor movement Solidarity in Poland, which successfully negotiated increasing rights.

China during this period continued the economic reform initiatives instituted by Deng Xiaoping, which called for constrained military spending and greater investment in the economy. In 1979, the United States had recognized the People's Republic of China (replacing Taiwan), and economic ties between the two countries began to grow. Consequently, China did not figure prominently in U.S. defense planning, although Congressional pressures meant that U.S. arms sales to Taiwan continued, as well as other less visible means of support.

In the Middle East, the Iraqi invasion of Iran in 1980 led to nearly ten years of war between those two countries, which the U.S. observed carefully but took no action. Because both Iraq and Iran were hostile to Israel, the lengthy war offered a respite from conventional threats to the major U.S. ally in the region. A 1981 Israeli air attack destroying Iraq's nuclear reactor at Osirak pushed any nuclear threats to Israel from that country much further into the future. Israel's 1982 invasion of Lebanon was intended to punish the Syrian-supported Palestinian Liberation Organization for multiple attacks over the years and push it back from the border; and while it achieved the immediate, near-term tactical goals, it did not produce a long-term solution.

Outline for 26 July 1994 Talk

[Note: Italicized items were not discussed during the interview.]

1981–1984 Themes

1. Analysis of Soviet weaknesses
   — [Discussions with Soviet specialists]
     • Shlapentokh/Pipes
     • Birman
     • Jim Billington/stories
     • Peter Reddaway
     • *Vlad Treml*
   — ASW/Maritime Strategy—>Iklé/Adm. Small/Weinberger
   — Long-term competition ideas/cost-imposing
     • Studies—Chip Pickett, George Pugh, Reckmeyer
     • Section inserted in Defense Planning Guidance/Iklé—first one (fall '81/spring '82)
       –Impact/Iklé statement at meeting with labor representatives
   — '84 Roche/AWM paper for Fortier
   — Estimate of Soviet military burden (Wolf "cost of empire"/failure of CIA effort)
   — Weinberger rejection of Soviet weaknesses
2. CIA in Net Assessment
   — Casey/Weinberger agreement
     • Share Blue data
     • Joint assessments
       –'82 Strategic Balance assessment/Rowen, Larry Gershwin, Charlie Pease, AWM
       –Meeting with Weinberger: {…}, B-2, ASW
3. Assessments
   — Asia/Rosen
   — Power Projection
   — Northern Flank
   — NATO/Warsaw Pact
   — Investment Balance
4. Restart of Israeli talks
   — Perle/Iklé OK
   — Trip/Osirak attack/washout
   — Briefing of Bud McFarlane—>strategic dialogue—State/DoD/Joint Staff transition to planning
5. Start of Defense Policy Board—experiment/Sy Weiss
   — Soviet weaknesses
   — Warsaw Pact cracks/John Vogt
6. Start of Summer Studies—Los Angeles/Dennis Ross
7. Negotiations with Israel post-'82 war/Sharon/*Weinberger reaction/undermined by Secord/Feckless support of Bing West*
8. *Context/Paul Wolfowitz/Jim Roche/Dennis Ross at State/Gene Durbin my deputy*
   — Space?
9. Change in orientation of office due to Weinberger's lack of interest

1981–1984

Marshall: I made some notes and they're organized more along kind of themes or areas in this period. The first one is this analysis of Soviet weakness. That was one of the major things I was working on. A second one is the Central Intelligence Agency's involvement in net assessment, because there was a period of collaboration with them. Then I thought I'd review several of the assessments that were written in this period. Then there was the abortive restart of the Israeli talks that took place in this period. There's the start-up of the Defense Policy Board; the start of Summer Studies [in 1985], and there was an episode of negotiation with the Israelis over the post-'82 war data.

Another thing to talk about is the gradual change in the orientation of the office, which was driven, in part, by Secretary of Defense Caspar Weinberger's kind of total lack of interest in anything. There's one other thing which I hadn't put down in my notes, which was an effort to continue with this development of the idea of strategic planning that was pursued during this time. That's really, in some ways, part of the analysis of Soviet weakness.

Let me get back to this matter of the analysis of Soviet weakness. When the Reagan Administration started, it coincided with a growing feeling on my part that the Soviets really were in a lot of trouble. Now, that had started back, in a small way, in the very late '70s, with an increasing sense that Soviet military expenditures were being very much underestimated by the CIA in all kinds of ways. One way was that they were simply leaving out a lot of Soviet activities that were relevant to their military, but weren't reflected in army units or people wearing uniforms and so on.

A particular thing that was of that sort was in the whole naval area. The Soviet navy never acquired all of the oilers and other support ships, but the whole merchant marine was designed to convert in wartime to support functions. Yet none of that was ever included in any of the CIA estimates. I had written something in '74 or '75 pointing a lot of this out—they never did anything about it. I don't know whether it was in the late '70s that I made still one more effort to get them to do something, and they didn't do a very good job. Some ridiculous small addition was added. So I finally got Charlie Wolf started on some of this, although his part was the other part of it. It had to do with the cost of empire, which is another part of it.[5]

Q3: You got John Jederlinic, as I recall, to go out and look at some Soviet merchant ships.

Marshall: Yeah, right. And there was a Russian naval architect, an emigre—but that probably was in the early '80s—who came and told this story about the design of the support ships for the fishing fleet. This guy had worked in one of the Leningrad yards, at kind of a design bureau there, and the fishing ministry had come to them asking for a set of refueling ships to support the worldwide operation of the fishing fleet. And they'd come up with a preferred design for the ship and the number of ships. When the fishing fleet people put it through the administrative mechanism, it ended up in the hands of the Navy, who came back with a counter-proposal in which they doubled the total capacity and came up with three different ship designs, a mix for the whole thing, and that was the program they actually went ahead with.[6]

And you could never get these Agency people to reflect any of that.[7] Now, subsequently, all of this looks very plausible. But the Agency's performance was absolutely abysmal in all of this. The things that made a big impact on me were, first, there was this clear underestimate of the defense expenditures and defense-related expenditures. Then, it was only really probably in 1980 or '81, '82, that it became increasingly clear to me that the estimates of the Soviet gross national product were too big, in addition.[8] So that the total effect of these two errors was far more dramatic. It became credible to begin to think that about 25, 35 percent of the GNP was really going into the military and military-related things.[9]

That things were bad in the Soviet Union, I think, came first from Igor Birman, who appeared in the late '70s, and certainly had the view that things were bad. I remember also a few talks with Vladimir Bukovsky, and his views of things. Then Richard Pipes [Director of East European and Soviet Affairs for the National Security Council, 1981–82] brought Vladimir Shlapentokh to my attention, probably in '81, and urged that I support him, and so I did that. Also, about the same time, I had a meeting with Peter Reddaway [Professor at the London School of Economics], and he talked about, essentially, the despair, demoralization of the people quite high up in the Soviet establishment. This was based a lot on a set of interviews that Reddaway had with the son of a very high-level official. And what it reflected was the conversation around his father's dinner table among very high-level officials about the state of things, and a conviction that they were never going to be able to match the West. They had once thought that they could, but then they realized they couldn't. The reasons were, on the one hand, the inadequacy of the Russians in the population. The "materiel" they were given couldn't be compared to Germans or Swiss or so on. And there was also a sense that the whole system was nutty, including the people above them.

Q3: Was this the son of Roald Sagdeyev, or was that independent?

Marshall: No, no. I don't know who it was. No, it was not Sagdeyev's son. Now, Shlapentokh told me that when he left Russia, he sold a dacha that he had to Sagdeyev. And they met at the dacha to close the deal and so on. Supposedly, according to Shlapentokh, Sagdeyev said to him—this would have been '79—that, "In the past we could always find resources to meet anything that the U.S. did, but now [the Soviets] had come to a stage where that was impossible. Any new, really big challenge would be impossible [to meet]."

Q4: This was generally, or as it applies to the military?

Marshall: I don't know how directly military, but it was just a sense that, "We're stretched to the limit. We just can't mount any big programs."

Q4: Were these the key things that influenced you to conclude that the Soviet GNP was overstated?

Marshall: I guess the thing that convinced me most was, initially, not so much that I had a good idea that the thing was dramatically smaller, that it was probably overestimated to some degree, but the main thing was that they were just in a lot of economic trouble, that they were just stretched to the limit and were going to have trouble.

Q2: Did you have Murray Feshbach looking at some demographic things around this time, too?

Marshall: That had started earlier. As I'd mentioned in an earlier meeting, the Feshbach stuff had raised real questions about the whole demographic situation. So when he left the government, I had him do this work on the health area, because it seemed to me that one of the things that they were clearly underinvesting in was health care. And that, therefore, in the longer run, the consequences of that were going to be so pressing that they would have to begin to invest more, and that, therefore, that would be a competing requirement for investment on their part that would compete with the military eventually. So I was interested in that.

I had also started the stuff that Enders Wimbush and the people at RAND were doing on trying to understand the impact of the changing composition of the conscription cohort, the fact that rapidly a larger and larger portion would be coming from the Asian ethnics.[10]

The other person, I think, that had a big impact on my beliefs about it all was Jim Billington, who had started talking in the middle '70s about the religious revival that he saw in Russia, and the decline of the official doctrine's hold on people. He also told me of a discussion he'd had with a quite high-level Party official about conversations that were taking place—again, this probably was around '82, '83—about their concern about developments.

Q3: The Soviet GNP wasn't as big as many thought. In addition, the military was bigger than usual for the Soviet Union. I remember that you were mentioning this economist, who sort of in the late '50s, early '60s was talking about how centrally planned economies probably are not as good as everybody at the time in the U.S. was cracking them up to be. What was his name?

Marshall: Warren Nutter.

Q3: And you said that he was kind of downgraded…

Marshall: "Downgraded?" It's worse than that. He was kind of exiled. He could not get a position at a really major university and ended up at the University of Virginia. And he actually served in the [Defense Department in the] first Nixon Administration as Assistant Secretary for Policy Planning in ISA [International Security Affairs]. He and I were graduate students together at the University of Chicago. In fact, he was Frank Knight's last major PhD student.

And he was made to seem like a kind of right-wing, conservative nut by these people, which he definitely was not. As I've said on other occasions, I would see it as a sign of a return to some intellectual health of this field if people could stand up and say Warren Nutter was right. If you read what he wrote, the tone was more questioning, Why should we believe that these societies are going over the long term to grow at a significantly faster rate than market economies? And he did a lot of statistical work and so on.[11]

Q5: What about somebody like [Friedrich] Hayek in all this? Was he just sort of ignored through this whole period?

Marshall: He was far less ignored, and I don't really think of him as having played a big role. In fact, I'm unaware exactly what he was writing and saying during that period.

Q3: He was writing more about the political pathologies of the socialist system, *The Road to Serfdom*. Was Nutter sort of how you got started on this idea that, well, this conventional view of the Soviet economy just really isn't right and it's worth talking to lots and lots of people to find out how it might be screwed up?

Marshall: No. I talked with him, but we talked mainly about, I think, the defense budget underestimates. He'd also spent some time at CIA in, I believe, the middle-through-late '50s [1952–53], and then went away and did this work for the National Bureau of Economic Research on these questions.[12] And he published this book in the late '60s, I think [*The Strange World of Ivan Ivanov,* 1969].[13] He'd published a couple of articles.

Q4: I'm interested if you will talk, take this one notch deeper: Andy Marshall's sort of intellectual process. That is, you seemed, basically, to find individuals, sometimes outliers, and if they say things which seem to fit in an informational sense with what might be going on, then you begin to push studies along to try to probe and investigate.

Marshall: Yeah.

Q4: You also remarked that you let people walk in the office who are conventional, straight-line and talk stuff that you don't believe is the case. You say, "Thank you very much," and you don't talk to them anymore. So what's your mechanism for finding what the hell's going on in the world?

Q3: What's the filter?

Marshall: I'm not sure I can articulate it very well. I certainly have had this view, much like Nutter, that it's hard to think that the Soviet economy was going to grow that rapidly over the longer term. Some of that went back, really, to the '50s at RAND. Here were all these guys at RAND who were studying the Soviet economy, and I thought what they did was very unimpressive. Certainly when Joseph Loftus and I were doing all that work for Project Sovoy we never could get anything out of these guys that was of any use, of any sort whatever. And also-this sounds maybe primitive—but you can do rough checks. Compound rates of growth are just very impressive over any length of time, and the kind of things that people were saying about the rates of growth in the Soviet Union didn't jibe with assumptions you had to make, crude assumptions, about the relative size of, say, U.S. and Russian GNP in the '40s, early '50s. And over time, you just had to get more and more skeptical that these rates of growth are consistent with other information. Because if they're growing at that rate, they are going to be a lot bigger than you perceive them to be.

Q3: Right. Either they had to be very small then or very big now.

Marshall: That's right. So there was that. I suppose also that by the time some of the emigres started showing up in the late '70s and early '80s, I'd already lost any confidence in the CIA numbers. Of course, the CIA can't, as I mentioned in another connection, be singled out. It was the whole of the American enterprise set up to study this. Except for a few people who didn't agree, it was kind of 95 percent of the effort in this country, in academia and the CIA, was agreed about these things.

Q8: There's a retrospective paper about the economics profession in the '50s, written by Anne Krueger, who was the chief economist at the World Bank, saying that a strong, dominant wisdom in the '50s was that the Soviets have got it right, basically.[14] And the so-called "neo-orthodox turn" doesn't really happen until the '70s.

Marshall: That's right.

Q8: Is Frank Knight known for this? You mentioned that he was at Chicago, he was one of your professors. Was he for a [indistinct] neo-orthodox economist?

Marshall: Oh yes.

Q8: Did he suffer?

Marshall: No, not at Chicago.

Q8: He already had tenure, he somehow got tenure before...

Marshall: He was one of the outstanding people at Chicago and had been there for a long time. And yet he was not a person who was doctrinaire in the way that Milton Friedman is. He was a much broader person than somebody like Friedman.

Q8: Did he ever visit you here in the Pentagon?

Marshall: No. He was the most prestigious person in the [Chicago] Economics Department in the late '40s, probably throughout the '40s.

Earlier I was talking about Billington, the story he told me of having dinner with this high-level Party official. According to this source, there was a great argument going on within the Party about the state of affairs in the Soviet Union. The striking thing to me, though, was that there was a substantial group of people whose view was that the only solution to their problem was the restoration of the church. My feeling was that if there's a wide group of people in the Communist Party who sees the solution as being the restoration of the Orthodox Church, they are in trouble. He also told me other interesting stories about the private collections of religious art and so on that high Party people had, and that there was a separate institute run by some professor and supported by the Party that specialized in the restoration of all these works of art. So I think—it's hard to place it exactly—but certainly by '82, '83, I was very convinced they were in a lot of trouble. And I suppose that then there also was the beginning of the notion that we had seriously overestimated the GNP.

Those are two separate issues, in a way, but the stronger was the fact that they were in very significant economic trouble, and that had a lot of effect on my thinking about the situation. And so my conviction continued through the '80s to grow on that. So that's one thing that was going on.

Another dimension of this weakness was more specifically military. I talked last time about the stuff that James Roche and I had done on the antisubmarine warfare problem and its effects. One of the first things I did when the new people for the Reagan Administration came aboard was to talk to Fred Iklé [Under Secretary of Defense for Policy] about this. By that time the Navy was undertaking some very interesting war-gaming and analysis, all done under Admiral William Small, who was then the Vice Chief of Naval Operations.

Ron St. Martin was involved in a number of these things. A special briefing was arranged for Weinberger to apprise him of this opportunity. All of that, therefore, continued and, in fact, was picked up and reinforced when Admiral James Watkins became the CNO.[15] He picked up on this in a very major way, which I think, given most people's view of Watkins, they may be surprised. But he was one of the most innovative guys in this respect. And he organized an effort that Bill Manthorpe was very much involved in, which led to something called the "Blue Book," which was a whole set of things that could be done—a deception program and some other things. So that continued, and through Small I was kept informed on what the Navy was doing.

Q4: Would you speculate on why, say, the Navy got to this thing faster than, say, the Army or the Air Force? Was it because, for example, they spend more time among the Soviets?

Marshall: Well, there was that, and they had a very special intelligence window into things. Also, it was probably that the kind of diagnosis that Roche and I and maybe a few others began to see in them, beginning in the middle and late '70s, was convincing for a variety of reasons. And it was such a big shift, and suggested such high leverage that we had, that it was very compelling to people. There was nothing quite comparable, I think, in other areas. So it represented a really big shift in beliefs on the part of people once they saw this.

Another kind of thing that was going on in this general area of exploiting Soviet weakness was this return to, or further development of these ideas on long-term competition. It's about this time that I started some studies. I had three study efforts on this "cost-imposing strategy." Chip Pickett did one, George Pugh directed a set of studies, and Bill Reckmeyer picked up on that. When do you date that?[16]

Q4: The study was delivered in the summer of 1983, so we started the work in the winter, January of '83.[17] And it was very difficult to get people to work on it, to get people to take the right focus. Getting them to really think about the issues the way that you were trying to think about them.

Marshall: Yeah, right.

Q8: Was one of the studies done by Booz-Allen and it said on the cover "Soviet Vulnerabilities" or something like that? It was in '84? Something that I was always stunned with, your office, whoever, on this whole issue. I can't remember who did it. It was either Booz-Allen or BDM. And it was a conference you had hosted on Soviet vulnerabilities; that was in the title. And there were color tabs for each topic or vulnerability.

Marshall: Yeah, that was BDM, I think, that did that.

Q8: Same period? '83? Or '84?

Marshall: No, I think all those things happened about the same time. I think probably around '82, '83, something like that.

Q8: It went beyond imposing costs. It was laying out a blueprint almost of what could be done.

Marshall: That's related, really, to one of the things that happened as part of this start-up of the Defense Policy Board. I'd actually forgotten that had happened, that BDM thing.

Q2: Was that Phillip Karber?

Marshall: Yeah, he ran that. But the most systematic thing, in some respects, is probably a series of things done by George Pugh, I think. They were the most elaborated things.[18]

Q2: On cost-imposing?

Marshall: Yeah.

Q2: And where was he?

Marshall: He ran his own company.

Q2: "DSCA" or one of those?

Marshall: Yes. "DS-something" [DSA—Decision-Science Applications, Inc.]

Q2: Did he do that cruise missile one, as a cost-imposing strategy? Did he do that?

Marshall: I think so [*Cruise Missile Cost-Imposing Effectiveness: A Case Study of the Impact of the U.S. Air Launched Cruise Missile (ALCM) on Soviet Air Defense Costs*].

So there were a series of studies that were done. This began probably in early '82 or something like that, maybe even late '81. Some of that was a response to something that happened in either late '81 or early '82, and that was the Administration putting together its first defense policy guidance [Defense Guidance]. I don't know who Iklé had really write that, but he had me look at it, and asked whether there was any other thing that ought to be in it. I said, "Well, there is this idea of exploiting Soviet weaknesses." And he said, "Oh yes, right. That's a reasonable idea. So why don't you write a section?" So I wrote about two pages that said we're engaged in this long-term competition with the Soviets, they have resource constraints and so on, and, therefore, in your programs that you propose, there should be a discussion of how this program contributes to or directly exploits Soviet weaknesses. In particular, there's this idea of cost-imposing strategies and, if possible, estimates ought to be made of the costs that are imposed. That was in there [the guidance] the first time, and then the next time it was largely just reproduced, but then it began to be negotiated away by the bureaucracy as just too hard, too difficult, with people essentially saying, "We don't know how to do that," and it gradually became neutered.[19] One interesting aspect of that was that when Frank Carlucci left as Deputy Secretary [of Defense], and was replaced by this businessman—I forget his name…"Norton" or something like that. But at any rate, one of the interesting things was that he happened to read this thing and said, "That's a terrific idea! What have people actually done about it?" And he sent out a memorandum to each of the Services asking them, "What have you done about this?"[20]

Q3: Thayer, wasn't it? Thayer. Paul Thayer.

Marshall: Thayer, yeah right. And we got back these ridiculous answers from all the Services. From the Navy: "We do this all the time," and a bunch of examples that were all baloney, and not really reporting on the relevant things that it was really doing.[21] I think the Army never responded, and the Air Force came back with some feeble set of answers.[22]

Q2: They did this all again when we did competitive strategies.

Marshall: That's right. Same thing.

Q5: The one that's particularly puzzling is the Navy, because they actually had something to report.

Marshall: That's right, yeah. They didn't understand it in those terms, whatever. In fact, I've been trying to get a copy of this first DPG [Defense Guidance].

One other thing to report on about this general theme and topic is that fairly early in the Administration, Carlucci asked me to come see him. This was probably March of '82, although it might have been earlier than that [Apr '81]. Anyway, the Administration was being beaten up a bit in the Congress and elsewhere for not having a strategy, which in retrospect, is a little crazy. The criticism was that its defense policy was just a big budget, you know, spend more on everything, and that there wasn't any strategy.

Q8: In this case you have to conclude that Congress was correct.

Marshall: Yes, in that respect, probably. And Carlucci wanted to know, what would people accept as a strategy? What would a strategy really be? And if they ought have one, what could they do? So I wrote him two memos, which I do have, and I sent him a lot of the stuff that had been done earlier on strategic planning and about strategy. And I told him, "This is what a strategy might be and here are some examples of it." And I sent him some papers. I remember one thing, a paper that Roche and I had written for [Secretary of Defense] Donald Rumsfeld that focused on the Navy. I sent him things of that sort.

Then I also wrote him a suggestion about how you might go about putting together some group or process for developing such things. Nothing ever came of any of that, I mean, the one thing that came of it was the creation of this thing that Karber headed over at the National Defense University, where…

Q2: SCDC, Strategic Concepts Development Center.

Marshall: Yeah…where Weinberger kind of added on to something that— I guess, it was General David Jones [Chairman of the Joint Chiefs of Staff] who was starting a couple of things over there. One was to look at mobilization, and I forget the second one, some sort of gaming activity.

Q2: There were three of them. The first one was the Strategic Concepts Development Center that Karber headed up.

Marshall: Yeah, but that was the add-on, that was the thing that…

Q2: Then the Mobilization Concepts Development Center.

Marshall: Right, right.

Q2: And they had the war-gaming center [Wargaming and Simulation Center]. And then, eventually, those three became the Institute for National Strategic Studies.

Marshall: Yeah, right. But initially, Jones had the two and three of that, and Weinberger then added this other thing. And that was the whole response to the strategy notion. Then the way they actually used Karber and company was not to really focus on these strategic kind of issues at all.

The one other thing worth mentioning is that Weinberger totally rejected the notion of Soviet weakness, despite this book that's out called *Victory* [by Peter Schweizer], which claims that there was this comprehensive strategy, at least for some members of the Reagan Administration.[23] I think in Reagan's case that might well be true, in William Casey's [Director of Central Intelligence] case it might well be true, but was definitely not, as far as I can tell, true in the case of Weinberger…

Q8: Or George Shultz [Secretary of State].

Marshall: …or Shultz, right.

Q2: Did you ever talk to Bob Osgood? Osgood had a Policy Planning staff over there [at State]. He was a former Dean, Johns Hopkins School of Advanced International Studies. Initially, Paul Wolfowitz went over there, didn't he?

Marshall: Paul was over there.

Q2: Alexander Haig [Secretary of State] brought him over.

Marshall: Yes.

Q2: So that was later on when Osgood was over there. But Osgood had this view about they had these weaknesses and we should try to roll them back, but that would have been later on in the Reagan years.

Marshall: Yeah.

Q8: Weinberger's congressional testimony on the Soviet Union frequently used the word "relentless." And there's no notion of weakness, or vulnerability, or these people are spread too thin. [Indistinct] They were not in the congressional presentations at all.

Marshall: Oh no. And I know that a little later on—I think it really would be in the early part of the second Reagan Administration—we had a briefing to Weinberger of stuff that we had been doing, and he was very much focused on this underestimate of the Soviet defense expenditures and, therefore, they have this burden and it's a problem and so on—which he dismissed entirely. He had various arguments about why it couldn't be true because their labor costs were so low and they hardly paid their servicemen anything. So there was no sense there of any of this. I don't know the exact timing of that. It might have been at the very end of this period that we're talking about, up through the end of '84, or it may have been just after it.

On the other hand, early on in the administration, probably late '81, early '82, I remember a meeting where there was a substantial delegation of labor leaders who came, a meeting in the Secretary's conference room, and Fred Iklé gave a talk, which was very much of this. Not so much stressing their weakness, pointing to that and that they don't have unlimited resources, but talking about, on the one hand, we're building up and it's going to create demands on them to increase some of their expenditures to counter what we're doing; on the other hand, we're instituting new measures to prevent access to technology, and therefore, it makes it more costly for them to respond to us and so on. And there was this whole sense of a strategy that he was articulating. But most of the time I saw no evidence of it.

Although Weinberger was quite enthusiastic about the ASW [Eds: anti-submarine warfare] things the Navy was doing, and he certainly was not in favor of being friendly or helpful to the Soviets, and if you could do them in or do something that caused them a lot of troubles, that was fine.[24] But he did not have a sense, I think, of the economics of the situation very clear in his mind. In fact, very few people did, as far as I'm concerned. The thing that I'm struck by now is the emergence of all these people who supposedly had these views. There's an increasing number who saw the light early, I guess.

Let's get on to topic two, the CIA and net assessment. An interesting aspect of this period was that early on, Casey and Weinberger had a meeting.[25] In the Harold Brown period, the effort of the CIA to get into the net assessment business had been strongly rebuffed.[26] But this [the Casey-Weinberger meeting] was an attempt at an agreement to stop the conflict between the CIA and Defense. There was an agreement laid down that Defense was going to supply more Blue data, and there were to be these jointly done net assessments, rather than the Agency going ahead and doing its own.[27] Given that Harry Rowen had become the head of the National Intelligence Council that worked very well. Not very many were done.

In fact, only one was really done, which was a really big effort, on the strategic balance. And that succeeded very well, as I say, because Harry was there and I was here. Charlie Pease [a Navy Captain in Net Assessment] and Lawrence Gershwin [National Intelligence Officer for Strategic Forces], who had just gone over there [to the CIA], got together, and it's probably the most satisfactory assessment of the strategic balance that we ever did. We didn't really do very much after that, and the ones before were not adequate in lots of ways. Even then, there were things that couldn't be in it because they were too classified.

So I had a short meeting with Weinberger, and I had a meeting with Iklé, to remind them that, here was this assessment, which on the whole I thought was good, but that there were three areas where we hadn't been able to adequately account for things that were largely a plus on our side. So that the real balance was probably more favorable than the one expressed in this thing. One of these was the B-2, that was going to be coming along—stealth; [one was] this ASW advantage and all the implications it had, the fact that very likely in the early phases of any conflict, during any conventional phase, we would attrit a lot of the nuclear-powered ballistic missile submarines; and a third thing, which I still don't think I can talk about, but which was going on.

The other kind of balances that were being done in the period included Asia.[28] It's the most comprehensive Asian thing that we've done. We haven't really done one since then. We're working our way to maybe do another one. We did a power projection balance which was pretty good. Early in the period, we finished off this

Northern Flank assessment, which reflected some of the stuff that we had done before.

Q2: Who did these? Who did power projection? Northern Flank?

Marshall: The Flank one was done by Jerry Burke.

Q2: And who did power projection?

Q3: I think Steve Rosen did that. But there was an older Jim Roche Power Projection Balance that Rosen sort of updated and Dennis Ross "blessed" it.

Q8: Who wrote all the parts about how to reorganize the U.S. government to do better at power projection? It's in the recommendation section at the end of it.

Q3: There were parts of it, because we'd talked about how, you know, we didn't think about power projection as a separate category. Lots of parts of the American military do that. But Dennis was certainly [indistinct].

Q8: That was a very rich assessment, actually. It ought to have had some impact.

Marshall: Yeah. Well, Dennis…

Q3: Well, as I recall, the thing was that some of the earlier projections about Soviet power projection capabilities having [indistinct].

Marshall: Yeah. But Dennis, in the early part, was over with Paul Wolfowitz [at State]. He came back in '83, '84 and became my deputy and then he went away to academia. He returned to the National Security Council staff in '86 or something.

Then there was the NATO-Warsaw Pact thing at that period that Barry Watts was involved in.

Q5: Yeah. And Dave Andre got it. John Hines did the Soviet assessment stuff. Eventually Jeff McKitrick inherited it.

Marshall: And then there was the investment balance stuff that we were doing, that David Epstein did. And those are the main things that were going on in that period.

But by the end of that time, because Weinberger really wasn't that interested and Iklé kept adding other kinds of things for us to do, we began to limit the number of assessments that we were doing. And in the second Reagan Administration, the NATO-Warsaw Pact thing was carried out under Jeff McKitrick, and we continued to look at power projection, but I don't know that we completed anything. There weren't too many assessments that were really finished in the second Administration.

Q2: Probably the Middle East one under Ned Cabaniss.

Marshall: Yeah, right. But we'll get to that, and the other diversions, later. Going back to the beginning of this Administration, there was an attempt to restart the strategic discussions with Israel.

Q4: Before you switch out of that, what happens to the CIA role in net assessment? Is it just basically on the horizon for a couple of years and then dissipates?

Marshall: We did the one big thing, and then there was an update, I think the year or so afterwards. Then, they were so resource-demanding on both sides that nobody really wanted to do more. Everybody was so busy. So it was there, but nothing more was ever done. We've done a couple of smaller efforts which never quite came to completion, to do further updating of the strategic assessment, but it's just petered out.

Q3: Before we move on here, my memory is that of one of the things which took up a lot of time and money in the office was the RAND Strategy Assessment Center effort, and the effort, in general, to improve the modeling capabilities of the American government with regard to nuclear war.

Marshall: Yes, yes. That's true, I'd forgotten.

Q3: On which we spent a lot of money, we spent a lot of time.

Marshall: A lot of time, particularly initially.

Q2: We had a whole series of workshops, and walked through the whole thing on the European model with all kinds of people reviewing those things.

Marshall: Of course, the whole thing had this history. There was this Defense Science Board study of '78. And then there was consideration of what to do about the recommendation to try to develop a more war-gaming style of analysis. It was all tied, initially, to this sense that our analysis of strategic warfare was…[gap for tape change]…on missile exchanges and so on. In fact…

Q3: It was a strictly economic model for optimum values, which had nothing to do with the way people would really fight the war.

Marshall: Well, every possible defect imaginable. If you want to look at an area that is intellectually awful, it's that area. It's as though it got frozen in a middle '50s framework or something. It has a whole bunch of problems. Anyway, it's an area of analysis somebody ought to just go and do a history of and the accumulating defects.

So the idea was that a more war-gaming style of analysis would be better.[29] First, it would put this strategic exchange in the context of an ongoing war, which was probably the way in which it would happen. So it would be simply a part of a war. With the standard analyses of strategic nuclear war it's as though you just had taken them [U.S. and Soviet nuclear forces] off to one side and it's like a weight-lifting contest or something like that happening on the sidelines of a football game. Then, it had this focus so exclusively on the missile exchange, as contrasted with the bombers, let alone the submarines, which were simply treated as invulnerable—everybody kept theirs. We knew from this other work that some of the submarines might well be sunk. This possibility was never treated, and so the whole interrelationship of the ASW picture with the strategic balance was never looked at in any of these models.

Q3: I remember [indistinct] it kind of totally ignored the logic of the Soviet campaign in the northern area which was designed to increase their submarines' survival.

Marshall: Yes, right.

Then Brown agreed that something should be done. So probably in '80 [1979], we let an initial contract. I think we put out a request for proposals. We got somewhere from five to eight. We winnowed it down to two, and then we gave money to both SAIC and to RAND.[30] And we also created a board of people who would make the decision. After a year's work, there was a briefing by each one of them, and a choice was made. And the [indistinct] was very much in full tilt during these early years of the Reagan Administration.

Let me interject one other thing which I have a note on, which I had mentioned. This is something that was kept very quiet at the time, and nothing has been said about very much since. In, I think it was '84, Don Fortier [Senior Director for Policy Development on the NSC staff] approached Roche and I to talk about something that the people on the NSC were going to do. The idea was that they wanted to have a more comprehensive, strategic approach, strategy for the second term.[31] He came and talked with Jim and me. So Jim and I wrote a paper for him and sent it over. I don't know whether Rosen [was working for Fortier at the time] saw it? The paper was very much along this line of exploiting our advantages and their weaknesses. I have a copy of it someplace ["Stressing Soviet Defense Planning"].

Q3: And there was finally a meeting with you and Admiral John Poindexter [deputy to the Assistant to the President for National Security Affairs]?

Marshall: Yeah.

Q3: You, Jim, Don, Poindexter, and Rosen, as I recall.

Marshall: Yes, right. Robert McFarlane [Assistant to the President for National Security Affairs] was also at this...

Q3: Maybe there was another meeting with McFarlane?

Marshall: [Indistinct] meeting with McFarlane. Which took place, I don't know, early in, probably, '85.

Q2: And what came out of that?

Marshall: Nothing. I guess Fortier had asked other people for advice as well. I think Albert Wohlstetter wrote something, or did he?

Q3: Yeah. Bill Odom. Sam Huntington.[32]

Marshall: Right.

Q3: Good idea, but nothing happened [33]

Marshall: Yeah, right.

Q2: When did Fortier pass away?

Q3: Summer of '86, August of '86.

Q2: Was that a reason nothing happened, or was it dead before that?

Q3: It was dead before that. As I recall, in the meeting with Poindexter, Jim Roche finished, after all the presentation had been made, and all the steps that had been prepared [indistinct], Jim turns to Poindexter and says, "John, we've brought you [some ideas]. You've gotta do something now."

Q2: Was Phil Dur up there, too?

Q3: Part of the time. He left the summer of '84 to go off to Lebanon for something.

Q8: The context of this debate, I should add, in '85 was the so-called "Reagan Doctrine" had been announced by the press. But the NSC staff, and State, and Defense were denying that there was a Reagan Doctrine. And part of it was coming from Don Fortier himself, who sympathized with some aspects of it, but I guess McFarlane and others kept saying, "Well, there's not enough consistency among these different issues to call it a doctrine." And so... I notice even now in memoirs about the Reagan Administration, people on one side want there to have been a Reagan Doctrine, but others say there never was.

Q3: It turned out that we didn't have a National Security Decision Directive on Afghanistan. So we decided to write one.[34] Essentially [it said], we should have a cost-imposing strategy for our activities against the Soviets in Afghanistan. And in order to do that, we had to be able to measure how the Soviets assessed their own costs in Afghanistan, figure out what kind of losses in Afghanistan meant the most to them, and then skew our effort to driving up their indices or their measures of the costs that they were suffering. We might be measuring it all wrong. They might not care about infantry casualties. They might care about officer casualties, or they might care about technology compromises. And that was written into the NSDD and that became part of— That's the only example—

Q8: The whole topic of Afghanistan... It's not clear how much you [Marshall] were personally involved in it, but lots of the work became very operational nitty-gritty stuff, both [indistinct] Steve Rosen and Michael Pillsbury, and others had copies of things you had produced and were using them in the persuasive part of it. And the opposition was just enormous, not only to the NSDD framework, but later on to what could or should be done. And it got into the whole issue of how strong is the Soviet Union. And there was a "fear" on the part of every single member of the Joint Chiefs of Staff, that if you provoke these guys—

Marshall: Yes, right.

Q8: —they're going to invade Pakistan, and we cannot defend Pakistan under any circumstances, so it's just foolish for you to be toying with this monster. There would be images like this created in the meetings—"Toying with the monster," and things like that. The whole notion of weakness was not...

Q3: And the flip side would be that there were people that just had these two contradictory notions at the same time. First of all, you would so hurt them that they would go mad and nuke you. On the other hand, they were so strong that there was no way that you could hurt them enough in Afghanistan to stop them from doing whatever they wanted to do. So it was both...

Q8: This was coming out of Robert Gates's [Deputy Director for Intelligence] office, too. The DI under Gates, remember, had a big Afghan task force. They were writing these papers that they would give to us at meetings, along the lines that, "nothing you could do—"

Marshall: Yes, right.

Q8: —will do any more than cause the death of a few more mujahedin. Our heroes will die if you do anything more.

Marshall: Yeah.

Q8: So the thinking about Soviet strategy was…

Marshall: Well, right. I should preempt something that we can talk about at the next meeting. I remember when the competitive strategies thing came out I had a discussion with the then-head of the Air Force [General Lew Allen], who I knew, and his argument was, "Well, yes, this is a good idea in a way, but, first, whatever we do, the Soviets will match us." [But] there's no sense in his argument. The Soviets have resource constraints; there was not just an infinite supply of resources available to them. And [his] second thing was, "We're so inconsistent that if we start something we'll weaken and fall off, and they'll keep going, and then we'll be in worse shape than if you'd never done anything."

Q8: Does he [Eds: Guthe] have a research assistant or someone who can dig up a lot of quotes? Because these were… I've noticed in the memoir literature now, everyone is simply ignoring the quotes of the mid-'80s about the power of the Soviet juggernaut, not by any professors, but by heads of the Air Force, Army, Navy, Weinberger himself.

Marshall: Yeah.

Q8: And it's just being laughed away now: "Oh, we never thought that. We always knew we were just three years away from the Berlin Wall being cracked open."

Marshall: Yeah, right.

Q3: The only other thing on the Reagan Doctrine, was, following on from Andy's studies of Soviet power projection, Fortier ,was saying, "OK, how can we drive up Soviet costs of power projection?" Because they were active in Angola, and doing all that other stuff. And I recall {…}

Q8: Are you going to do different assessments at different times [during this project]? Because relevant to this, also, is the ADCAP [advanced capability] torpedo. The whole issue of Weinberger and strategy, in some sense, came down at one point to one of your…I think it was the Navy balance, wasn't it? I remember reading it from… It was an official balance…

Marshall: Yeah.

Q8: You're leaving out the effort you made, you had to get Will Taft [Deputy Secretary of Defense] to…Taft said, "Why don't you…"—it was Dennis Ross's idea, actually—"Why don't you get together every single formal assessment Marshall has done, put all the recommendations together, and check which ones have been implemented and which ones haven't, and then report to Fred Iklé, [Undersecretary of Defense for Policy] and Taft about what more needs to be done, and I, Will Taft, will sort of oversee this process." And Marshall drafted one of those memorandums to all Services, everybody in the Pentagon, describing this. Then he gave it to Fred And remember Fred? "No!"—he didn't want to do that. But we started doing it anyway, informally, to see what had been done.

And one of the most vivid things was your Naval Balance, said, at the very, very end, it had like 12, a certain number of recommendations, and one of them was the ADCAP torpedo. That for reinforcements to NATO, if you can get the probability-of-kill way up against Soviet submarines that were going to otherwise choke off U.S. supplies—

Marshall: Yeah, yeah.

Q8: —it made a big difference in whether or not Europe survived long enough to continue resupply. And all of a sudden the ADCAP torpedo, which had a Pk [probability of kill] [that] was very, very high. You know, most torpedoes were considerably lower...Well, the Senate Armed Services Committee staff had put the ADCAP torpedo as— They'd zeroed it out that year, because there was testimony from Litton that it wasn't ready yet or something like that. And here was a concrete example of where you'd think the Secretary of Defense would have said, "NATO, Europe—they're important. This is a strategy against a known Soviet strategy to sink our resupply effort. This is very much needed. It's crucial." When asked, ["Has] Weinberger or anybody, even the Navy, singled out the ADCAP torpedo?" [the SASC staff] said, "No, there's not even any testimony." It was just buried away in the munitions request from the Navy. So that got plussed up, once it was explained what the importance of it was. But that was an example of what would have happened if Taft had gone through with the—

Marshall: Yeah.

Q8: —[indistinct] what he thought was right, and said, "No, I really think Andy Marshall's recommendations ought to be inventoried and we should see what's been done and what's not been done." This little anecdote, are you going to have a session on this? Because this went into "Reagan II," later on. Your operating environment, here in the Building.

Marshall: Yeah. Well, we can try to talk about some of these things.

Q2: I think on that one I agree, with Fred [Iklé].

Marshall: On the whole, we haven't been in the recommending business. We've sort of....

Q2: Diagnosis.

Q8: [Indistinct] your recommendations in some of these.

Marshall: Yeah, there are.

Q8: Your name could be stenciled on every ADCAP torpedo. The chairman at Litton should be in here thanking you.

Marshall: I see. Yes, right, right.

Let me get back to the restart of the Israeli talks. When Richard Perle [Assistant Secretary of Defense for International Security Policy] and Iklé came on board and—Perle in particular knew about the fact that we'd had those earlier discussions with the Israelis—asked about them, I told him about their demise in the last year of the Carter Administration. Perle and Iklé wanted to restart them. So I went over to Israel to make arrangements and talk to them about starting the talks. But then the thing got squashed by Weinberger, because a few days after I left Israel, the Israelis attacked [on 7 Jun 1981] the Iraqi [Osirak] nuclear reactor. Then later, I talked with Bud McFarlane, who I guess during the first year or so of the Reagan Administration was an ambassador-at-large [Counselor] in the State Department. And it was then they were transiting to this new form of these talks with the Israelis, in which State and, really, the Joint Staff, I think, were the main participants, because it was the direction which the Israelis wanted to go, in any case, of having these talks as real planning sessions and so on about various things.[35]

A next thing to talk about is the start of the Defense Policy Board. I guess fairly soon after Iklé was on board, Sy Weiss went to him with the idea of establishing a Defense Policy Board modeled after the CNO Executive Panel. Fred decided to go ahead with it, or to do an experiment with it. And there was a period when, in some way, we probably were breaking the law by having these meetings and it wasn't fully established, all the procedures hadn't been gone through.

Q2: I'm sure no laws were broken.

Marshall: In any case, it was run out of my office at this point. I think it probably started in '83.

Q3: I think it was earlier, '82.

Marshall: It might have been. I suppose so, I don't know. The reason I place it in '83 is that I had the impression that from the beginning Harry Rowen had been a member of it, and he spent the first two years of '81 and '82 out at the …

Q2: National Intelligence Council [NIC].

Marshall: Yeah.

Q3: He came over while he was still head of the NIC. And Paul Wolfowitz was part of it when he was still at Policy Planning in the State Department.

Marshall: Yeah. One of the topics early on was another dimension of the Soviet weakness theme that began to be explored, particularly the possibility that in Europe, that the Warsaw Pact might have some serious internal problems. And retired General John Vogt was hired to do a study and exploration of that.

Q2: Is this the one that was locked up, the Vogt study—

Marshall: Yeah, right.

Q2: —was locked up. And all the copies collected.

Q3: Rosen also wrote a paper for you on the character of mutinies in the Warsaw Pact that might be induced.[36]

Marshall: Yeah.

Q3: No one was allowed to take any copies of that paper anywhere, either.

Marshall: Right. There must be some copies of the Vogt study around here yet, I don't know.

Q2: Who was it that was really against the Vogt study? Was that Weinberger?

Marshall: No, I don't think Weinberger was against it. I don't know where he really was [on that] [indistinct].

Q2: There was a lot of opposition to that study. I mean, it was like don't…Somebody said this study is not to be circulated

Marshall: Yeah.

Q3: But then SCDC went and sort of tried to replicate a similar kind of analysis about how to get the East European nations to defect in time of war. I remember being sent over to listen to their brief.

Q2: A Jeff Simon study?

Q3: I can't remember who did it.

Q2: But I remember the Vogt study. It was classified TS, and when I came here, I think you said, "Read the Vogt study." And there's only one copy, and it's locked in a vault, and all the rest have been destroyed. It was the only copy left of the study, which also was supposed to have been destroyed, too, but wasn't. Do you remember who was opposed to that or why?

Marshall: I know that Vogt went and talked with General John Vessey [Chairman of the Joint Chiefs of Staff] and some of these other people about it and got some response. It was a very mixed reaction, no doubt about that. And there was some discussion with Bill Casey [Director, CIA] about intelligence activities, operations that might have to be taken. And again, there was no enthusiasm for doing that. I mean Casey probably was interested, but the establishment as a whole was rather uninterested.

Another thing to mention from this period is that probably the first of the Summer Studies was done, during, I think, '84.

Q4: Wait a minute. Before you get out of [indistinct] the Policy Board. Are you going to come back to the Defense Policy Board in the subsequent discussions and talk about how it has grown and changed, and whether you think, in reality, it has been useful?

Marshall: We can do that, but I think probably we'll do it next time. But I'll say a few words now. I think that the original notion that Sy Weiss pushed made it not very useful. It was the idea that here is this set of wiser people who would listen to things and then make judgments and give advice, but not really do any work. I think that was not a good way of functioning. I don't know how it's going to function now, but it was evolving toward something a bit more like the Defense Science Board. Not quite, but it is more helpful where there are task forces, people really do some sort of work, and it's not just off-the-top-of-the-head, seat-of-the-pants reactions.

Q3: John Deutch [Deputy Secretary of Defense at the time of the talk] was involved in it in the early days. And I remember he was pretty disgusted with the whole business. That may affect his attitude.

Q2: Who was on it the first year? You said Rowen, Paul [Wolfowitz]...

Q3: Paul, John Deutch, Albert Wohlstetter, Fred Hoffman. I remember Sy Weiss sitting Fred Hoffman down, and he said, "We'll let you come to the meetings, but you can't vote the way Albert does all the time." Sort of like Outer Mongolia and the Soviet Union [both being in] in the [United Nations] General Assembly. Fred Hoffman, being such a sweet guy, just sat there and said, "Right."

Marshall: Bob Long [Retired Navy Admiral] was on it.

Q3: John Deutch, Larry Silberman…

Marshall: Oh, and Charles Herzfeld [former Director, Advanced Research Project Agency]. OK, the Summer Studies. The first Summer Study was run out of RDA in L.A. in '85.

Q3: And then in '87 we did the Monterey business for the *Discriminate Deterrence*…

Marshall: Yeah, right.

OK, another item: This negotiation with the Israelis about the "lessons learned" activities that were to use their experience, any materiel they had gotten out of the '82 War, turned into a fiasco. Dennis Ross and I and some other people went over [in Nov 1982] to try to negotiate something with them.[37] Anyway, I say this just for the humor of it. We negotiated this long agreement. Weinberger thought there should be just one page, some minimal kind of agreement. And they wanted this long, elaborate document, giving this or that access to the results that we would have and so on. So we negotiated it and came to sort of an agreement that was enough to take back, although I was doubtful that the people back here were going to agree to all of it. Then I guess Ariel Sharon [Minister of Defense] because of his own political interest decided to get some publicity out of the fact that this agreement had been reached. [gap for tape change]

Q2: You were on TV? In Israel?

Marshall: In Israel. Just at this meeting. And it was leaked to the papers later and so on. At any rate, I started off the meeting by telling him that we'd signed this thing, but very likely people back in Washington were not going to agree. In particular, I mentioned that they [the Israelis] had insisted on, throughout this document, the name or the phrase that they had used to designate this war—something like "Peace in Galilee" [Operation Peace for Galilee] or something, I forget what it was—and I had been instructed that we would not sign any document of this sort. So I told him that. I said, "Look, I'm taking this back. We've gotten as far as we can with this thing. But I just want to tell you that very likely we're not going to sign any document that has this in it." He then went off for about a 20-minute tirade on, "'We don't tell you what to name your wars,' and went on and on and on. And then at the end of it all, he asked my help in getting Weinberger to ship off another $200 million that had been held up for something.

Q3: Before we move on…You had this sort of discussions with the Israelis, but you also had a set of ongoing discussions with the French and the Germans.

Marshall: Yes, right.

Q3: The only reason I mention that is that my memory of the sessions that I went to, a lot of the stuff that now gets referred to as the "military-technical revolution" and the "revolution in military affairs" was presented in one guise or another, both by the Germans and the French, and a little bit by us.

Marshall: Yes.

Q3: This was also following on some of the "lessons learned" from these Middle Eastern wars, but also some of the early work on the impact of precision guided munitions.

Marshall: Yeah, right.

Q3: Again, I wouldn't [indistinct] seems like a lot of your interest in what we now call the "revolution in military affairs" stems from this period, and you used that in discussions with other governments. Is that right?

Marshall: Well, I would locate my interest in this really from the Russian writings which we became aware of—their internal, still-secret writings—in the late '70s, which became more public in the early '80s, '83 or so, in particular, and also the analysis of exercises that we were aware of. In, say, '79, they were already beginning to look at how to counter reconnaissance-strike complexes and so on. So we were interested in that set of ideas increasingly through that period. And then at the time of the Iklé-Wohlstetter thing [the Commission on Integrated Long-Term Strategy],which began in the summer of '86, we did this assessment of the next 20 years in which that was one of the major themes. [38]

Just to leap ahead: It's kind of symptomatic of this denial of Soviet weakness that Iklé as late as '88 required that Charlie Wolf and I use, in effect, the CIA estimates of Soviet GNP as the base case in our projections, which was not what we believed, and we put in the variants.[39] If it had been left to us, we would have put a much lower estimate of Soviet GNP as the base case and the OA [CIA estimates] as a nutty variant.

Q8: Did you hit Iklé on that, too?

Marshall: He had two objections to things that we wanted to do. One was that we wanted to put a little emphasis on AIDS as a problem…And secondly, any portrayal of the Soviets as weak or smaller as opponents was not to be highlighted.

But certainly by the middle '80s, I very much saw this as a major theme, and certainly included it in the NATO-Warsaw Pact assessments that were done. And I'm sure that we talked some about that with the French and the Germans and others.

Q3: What sticks in my mind is that both general staff officers from the German army and from the French army talked about their ideas about the nature of battlefield tactics—

Marshall: Yeah, right.

Q3: —and operational behavior in a battlefield in which information was available to everybody, where everybody else was, and the importance of stealth on the battlefield, of tactics which mingled your forces with those of the enemy so they couldn't be sorted out. I was just kind of struck by—

Marshall: Yeah, right. The French in particular…

Q3: Colonel [indistinct] was a very impressive guy.

Marshall: Yeah, right. The discussions with the French, which did start in this period, began in '81. The approach had been made to Robert Komer [Under Secretary of Defense for Policy] in the middle of '80, and Komer had assigned us to undertake those. The first meeting, I think, was in Paris in the spring of '81.

Q8: And the French—I misunderstood—the French did or didn't have the idea that the Soviets are weak?

Q3: They may or may not. I don't recall them discussing that specifically, but they did— We, the French and American delegations, seemed to be engaged in a dialogue about the nature of what we now call the "military-technical revolution."

Marshall: Yeah, that's true.

Q3: But what was the purpose of the discussions with the French? Was it just to…

Marshall: It came on the French initiative, through the attaché, who they approached wanting to create another channel for discussion of particular problems, sharing of information, and so on.

Q2: I think the French were doing that because of the Israeli talks?

Marshall: No, I don't think so. I don't think so. I don't think they particularly knew about the Israeli talks prior to this.

Well, I don't really have anything much more [to say]. Maybe some of you have…

Q3: Well, how about the changes in the orientation of the office? You didn't kind of take the office as a topic.

Marshall: Well, as I say, we came into this particular period from what we were doing in the years with Brown, during which things were going very well, and so we had a bunch of things, and so in the first few years of the Reagan Administration we were finishing off various things and continuing on. But then, it was, I thought, very clear that Weinberger just didn't have any interest. And so the number of assessments, I think, began to trickle off, and there were other things that Iklé had us involved in. It got worse in '85, with the competitive strategies thing being added on. And there was also the start up, then, of the Wohlstetter commission [Commission on Integrated Long-Term Strategy].

But before that, there had been something which I had forgotten entirely about, which was a big set of activities which had to do with this effort to look at the strategic forces—the whole set of games that Charlie Pease did.

Q2: Oh, the NSDG [Nuclear Strategy Development Group] games?

Marshall: Yeah, the NSDG effort, which was an initiative that Fred Iklé started, I suppose, given that this was probably in '83, and just about the time of the Strategic Defense Initiative. It wasn't triggered by the SDI, but it sort of became part of it. The whole idea was that we have this elaborate Master Plan [Nuclear Weapons Employment and Acquisition Master Plan] for our forces that runs through the early '80s, or up to '85—I forget the exact dates of it now.

There was this Master Plan for fixing up the strategic forces, and Iklé thought we needed to look at the period beyond there, because, first, the target system was changing, becoming much harder. As a result of the change in targeting policy we've already touched on, they began looking for the leadership things. We find all this whole new range of hardened targets. You look at what's happening on the silo upgrades, changes and so on. In other words, just a whole set of things are happening on the Soviet side so the target system…

Q2: Started going to mobiles.

Marshall: Yeah…is changing. Then along comes SDI, or the possibility of missile defenses, and the questions arises, "Will you want to shift to a defense-dominated strategy? How should you be thinking about all of this?" So he wanted to get started with the studies to look at the period beyond the Master Plan. So we ran a lot of the activities for him. One, there was this set of games. But there were other studies that were done that we were responsible for. There was an analysis working group that I chaired for him.

All of these activities began to displace the assessments, and we were down, I think, to continuing— We were doing the NATO-Warsaw Pact by the end. We were doing the Investment Balance. I think we continued to try to do a few things in power projection. But we were narrowed down quite a bit, and only began to expand back into this net assessment business when the Bush Administration came to power.

Q4: At this point in time, you've been in Washington a little over 10 years. You've been in [the Office of the Secretary of Defense] for about 10 years.

Marshall: Yeah.

Q4: What are you thinking about in terms of going back to California? Is this analysis effort succeeding or failing? What were your sentiments?

Q3: You were pretty grumpy at the time as I recall.

Marshall: Yes. Right. I don't remember exactly, but, yes, I think that's right. And I was probably thinking a lot more about going back to California.

Q2: Now, why do you think Weinberger never picked up on this Soviet weakness stuff? He clearly was very interested in doing as much as he could do across the board on defense spending to get forces up and everything to oppose them…

Marshall: Yeah, right.

Q2: A big proponent of SDI. And in many ways, that's just the culmination of the whole cost-imposing thing. So, why not? Beyond political reasons. One could understand why you wouldn't want to give the political opposition any notion at all that they could do less because the other guys were weak, as opposed to doing more and driving them into the grave. If you said too much, "Aw, they're weak over there," then everybody says, "Well hell, we can cut the budget then, if they're weak." But setting that aside, was there some really fundamental problem? He wanted to do all these things, but he didn't have an objective? Is that….

Marshall: I don't know. I don't know how he felt about me. I saw him only two or three times or so. So I don't know what he thought. But I see no signs in the discussions that I know of, or heard anything about, that he had any understanding of the economics of the situation. And obviously tactically bad. Whether, as this book *Victory* alleges, he was part of an inner group that understood this all and was pursuing this strategy…Maybe he kept it to himself. I don't believe that, but it's possible.

Q8: On the pipeline issue, Weinberger totally was with hawks.

Marshall: Yeah, well so was Perle and everybody. There was a whole set of people here that were very much against that.

Q8: Was there anything in '84 about Weinberger giving the order to shut down the Office of Net Assessment? Did you hear anything about how the Joint Staff was going to create its own office of net assessment? And get public release rights for every assessment wherever it would be done? That this office [ONA] would not ever be allowed again to testify in public?

Q3: Yeah. There was gossip that the JCS wanted to take over the net assessment function entirely.

Q2: It was more than gossip. They tried to get that under the Goldwater-Nichols [Defense Reorganization Act of 1986).[40]

Marshall: Yeah.

Q2: They got it insinuated in some way in the Packard Commission [President's Blue Ribbon Commission on Defense Management],[41] and then they tried to get the language in Goldwater-Nichols structured in such a way that, well, they did get it structured in such a way that they were given the charge by Congress for conducting net assessments.

Marshall: Yeah.

Q2: And they tried to argue that they gave them sort of a higher moral authority because theirs came from Congress, whereas this office's only came from the Secretary of Defense. And so therefore, they should have the lead on doing all that. I mean, this is looking forward to the next session.

Q8: I was told that by the special assistant to Weinberger in roughly May '84. When I said I really wanted to go over there and work with Andy Marshall, she said, "Don't you know he's being shut down? Weinberger has decided to shut down that office." I remember being quite startled. And as I remember, there were a number of steps taken involving senators calling Weinberger, saying, "You shouldn't do that," in June and July, I think it was, of '84. Phil Merrill [publisher of the *Washingtonian* magazine] also got into it. He was quite panicked by this concept. And when given…When asked why, Weinberger's special assistant said there are two reasons. The office never does anything, number one. They never produce any assessments. This is not her opinion; she's reflecting what she'd heard discussed around Weinberger. Number two, the Joint Staff—or a word like that, I don't know the exact…whether it was J-5 or JCS, or the Joint…the word "Joint" is in there—was real angry about the way the assessments come out.

And looking back—this is not something that happened at the time—but looking back, I think the Soviet weakness argument, perception, or whatever you want to call it, brilliant analysis, was causing real political enemies to be formed, that you cannot— This office was being misperceived as a liberal, sort of left-wing operation. Because at the time, the Federation of American Scientists…there's a unit of public opinion in America at the time saying the Russians are weak, they want peace, we should extend the hand of friendship. And some of Marshall's arguments were being scrambled with the American left-wing view, and therefore seen as mutiny or treasonous against Weinberger's patriotic efforts to sell the Services' budgets on the Hill.

Q3: Yeah. It sort of feeds back into the idea that by talking about Soviet vulnerabilities, you sort of gave ammunition to the left wing in the United States. I think it came back in what was one of Weinberger's few marginalia on the Investment Balance, which contains some idea, "Look, the Soviets have this big advantage over us in invested military capital, but one response the United States might take would be to undertake new military programs to render obsolete some portion of the Soviet military capital stock." Which I believe Weinberger's response to Mr. Marshall—obviously he remembers better than I do—[was], "Is he one of those liberals who thinks the Soviet military is full of obsolete equipment and therefore is no good?"

Q8: So that's around here somewhere? That handwritten note is around here?

Q3: I think it was in marginalia on the short version that was sent up. David Epstein perhaps knows something about that? So that's the only concrete piece of evidence that I know of.

Q8: When Senator Orrin Hatch came over to see Weinberger about Pillsbury joining the Defense Department, that was one of the topics, that is, his decision to shut down…

Q3: [Indistinct] confined to Weinberger, because Andy, as always, had many adherents in the senior leadership of the military. And I don't think the outcome of the Balances that I was aware of annoyed— For example, remember that famous occasion when we all went out to Pearl Harbor to brief Admiral William Crowe [Commander in Chief, Pacific Command, 1983–1985] on the East Asia Balance? A lot of people on Admiral Crowe's staff didn't like the Balance, but Admiral Crowe looked at the thing and said, "Yeah, yeah, looks good. Count on Andy to come up with something sensible," you know. And Admiral Long [Crowe's predecessor as CINCPAC] was a fan of Andy's. So I think Weinberger had one set of perceptions and the senior leadership of the military was different.

Q8: I never got a name of an individual, general or admiral, who was sort of out to abolish Net Assessment…Somehow this thing called "Joint Military [Net] Assessment" …I lost track of it. The next thing I know people are coming to sit in offices, handing people copies of this thing called "Joint Military [Net] Assessment," JCS on the cover.

Marshall: Yeah, well that's later.

Q2: That comes after Goldwater-Nichols and that's their first thing. J-5 did something on it, and we have a whole process set up to do these things on it. Coordinated, Net Assessment Coordinating Committee we set up—

Marshall: Right.

Q2: —and all that in response to this. And, you know, J-5 tried to use that to capture the flag, the whole net assessment flag. It was all pressure from the Hill, by the way, for net assessments. This was just being written into legislation, "You will do net assessment on this, you'll do net assessment on that." They use the phrase "net assessment" in ways that this office doesn't. So, you, "You'll do a net assessment on which way rivers flow, and why they do that," things like that, I mean everything, partly because Tony Cordesman was over there. He was kind of writing [that] everything had to have a net assessment to it.

Marshall: Right.

Q2: It's his [Cordesman's] way, I think, of bolstering the importance of the office. But then there was this sense that these were becoming increasingly important and demanded, etc. So the JCS thought they must be good and they didn't want this office doing it. So they wanted to capture the thing for themselves. The only general officer I know—and again this is a feedback to Weinberger—Arnie Schlossberg, who was the head of the Chief of Staff of the Army's staff group as a colonel, and the Chief was General John Wickham, and then Schlossberg got his first star and went down to J-5, so he was involved in this kind of stuff, although I met him as a colonel. He once pulled me aside at a conference up at Harvard, when, I think, he was the deputy J-5, and said, "You know, you really ought to get out of that office because, you know, Weinberger just doesn't think very highly of it all [indistinct] Marshall."

Q8: '85 this was?

Q2: Yeah, somewhere thereabouts.

Q3: Did he go to MacDill Air Force Base [Rapid Deployment Joint Task Force and Central Command headquarters] after that?

Q2: Well, he got his second star and went off and did something else.

Q3: What was it, like RDJTF [Eds: Rapid Deployment Joint Task Force]?

Q2: Well no, it was CENTCOM.

Q8: One of the things that Senator Hatch said to Weinberger was about China. He praised the Asia assessment; how the Senator knows about this, who knows? Weinberger perked up—I was in the meeting—Weinberger perked up and he did not seem to know that Net Assessment had said that there's a big payoff if we do things with China on [indistinct] used the term "imposing costs." That's why I wanted to underline the part in the assessment about the tremendous costs of doing things in Siberia. There's about ten pages in there—God knows how you dug all that out—but it's tractors and trucks, and building roads, and permafrost.

Q3: There were studies of oil exploration in Siberia and that was the basis for what we had.

Q8: And Weinberger, of course, had just been to China for the first time in '83. We hadn't sold anything in terms of weapons yet. We did that in '86. Then they came…I noticed in his memoirs, he really plays up his China trips and going to upgrade the Chinese MiG-23s with American avionics and— He's very proud of his China role.

Q2: That wasn't a great idea.

Marshall: Yeah, right.

Q8: Through the second term and Bush and later comes more the Asia stuff, but the seeds begin in this 1980 100-page-plus Asia assessment, which was not offensive, I would think, to the Joint Staff, if they ever saw it. Of course, [indistinct] none of your studies, apparently, so they really didn't know what you were saying. They would just know from conferences or meetings.

Marshall: No, we sent copies of that all over the place.

Q3: It was briefed to the CNO Executive Panel.

Q8: No, I mean all assessments. A J-5 colonel could not lay his hands on all of Mr. Marshall's assessments.

Marshall: No, no. I mean we send them there and they systematically destroy them. They have no memory…

Q2: In fact, one of the first things they did when they tried to capture the net assessment flag was they wanted copies of all assessments.

Q3: One of the, I think, more successful stories was something which you didn't talk about at all, which is the whole effort to develop Soviet-style assessments—

Marshall: Yes.

Q3: —which I think actually was useful in itself, and also kind of caught on in the larger community.

Marshall: Yeah, that's right. That continued on through this period.

Q5: Which is progress. A lot of progress was made on that.

Marshall: Yeah, right.

Q3: Because it was something that you could— Remember [indistinct] had the very intelligent idea from you about setting up schools to teach people how to do Soviet-style— So you could propagate it throughout the bureaucracy and it wasn't the exclusive possession of—

Marshall: Yeah, right. That's true. I don't know the timing of all that. It certainly was continuing on during this period. And I remember—I guess I'd have to look at the records—we had a number of meetings with people like Fritz Ermarth [National Intelligence Officer for the USSR and Eastern Europe], and General William Odom [Army Assistant Chief of Staff for Intelligence], and a guy from [the Defense Intelligence Agency]…

Q3: Gordon Negus.

Marshall: Right. And I forget when John Hines first came to the office. And Steve Rosen came during this period.

Q2: And you had started the Foreign Systems Research Center at SAIC in Denver, John Battilega and Judy Grange and Project Eager out there.

Marshall: Yeah.

Q2: In fact, they taught a week-long course on Soviet-style assessments. And you got the intelligence community to actually start looking at that.

Marshall: Yeah, right. And Odom started to do stuff with the Army.

Q2: And actually started getting some information that, you know, was useful.

Marshall: Right.

Q5: The first balance in which that was really well laid out was the strategic net assessment that you were talking about during this period.

Marshall: Yeah. It came out in '82, '83 [1983].

Q5: Going back to these issues about how things were perceived outside the office… I seem to remember initially on that NATO-Pact Balance that Dave Andre began that we really got around to a clear understanding that their view of the balance was really much more pessimistic than what we said, you know, in *Soviet Military Power,* I mean. There were a lot of unhappy campers about that whole story, although I think it was well-documented.

Marshall: Yeah, right.

Q3: You're right. There were some very specific technical aspects of the strategic nuclear balance where it was just clear that the Soviets analyzed damage probabilities in a completely different way than we did.

Marshall: Right.

Q3: In ways which sometimes made us look better. [gap for tape change]

Q5: It was difficult to get almost anything out of you on subjects like "what do you think about the Soviet defense burden" back in '83, '84, '85. I just say that in the sense that you're much more open about this stuff the last few years, it seems to me—

Marshall: Yeah.

Q5: —than you ever were. And that, I think, has some consequences, from your standpoint, for how clear messages maybe were communicated, even within the office. Now, I assume you were completely forthcoming in private discussions, you know, in appropriate…

Q2: When did the German talks start?

Marshall: They started back in the '70s. They were the first ones we started. Mainly because this group was headed by this General [von Orr/von Knorr?]…who was a close friend of Peter Haas [Deputy Director for Science and Technology, Defense Nuclear Agency] and had come to me. So they started, I think, probably around '77, '78, something in there. The first actual meeting, I forget, whether it took place here or there. And they did some interesting things for that. And then the French thing, as I say, started in '81. And we had this whole other set of discussions of a different sort with the Israelis, beginning in '76.

Endnotes

1 Eds: This chapter is based on an interview conducted 26 July 1994.
2 Eds: Jimmy Carter, Nuclear Weapons Employment Policy, Presidential Directive 59, July 25, 1980, formerly Top Secret//Sensitive, partially declassified 8/20/96, 1.
3 Eds: National Security Decision Directive 13, Nuclear Weapons Employment Policy, The White House, Washington, October 19, 1981, formerly Top Secret//Sensitive, declassified 12/09/08, 1.
4 Eds: Richard Allen, Interview for the Ronald Reagan Oral History Project, Miller Center, University of Virginia, May 28, 2002, accessed March 8, 2013.
5 Charles Wolf, Jr., K.C. Yeh, Edmund Brunner, Jr., Aaron Gurwitz, and Marilee Lawrence, *The Costs of the Soviet Empire,* R-3073/1-NA (Santa Monica: RAND Corp., Sep 1983); Charles Wolf, Jr., Keith Crane, K.C. Yeh, Susan Anderson, and Ed Brunner, *The Costs and Benefits of the Soviet Empire, 1981-1983,* R-3419-NA (Santa Monica: RAND Corp., Aug 1986); Charles Wolf, Jr., "The Costs and Benefits of the Soviet Empire," in Henry S. Rowen and Charles Wolf, Jr., eds., *The Future of the Soviet Empire* (New York: St. Martin's Press, 1987), 121–140.
6 Andrew W. Marshall, "Commentary," in Joint Economic Committee, *Gorbachev's Economic Plans,* Vol. I, 100th Cong., 1st sess. (Washington: GPO, 1987), 483.
7 Ibid., 481–484; David F. Epstein, 'The Economic Cost of Soviet Security and Empire," in Henry S. Rowen and Charles Wolf, Jr., eds., *The Impoverished Superpower: Perestroika and the Soviet Military Burden* (San Francisco: Institute for Contemporary Studies, 1990), 127–154.
8 For criticisms of CIA estimates of Soviet GNP, see Anders Åslund, "How Small Is Soviet National Income?" in *The Impoverished Superpower: Perestroika and the Soviet Military Burden,* 11–50; Senate Foreign Relations Committee, *Estimating the Size and Growth of the Soviet Economy,* 101st Cong., 2nd sess. (Washington: GPO, 1991); Nicholas Eberstadt and Jonathan Tombes, 'The Soviet Economy: How Big?" *The American Enterprise* 1, no. 4 (Jul/Aug 1990): 76–80.
9 "Commentary," 484; Andrew W. Marshall, remarks at 11 Jun 1987 seminar of the International Security Studies Program of the Wilson Center, in *Security Digest #20,* Jul/Aug 87; 'The Economic Cost of Soviet Security and Empire," 153.
10 Alexander R. Alexiev and S. Enders Wimbush, 'The Ethnic Factor in the Soviet Armed Forces," and Edmund Brunner, Jr., "Soviet Demographic Trends and the Ethnic Composition of Draft-Age Males, 1980-1995," in Alexiev and Wimbush, eds., *Ethnic Minorities in the Red Army: Asset or Liability?* (Boulder: Westview Press, 1988).
11 G. Warren Nutter, *Some Observations on Soviet Industrial Growth,* Occasional Paper 55 (New York: National Bureau of Economic Research, 1957); G. Warren Nutter, *Growth of Industrial Production in the Soviet Union* (Princeton: Princeton University Press, 1962).
12 See biographical sketch of Nutter in House Foreign Affairs Committee, *Strategy and Science: Toward a National Security; Policy for the 1970s,* 91st Cong., 1st sess. (Washington: GPO, 1969), 241.
13 G. Warren Nutter, *The Strange World of Ivan Ivanov* (New York: World Publishing Co., 1969), 121–132.
14 Anne O. Krueger, *Economists' Changing Perceptions of Government* (Kiel: Institut für Weltwirtschaft an der Universität Kiel, 1990).
15 See statements by Watkins in Edgar Ulsamer, "Bobbing, Weaving, and Fighting Smart," *Air Force Magazine* 66, no. 8 (Aug 1983): 88–89, 91, 94; Richard Halloran, "Navy Trains to Battle Soviet Submarines in Arctic," *New York Times,* 19 May 1983, A17; Senate Armed Services Committee, *Department of Defense Authorization for Appropriations for Fiscal Year 1985,* Part 8, 98th Cong., 2nd sess. (Washington: GPO, 1985), 3887–888, 3893; House Appropriations Committee, *Department of Defense Appropriations for 1986,* Part 2, 99th Cong., 1st sess. (Washington: GPO, 1985), 927–928; "The Maritime Strategy," in *The Maritime Strategy,* Supplement to the U.S. Naval Institute *Proceedings,* Jan 1986, 7, 9, 11, 13, 14. See also Ronald O'Rourke, *Nuclear Escalation, Strategic Anti-Submarine Warfare, and the Navy's Forward Maritime Strategy,* Report No. 87-138F (Washington: Congressional Research Service, 27 Feb 1987), which draws on official statements, including those of Watkins.

16 W. Reckmeyer, *Cost-Imposing Strategy Development* (Bethesda: Booz-Allen and Hamilton, 7 Feb 1983), For DoD Use Only [Available from Defense Technical Information Center, ADB 158346].

17 Booz-Allen and Hamilton, *Analysis of Alternative Techniques for Using Cost Imposing Strategies in Defense Decision Making (U)*, DNA-001-82-C-0331, 21 Apr 1983, SECRET.

18 *Enhancement of the Efficiency of U.S. Military Programs Through the Application of "Cost-Imposing" Strategies (U)*, DSA Report 485 (Arlington: Decision-Science Applications, 24 Jun 1983), SECRET (executive summary only); *Cruise Missile Cost-Imposing Effectiveness: A Case Study of the Impact of U.S. Air Launched Cruise missile (ALCM) on Soviet Air Defense Costs (U)* (Arlington: Decision-Science Applications, date?), SECRET; G.E. Pugh and R.R. Hibbard, *Cost Imposing Effectiveness of an Attack RPV: A Case Study of a Remotely Piloted Airframe for Ground Combat Support*, Partial Draft (includes summary of findings and conclusions only), DSA Report 521 (Arlington: Decision-Science Applications, Jan 1985).

19 Memo, Caspar W. Weinberger to Secretaries of the Military Departments, Chairman of the Joint Chiefs of Staff, Under Secretaries of Defense, Assistant Secretaries of Defense, General Counsel, Assistants to the Secretary of Defense, Directors of Defense Agencies, Commanders-in-Chief of Unified and Specified Commands, subj: FY 1983-87 Defense Guidance (U), 10 Jun 1981, with attachment, "Competing with the Soviet Union in Peacetime (U)," SECRET, 13; Memo, Caspar W. Weinberger to Secretaries of the Military Departments, Chairman of the Joint Chiefs of Staff, Under Secretaries of Defense, Assistant Secretaries of Defense, General Counsel, Assistants to the Secretary of Defense, Directors of Defense Agencies, Commanders-in-Chief of Unified and Specified Commands, subj: FY 1984-88 Defense Guidance (U), 22 Mar 1982, with attachments, "D. U.S. Strengths and Opportunities (U)," 5, and "F. Competing with the Soviet Union in Peacetime (U)," SECRET, 24–26. See also Richard Halloran, "Pentagon Draws Up First Strategy For Fighting a Long Nuclear War," *New York Times*, 30 May 1982, 1.

20 Memo, Paul Thayer to Secretaries of the Military Departments; Chairman, Joint Chiefs of Staff, subj: U.S. Weapons Development Policy (U), 10 May 1983, CONFIDENTIAL.

21 Memo, John Lehman to Deputy Secretary of Defense, subj: U.S. Weapons Development Policy (U) - Info Memorandum, 18 Jul 1983, SECRET.

22 Memo, Verne Orr to Deputy Secretary of Defense, subj: U.S. Weapons Development Policy (U) - Information Memorandum, 20 Jul 1983, SECRET/

23 Peter Schweizer, *Victory: The Reagan Administration's Secret Strategy That Hastened the Collapse of the Soviet Union* (New York: The Atlantic Monthly Press, 1994).

24 "I think that [attacking Soviet ballistic missile submarines in a conventional war] is a vital part [of our strategy], that we should try to eliminate the Soviet means of warfare as quickly as possible. You have with these submarines always the question as to precisely what they are carrying at any given moment despite their capability. It certainly is an isolated part of the strategy, of the way we would go about dealing with the situation that we hope to deter. One of the ultimates you have to bear in mind is that the Soviets know we have this capability. It adds to the discouragement to them to start it." Weinberger, in Senate Armed Services Committee, *Department of Defense Authorization for Appropriations for Fiscal Year 1987*, Part 1, 99th Cong., 2nd sess. (Washington: GPO, 1986), 82.
"Our ASW capability has reinforced the Soviet navy's defensive orientation, keeping them close to their homeland in order to protect their ballistic missile submarine fleet. This limits the Soviet navy's threat to our sea lines of communications with our European and Asian allies in the early period of war." Caspar W. Weinberger, *Department of Defense Annual Report to the Congress for Fiscal Year 1987* (Washington: GPO, 1986), 86.

25 According to Bobby Ray Inman, "Mr. Casey and I struck an agreement with Secretary Weinberger and Deputy Secretary Carlucci in the second week of the Reagan Administration that, after the intelligence community had finished its national intelligence estimate on strategic forces, the intelligence community and the Defense Department would sit down together and prepare a net assessment." Bobby R. Inman, "The Military Perspective," in Thomas J. Hirschfeld, ed., *Intelligence and Arms Control: A Marriage of Convenience* (Austin: Lyndon B. Johnson School of Public Affairs, 1987), 55.

26 For Turner's comments on net assessment, see Stansfield Turner, *Secrecy and Democracy: The CIA in Transition* (Boston: Houghton Mifflin Co., 1985), 249–251.
27 Memo, Frank C. Carlucci to Secretaries of the Military Departments, Chairman of the Joint Chiefs of Staff, Under Secretaries of Defense (cc: Director, Net Assessment), subj: Preparation of Net Assessments and the Provisions of U.S. Force Data to the CIA (U), 12 Jun 1981; Frank Carlucci to Director of Central Intelligence, subj: Selected Net Assessments, 19 Jul 88, with attachment, Memo, Richard L. Armitage [to?], subj: Selected Net Assessments - Action Memorandum, 14 Jul 1988.
28 Ltr, Secretary of Defense Caspar W. Weinberger to Fred Iklé, 9 Sep 1983, SECRET with attachments (no attachments) (mentions White House use of OSD/NA analysis of Asian military balance).
29 A.W. Marshall, "A Program to Improve Analytic Methods related to Strategic Forces," *Policy Sciences* 15, no. 1 (Nov 1982): 47–50.
30 J.J. Martin and Douglas C. Olin, "Improved Methods for Strategic Analysis," and Marlie H. Graubard and Carl H. Builder, "New Methods for Strategic Analysis: Automating the Wargame," in *Policy Sciences* 15, no. 1 (Nov 1982): 51–69, 71–84.
31 This planning effort is discussed in Robert C. McFarlane and Zofia Smardz, *Special Trust* (New York: Cadell & Davies, 1994), 276–277, 285–289.
32 McFarlane lists the following individuals as among the participants in the planning effort: James Billington, Samuel Huntington, Irving Kristol, Richard Solomon, and Adam Ulam. *Special Trust,* 277.
33 According to McFarlane, the purpose of the planning effort was to produce "a series of studies, to be conducted by experts both inside and outside the government, of twelve issues of strategic interest to the United States, to identify the top two priority foreign policy items on which the Reagan administration should concentrate in the second term." Over the Thanksgiving holidays in 1984, McFarlane gave President Reagan 12 NSC-sponsored papers bound in a notebook and asked Reagan "to select the most important issues for us to focus on." Sometime later, Reagan handed the notebook back to McFarlane and said, "Gee, I've looked this over, Bud, and I think these are just terrific ideas. Let's do them all!" *Special Trust,* 287–289.
34 National Security Decision Directive 166, Mar 1985. *Victory: The Reagan Administration's Secret Strategy That Hastened the Collapse of the Soviet Union,* 213–214; Steve Coll, "Anatomy of a Victory: CIA's Covert Afghan War," *Washington Post,* 19 Jul 1992, A1.
35 *Special Trust,* 185–188; Karen L. Puschel, *US-Israeli Strategic Cooperation in the Post-Cold War Era: An American Perspective* (Boulder: Westview Press, 1992), 35–99.
36 Stephen Peter Rosen, "Mutiny and the Warsaw Pact," *The National Interest,* no. 2 (Winter 1986): 74–82.
37 Foreign Broadcast Information Service, *Daily Report: Middle East and Africa,* 5 Nov 1982, 113; 13 Dec 1982, 17–18; 15 Dec 1982, 16; 2 Mar 1983, 16; 22 Mar 1983, 13; 23 Mar 1983, 13–14; 14 Apr 1983, 110; 23 May 1983, 17; 6 Jun 1983, 14–15.
38 Notra Trulock, III, "Emerging Technologies and Future War: A Soviet View," Appendix B in *The Future Security Environment,* Report of the Future Security Environment Working Group, submitted to the Commission on Integrated Long-Term Strategy (Washington: Department of Defense, Oct 1988), 97–163.
39 *The Future Security Environment,* 6–7, 92–95.
40 Goldwater-Nichols Department of Defense Reorganization Act of 1986, Public Law 99-433, 1 Oct 1986, Title II, Part A, Section 201.
41 President's Blue Ribbon Commission on Defense Management, *A Quest for Excellence,* Final Report to the President (Washington: GPO, Jun 1986), xix, 12–13, 15, 18–20; *Appendix,* 5, 7.

6 The Second Reagan Administration: 1985–1988[1]

Introduction

Presidents Reagan's second term manifested significant continuity with the first term. Nuclear and conventional force modernization continued, as did research into strategic defense. Congressional efforts to dampen inter-service rivalry in the Defense Department seized on the idea of strengthening the Chairman of the Joint Chiefs of Staff and the Joint Staff in the Goldwater-Nichols DoD Reorganization Act of 1986. One of the new responsibilities for the Chairman was to produce an annual military net assessment.

In the Soviet Union, Mikhail Gorbachev's ascent to the leadership in 1985 produced a major shift in Soviet domestic and foreign policy. He cut military spending to free up resources to invest in the economy and promoted the concepts of *glasnost* (openness) and *perestroika* (restructuring) in a futile effort to save the Soviet Union from economic and political collapse. In an effort to maintain some sort of parity with the United States while cutting defense spending, Gorbachev agreed to negotiations on two classes of nuclear weapons, resulting in two arms controls treaties with the United States—the 1987 Intermediate Range Nuclear Forces (INF) treaty and the 1991 Strategic Arms Reductions Talks (START) treaty. Soviet armed intervention in Afghanistan continued until 1989 when Gorbachev finally withdrew the Red Army without having achieved the Soviet Union's strategic objectives.

Meanwhile, the Soviet Union itself began to dissolve slowly, starting with violence in the Baltic Republics and continuing through the Caucasus and Central Asian Republics. The United States continued to support evolving East European freedom movements (particularly in Poland, Czechoslovakia, and Hungary) with public statements and covert funding. Unlike Hungary in 1956 and Czechoslovakia in 1968, the Soviet leadership decided not suppress these movements with the Red Army, nor those in the Republics of the Soviet Union. Eventually those countries and republics won their freedom and the Soviet Union and Warsaw Pact collapsed in 1991.

In China, Deng Xiaoping continued to focus inwards. He maintained his efforts to reform the economy and the nation's industry while also downsizing and modernizing China's armed forces and rooting out corruption in the senior officer ranks. He negotiated the return of Hong Kong from the United Kingdom and the return of Macau from Portugal, demonstrating to the Chinese people and the world that the expansion of China could be peaceful.

Outline for 16 December 1994 Talk

[Note: Italicized items were not discussed during the interview.]

Assessments
1. NATO/Warsaw Pact-biggest success
2. Space

Competitive Strategies

Commission on Integrated Long-Term Strategy/Future Security Environment
1. Vision
2. [?] of (Backward look)
3. Vessey men's room conversation
4. Pol [Policy/Political?] censor
   — AIDS
   — Soviet economy estimate

Nuclear strategy
1. General
2. Games/SDI—insights

USSR endgame/AWM general view
1. Discussions with Dennis Ross at NSC
   — Louis XVI
   — Time on our side
2. Meeting on U.S. demands

Goldwater-Nichols
1. Coordinating Committee
2. Crowe/dubious
3. Joint Staff efforts/Marine colonel

Summer Studies

1985–1988[2]

Marshall: The idea is to talk about the period '85 through the end of '88. As I thought about that time period—I didn't make a lot of detailed notes—I thought I'd talk about assessments, about competitive strategies, about the Fred Iklé-Albert Wohlstetter effort [the Commission on Integrated Long-Term Strategy], about the—actually it preceded it [CILTS]—the nuclear strategy effort that Iklé [Under Secretary of Defense for Policy] wanted…

Q2: NSDG [Nuclear Strategy Development Group]

Marshall: …NSDG, yeah, and about the USSR endgame kind of things. Maybe there are other topics that occur to you guys.

Q2: I'll throw one out and you can throw it away.

Marshall: Yeah?

Q2: Packard Commission [President's Blue Ribbon Commission on Defense Management], Goldwater-Nichols [Defense Reorganization Act of 1986], NSDD [National Security Decision Directive] 219…

Marshall: Oh, yeah.

Q2: The whole nonsense about the [Joint Chiefs of Staff] doing net assessments and the Net Assessment Coordinating Committee and…

Marshall: Yeah, we ought to talk about that. [Adds topic to his outline.] OK, that's a good addition.

On the assessment side, the only assessment I really have much memory of as being a big thing, really, is the NATO-Warsaw Pact one. Now, we continued also with the Investment Balances during that period, but we'll take them up under the USSR endgame. I can't think of any other really major assessment that got done during that period.

Q2: There was one that George Kraus did on space that was not really a net assessment… Well, it was a net assessment, but it wasn't a future net assessment so much as kind of a current thing. And as I recall, the NRO [National Reconnaissance Office]—

Marshall: Yeah.

Q2: —so much disliked the results that they confiscated all copies and asked that it not be distributed further and that sort of thing…

Marshall: Yeah.

Q2: —And I think the guy that did that was General Thomas Moorman.

Marshall: Well, there was something that wasn't really our assessment. It was an assessment that was done jointly with somebody on the Joint Staff. And, yes, the NRO didn't really like the outcome. It was one of these things which was asked for by Secretary of Defense Caspar Weinberger, I think. It was requested. And we have copies around. It was conducted as a joint activity with a variety of people, various representatives. It turned out to be an assessment which, even though the NRO didn't like it, was, by my judgment, extremely bland, because it was a consensus sort of thing and nobody got particularly hurt. The one thing I remember about it is a recommendation that a bad decision had been made to rely entirely on the Shuttle and that Defense needed to really get back in the business of putting stuff up with boosters.

Q5: You don't suppose the "soda straw" sensor business came up? You know, [did] that set of decisions come up?

Marshall: No, not particularly, I think.

Q2: There was one other initiative that was not really a net assessment, but it was this Interagency Technology Assessment Group. Do you remember that, the ITAG?

Marshall: Yeah.

Q2: And you had some people over in the Office of the Under Secretary of Defense for Acquisition that were going to help. And I think maybe Ronald St. Martin was over there doing it for a while, then Jerry Murphy or somebody. And we tried to have this interagency group looking at what really key and critical technologies might look like, with this army colonel named Jack Wood, I think, running the thing…

Marshall: Yeah.

Q2: …out of Acquisition or something. The only reason I mention that is because we participated in the thing. You didn't lead it, but it sticks in the back of my mind that it was related in some way to net technical assessments that you were trying to get off the ground again. Nobody had really done one since Paul Berenson many years ago. And I think you actually tried to get, I think it was St. Martin, and then maybe Jerry…

Marshall: Well, St. Martin was over there. I think that comes later, though. I think it's maybe… You don't think so?

Q2: I don't think so, because I wasn't here after '88…

Marshall: Yeah.

Q2: …and that was going on, at least the ITAG…

Marshall: The ITAG thing was going on, but there was an effort to really get things going again in the early part of the Bush period when Admiral Bill Small came on, and I think that's when St. Martin was there.

Q2: Okay. Alright.

Marshall: But even then, because… I don't know, the timing of all that's unclear to me.

Q2: OK.

Q4: Going back up to your original menu, when does Newport start?

Q2: The Summer Studies.

Marshall: The first one is in LA, I think it was '85. And then in Newport in '86.

Q4: I think that ought to be something on your agenda. It certainly is a very sizable chunk of time.

Marshall: OK, we can talk about that. Right. Let me say one word more on the assessment side about the NATO-Warsaw Pact balance, which ended up being, I think, a very successful assessment.[3] It also was the framework within which, because things had matured, we were able to incorporate a lot more on the Soviet assessment stuff, the stuff that people had been doing and John Hines finally brought to fruition. And even after that—again, I don't know the date; maybe either towards the end of this period or the early Bush period, because I can't quite remember, again, when Hines finally left—

Q2: He left before this, because we got Ned Cabaniss in, and Ned was trying to do that Middle East assessment. So John left. He went over to the Army Intelligence Board.

Marshall: Yeah. But he did one very, very useful thing at the end, although Dmitry Ponomareff sort of had to finish it. He went down and tried to catalog all of the ways in which Soviet assessments might really be different than ours, in particular, some of these very surprising things, things that you would think objectively people could measure and so on, but they obviously had very, very different views of.

One of them was in terms of the size of the craters that nuclear weapons make and how the explosive energy and blast wave energy from a nuclear explosion couples to the ground and so on, which leads to very, very different expectations on their part about the vulnerability of silos and things of that sort.

The second one that struck me was just the very different estimate or belief in the vulnerability of human beings to at least the kind of overpressure regimes that are associated with fuel-air explosives, and may explain why they continued to go ahead with these weapons in a way that we did not. They just see them as much more lethal, and to this day one doesn't know whether they're right or we're right.

Anyway, that is still a very highly classified document, but it was a really very interesting piece to just show all the ways in which their calculations and assessments might be very different than ours. And some of the stuff that's come out in his [Hines] interviews with the Russians is like that, including the ramifications of that.

I guess the most striking story in that connection is that at least some of the Russians you talk to make the point that...which I think is a really very, very interesting one...The more general aspect is that whatever you're doing, you're communicating something to the other side. And because of the spacing that we chose for the Minuteman silos, supposedly wide enough apart that, under our belief you couldn't get "two with one" by putting a warhead in between, in their way of calculating, that wasn't true. Therefore, whatever we said, whatever we did in exercises, they always believed the Minuteman missiles were first-strike weapons [because they couldn't survive a Soviet first-strike, so the U.S. would use them first or lose them].

Q4: Have they subsequently confirmed that? Have we been able to talk to them since the Berlin Wall came down that, indeed, that was something of a root difference in perceptions?

Marshall: I don't know that anyone has probed it at the level of, you know, "Let's compare calculations" and so on, but it certainly comes out in some of John Hines' interviews with these Russians. They certainly believed that ground-based systems were much more vulnerable than we believed.

Q2: There was another area that I remember that was not quite as important, but, I think, exemplifies this whole thing, and that we got hold of some of their numbers for conventional systems. And they ranked the M60A2 tank very, very high. M60A2, that's the Shillelagh missile one. And we thought, of course, it was a dog, it wasn't worth a damn, and we stopped building them. Well, by their measures, you know, tanks that shot missiles through the tube were very, very effective. They were, I think, building on the lessons of the '73 Arab-Israeli war in terms of the vulnerability of tanks to anti-tank guided missiles. And so they went down this whole path of tanks; one series of the T-64, the T-72, the T-80—all have missile-firing main guns. We stopped it after the M60A2, because we just thought it was such a dog. Again, an asymmetric view of kind of the—in this case, more the technical than tactical characteristics of the system really led to very different kinds of weapon systems. And different views of their effectiveness on the battlefield. So when they're looking at what the correlation of forces are at a particular point they see, "Ah, M60A2s. Must be main attack" and we say, "M60A2s. National Guard. Got the dogs."

Q4: All 65 of them, or some modest number.

Q2: And then we have the Sheridans, of course, have the same kind of thing.

Q3: There was some talk or effort to get people to look at Soviet helicopters that had been shot down in Afghanistan, to see if we could get some insights into the differences in their design practices. Did that ever come through?

Marshall: I don't know. Right. OK, let's go on to item two, competitive strategies.

Q2: Can I say one more thing on the NATO-Warsaw Pact thing?

Marshall: Yeah.

Q2: I think that was the first place where this concept of the "revolution in military affairs" [RMA] was introduced in a formal assessment. Because we...

Marshall: That's probably true.

Q2: I know we introduced it in that assessment. I think it was the first assessment that introduced that notion.

Marshall: Well, of the assessments. It had first come up in connection with the CILTS stuff. But in terms of what we were including in one of the assessments, yes, that's right.

Q2: And then we also got some Marshal Nikolai Ogarkov stuff, and not just the open stuff like *Always in Defense of the Homeland* [*Always in Readiness to Defend the Fatherland,* 1982], but we had some human intelligence that was giving us stuff that, you know, really indicated that this was very, very important to the Soviets.

Marshall: Right.

Q2: And we briefed Major General John Shalikashvili, who was the Assistant DCSOPS [Deputy Chief of Staff for Operations and Plans] in the Army at the time, on this thing, as well as Major General Wilson [Dutch] Shoffner, who was his predecessor, on this particular thing, and both of them were very, very knowledgeable of that particular source of intelligence that was confirming this view that Ogarkov and the Soviet General Staff had about how there was going to be this revolution in warfare coming.

I guess the other thing I would say is that caused the net assessment to get a fairly wide distribution. Well, we did two versions. We had the codeword version, and then we did a Secret version, and the Secret version got a wider distribution.

Marshall: Right.

Q2: It was probably one of the first efforts by the office to have this RMA concept in it that then went out to people in the Department and said, "Here's something you ought to be looking at, because this is something the Soviets are taking very seriously."

Marshall: Yeah, right.

Q5: There's sort of one other tidbit about this. Earlier you and David Epstein were going off and worrying about the defense burden over there [in the Soviet Union], and it hadn't gotten connected into the balance in the way that it probably should have at that time—

Marshall: Yeah.

Q5: —if we'd been able to think the thing through a little more clearly. Did that occur further downstream?

Q2: Yes. In fact, the third section of the balance looked really at... You know, the first section was deterrence and how do the Soviets assess the balance. And the second section was, "What if deterrence fails? What do we think likely war outcomes might be?" And then a third section was, "What do we think is coming in the future?"

And here is where we introduced the RMA, the Soviet view that this RMA is coming. But also this notion that there's this increasing pessimism on their part that they're able to compete with us. They see that this RMA with the kinds of high-tech things that are associated with it, because they were really focusing on it as a technology kind of thing, like weapons based on new physical principles, and so forth. They were just very pessimistic that they were going to have the resource base, in its broadest sense [indistinct] S&T [scientific and technical] base [indistinct] that they were going to be able to compete with us. And I think that some of that is what came out of David Epstein's work on the investment balance.

Marshall: Yeah. And putting the two things together, it ended up with— And I remember raising an issue for the top-level people that, "You really ought to be thinking about how to invest in this stuff and to really take advantage of the..."

Q2: We specifically and explicitly also raised the competitive strategies notion—

Marshall: Yes, right.

Q2: —in this context. You know, looking to the future, what should we be doing? And we specifically said, "Well, we ought to think about competing with the Soviets in this competitive strategies framework, where they're having a very difficult time making the kinds of investments that they need to stay up with us, and we need to target our investments in such a way that it's just going to make it just that much more difficult for them, and get them to focus in areas that are not particularly dangerous to us, like defense instead of offense." So we were able to— Part of that is sort of the serendipity of taking many, many years and many, many authors to put this balance together.

Marshall: Right. OK, let's go on to the competitive strategies thing. As I mentioned last time, in some ways that story starts a long time ago, back at RAND.[4] That was in '69.[5] And then, in the first year of the Reagan Administration, when the time came to put together the policy guidance [Defense Guidance], Iklé asked me to look at the draft and then asked me whether there was anything additional that could be put in it. And I raised the point that there could be a section that tried to focus people's attention on the notion that in thinking about their programs they ought to be taking account of what the impact on the other guy was, and, in particular, cost-imposing strategies. So he asked me to write a section, which he put in there.[6] And then essentially nothing happened. Even though, when Paul Thayer came on board as Deputy Secretary of Defense, he looked at it, thought that was a good idea, and a query was put out to the Services about, "What have you done about this?" and various levels of nonsense came back.[7]

But then, what really happened was that in the second Reagan Administration, Weinberger asked Graham Allison [Dean, Harvard Kennedy School of Government] to come down and help him, come down a couple of days a week, I think it was, something like that. And one of the first things that Graham and I talked about was that Weinberger wanted Graham, for one thing, to write the next posture statement, and also wanted to cure this problem of these charges that Reagan Administration defense officials don't have a strategy and so on, and how should we do this. As I mentioned the last time, I'd already had a session and wrote a number of things to Frank Carlucci [Deputy Secretary of Defense] in the first year of the first Reagan term, but nothing came of that either. That work dealt with what you might do to really develop a strategy or have a strategy or what would that all mean. Because of Graham, we talked about it and one of the ideas was… the label picked was "competitive strategies." He talked to Weinberger, and even though it was something that had been really suggested before, this time he bought it, to my surprise. He bought it, at least in the sense that he was willing to see the thing kind of go along. I don't think he ever did very much one way or another about any of it. But it got started and therefore we got involved with Graham, helping him get that whole thing underway.[8] And then, both a good thing and a bad thing, they picked Dennis Kloske [Deputy Under Secretary of Defense for Planning and Resources] to be the one in charge. He was an energetic person and he tried to push it. He kind of screwed it up, but it wasn't like giving it to somebody who didn't do anything with it

He did all kinds of things with it. He was not bureaucratically very astute, and so in the end it didn't flourish very well. Although after the fact, there were people who came to me, guys who were younger and guys who were involved in it in the Services, and said, "That was a good idea. We ought to do things like that." But at the time it didn't flourish.

Q2: One thing in that regard, is that the O-5s and O-6s that were on the task forces were trying to figure what the strategies should be to make up this competitive strategy. They were really trying to make this thing work, as opposed to other efforts associated with the RMA, where people were, for bureaucratic reasons, trying to "block that kick." These guys were really trying to make the thing work. They were really analyzing things seriously and trying to figure stuff out. I've also talked to a number of people who were on competitive strategy task forces at the time that said it was one of the most broadening experiences that they ever had. Really the first chance they ever had in their military career to do some strategic thinking on something. They were sad that it just didn't go anywhere. So at one level, maybe it helped raise this notion for a group of officers that maybe now are general officers, that there was a broader, larger way of thinking about what their business really is, not just fighting the battle of the budget every week.

Q3: Can you think of any of the O-5s who are still around?

Q2: Well, they wouldn't be O-5s now.

Q3: No, I know.

Q2: One guy that just talked to me last week is Marty Spolarich, who is still up in OSD [the Office of the Secretary of Defense]. He was on the Navy Staff working competitive strategies at the time. On the Army Staff was this guy Monty Meigs. He wasn't one of the guys on the task force. He was an Army Staff guy watching all this. And he's now a general officer. He's a brigadier or major general now. And he probably will be a three-star or a four-star someday [he retired as a four-star general in 2002]. So he's at least been exposed to the "virus." And again, he was on the Army Staff, he wasn't on a task force, so he didn't get the intense exposure that people on the task forces did. And there was an Air Force guy Mike Hayden. Mike Hayden was an Air Force colonel that worked competitive strategies on the Air Staff. And Mike is now a brigadier general, J-2 in European Command, and in between there was on the National Security Council staff in the Bush White House. [Michael Hayden retired as a four-star general and served as Director of the National Security Agency and Director of the Central Intelligence Agency.]

Marshall: Oh, yeah.

Q2: We were able to use him as a vehicle to get competitive strategies introduced into a couple of the *National Security Strategy* [*of the United States*] documents, because he thought it was a good idea and stuff.[9] These are episodic things, but there are still some people that are at general officer level, that were exposed to this thing. Maybe someday they will come back to haunt everybody who was here.

Marshall: General Howell Estes, USAF, I think had something to do with it. Another aspect of the competitive strategies effort, though, was the really, again, abysmal performance of the intelligence people. They were told from the beginning, "If we're going to do this, we really need to know how the Soviets look at things, think about things, and so on." That message was repeated over and over again.

And they never seemed to get it. And then, sort of midway through the thing, I get feedback…I think there was a meeting with a bunch of them, and then it was as though, "Ah! We finally got it!" And then they did a few things…[10]

Q4: There was a briefing on arms control. Competitive strategies and arms control.[11]

Marshall: Jim Roche did that.

Q4: Barry Watts and Chip Pickett put it together and Jim gave it out in California. 24 hours later it was sitting on the desk in the Secretary of Defense's office. And as it was recounted, he sort of looked at the definition that we put up and said, "Yeah, that's it." And the study group served as a standing staff. I had a few windows into what they were doing from time to time. But my impression is they went off and said, "We're going to make our own definition," and six months later they were basically back to using the definition we had put together. We still have the charts.

Marshall: As I remember that meeting in L.A., I talked some about the general aspects of it. Then Jim had his thing. Then there was a third thing by somebody, which was one of these things where you just take the label and attach it to whatever it is you are doing.[12] So it was a whole list of things, which…

Q5: Yeah, it was the "everything we've always been doing" [indistinct] competitive strategies [indistinct].

Marshall: Yes, right. Stick a new label on it.

Q2: That was the initial kind of reaction to the Weinberger [competitive strategy] memos, when he said, "We're going to have a meeting, we're going to have this group and I'm going to chair it, and come in and tell me what you guys are doing that's associated with or even looks like this competitive strategies." The Navy gets in there and says, "Here is our competitive strategy. It's called the Maritime Strategy." And he says, "OK, fine." The Air Force stood up and said, "Here's our competitive strategy. These ten weapon systems here and technologies will allow us to compete effectively with anybody around the world, anytime anyplace, it doesn't matter who it is. Those systems are our competitive strategies." And then, Army General Max Thurman [Army Vice Chief of Staff] got up and gave a very interesting special-access-program briefing on some stuff that the Army was doing. That was closer to the mark. Obviously can't go into it. But generally the Service responses were—"We IS a competitive strategy."

Marshall: I must say that the more senior military did not cover themselves with the glory at all. General Colin Powell at some point said, "Well, we do this all the time. This is what we always do." Bullshit!

Q2: Jeff McKitrick had to go personally and brief Admiral Carlisle Trost [Chief of Naval Operations], because the Navy initially wasn't even going to play—

Marshall: That's right.

Q2: —and Jaymie Durnan took McKitrick in to see Trost, because Trost wanted to know "Who's the troublemaker down there doing this stuff?" And so Jaymie, said, "I think I should take McKitrick to see Trost." So McKitrick went up there and briefed him on what the initiative was and why and all that. And Trost said the same thing. He said, "Look, we do this all the time. This is the way we do business."

Marshall: Well, the Navy had more of a complaint because of what they were really doing. There was an area where they were really doing it.

Q2: That's right. [Indistinct]

Marshall: And the Air Force, in some people. Although I generally found that people who were by far more responsive were like Richard DeLauer [Under Secretary of Defense for Research and Engineering]. When that first guidance document came out that I mentioned, he was very interested. Guys from business saw it right away. In fact, he had some people do a quick-and-dirty analysis of, "What is the cost of the Soviet air defense system and how can we do things that really drive that up still further?"[13] Or William Perry [DeLauer's predecessor and Secretary of Defense at the time of the interview]. Guys like that just see it right away. But the military guys, on the whole, do not, despite Colin Powell's "We do this all the time" kind of thing. Maybe at a tactical level or operational level they do, but in this investment sense, it is rare, very rare. Although…

Q5: At a tactical level? I can't imagine what it would be.

Marshall: Well, you look at weaknesses. You look at exploiting weaknesses.

Q5: Yeah, you do try to put strength against weakness.

Marshall: Whereas most of the stuff that you see happen in the Building is characterized by countering new strengths that the other guy has.

Q2: Or, more typically, done irrespective of what the other guy is doing. You know, "We need a new year's supply of 'Y.'"

Q5: [Indistinct] the other ones are old. Tactically, trying to put your strengths against their weakness, in sort of a generic sense, was seen as producing tactical success. Cost imposition, in a long-term sense, was not part of that.

Marshall: Yeah.

Q2: I think typically you don't find at the tactical level this kind of series of moves. The, "I'm going to think about doing this—now, what might he do in response? So, what do I do then? What does that mean about my first move that I tried to get him in a position…" Tactically, you don't…

Q3: My understanding is that, in fact, some of the aspects of maneuver warfare do have that. You act in a way so as to force him to react, which forces him to respond in the way that you predict.

Q2: At the operational level, if you're doing it right. And certainly at the strategic level. But very seldom at the tactical level. You're looking at single engagements there, not…

Q3: Do you channelize the enemy when you force him to go to a certain way, but you know [indistinct]. The business I was thinking about [indistinct]. I thought commanders [indistinct].

Q2: Commander's Estimate.

Q3: Yeah.

Q5: I don't think they get much beyond the first step.

Q2: Yeah.

Q5: I would agree.

Q2: That's at the tactical level. At the operational level, you get some guys that are good. They will think about it as a series of moves and how they want to be positioned better at the end of the day or…

Q5: I don't know, but I can't remember any documentary or war story evidence from the Gulf War of us getting to a second round of steps and counters in anything that was done, in planning or executing that campaign.

Q2: Maybe through serendipity, with the Marines offshore. When it turned out that they couldn't land because of the mines, then we said, "Well, let's make a virtue out of a necessity and try to use them to hold forces down and then we would come in from the other direction."

Q5: But competitive strategies thinking is different than blind luck.

Marshall: Yeah, right.

Q4: [Indistinct] I wonder whether thinking second- or third-order sort of moves isn't actually fairly unusual, even in the business world. There was sort of a newsy article on learning in the *Times* about two months ago. And one of the points made was that even if you go look at experienced chess players, they don't, in fact, think "move, move, countermove, move, move, countermove," four or five iterations ahead. They have such a wealth of experience that they instantaneously can recognize the lay of the board; the pattern recognition is extraordinarily sophisticated. And their level of understanding the game is such that they can violate the early rules that you are taught in the game, because they can see a larger purpose. When the piece was written, it was really along the lines of "this is the way human thinking may be and function."

And I'm trying to think, maybe the reason why business people sort of latched onto it is not because they'd been through the move-countermove-move problem, but that because they had made so many mistakes in the past, and they'd had so much experience with how you deal with competitors, that they sort of instantly recognized the validity of that kind of approach. And the whole notion of thinking about moves and countermoves … in fact, they don't go much beyond the first iteration.

Marshall: Yeah.

Q3: I don't want to argue this, but it's an interesting point. I know some guys at Procter & Gamble. And they'd come up with this shampoo, like Pert. And what they do is they deliberately test-market it in an area where they knew that the competitors knew that Procter & Gamble test-marketed, also in an area where they knew that there was not going to be too much demand for this sort of big flop. So everybody says, "Nobody wants this. We were thinking about getting into this. We are not going to get into it." And P&G has this master plan for them going massively into it. And they just captured [?] percent of the market when they did. So at least there are a lot of anecdotes of people—

Q4: But the calculation there was to send the wrong signals to the [indistinct] competitor.

Q3: The idea that your action was a form of communication which shapes enemy behavior in ways which are favorable to your future competitive position.

Q4: Absolutely. I'm not contesting that, not at all. I was reacting to this discussion about whether we think two, three, or four moves beyond... [indistinct; multiple voices]

Q2: But in the Building, obviously, we generally have a hard time getting anybody to think one step vis-à-vis... [indistinct; multiple voices]

Q4: That example is an excellent one about just understanding about how to use the competitor against himself.

Marshall: Yes, right.

One other story about top-level people. I had a discussion with General Lew Allen [Air Force Chief of Staff, 1978–1982]. He said he really wasn't too sure about this competitive strategies concept. There were two problems, he felt. One was that we were so inconstant as a country that we could start something but not carry through on it, and that could cause the other person to react, but then we would fall off and we would end up in a worse position than we would otherwise be in. And that's a very real concern. There's no doubt about that. But the other thing was that we were up against the Soviets and they'll just spend the extra money. It was as though they had an endless supply of resources, that there was really no resource limit, no long-term consequences.

Q3: My memory is that was Weinberger's initial reaction to this. That the Soviets were essentially resource-unconstrained.

Marshall: Well, yes. That comes up actually a little later, when we'll talk about the USSR endgame, because it was around, I think, '85, '86 that we actually had a meeting and presentation for Weinberger. We'd done some things that Iklé thought he ought to hear, that came out of the Investment Balance, which was more along the line of, "They are having troubles, and these are..." Weinberger's view was that the cost to them of their military programs was very little, that they didn't have to spend a large percentage of their gross national product, because in one form or another, they had slave labor in their plants, and the amount paid to conscripts was very little. Which was true, overall true, but nonetheless represents a diversion of resources: this man is not doing anything else, he's over there in this army. So Weinberger just struck me as someone who had no understanding of economics at all.

Q2: Let me ask you what you know about actual Soviet responses to the competitive strategy initiative. A couple of articles, in journals like *Red Star,* complained about this as being another nasty move by the nasty Americans to drive the Soviet empire onto the ash heap of history, which it already was, but...

Marshall: Right, right.

Q2: One other story. There was a U.S. delegation meeting with a Soviet delegation, and it might have been a military...or it might have been military-to-military talks, I don't recall exactly now, but I know it was a Department of Defense-level talk to a Soviet, military, maybe political-military, like say a Soviet Ministry of Defense thing. And the U.S. delegation was asked by the Soviet delegation, "What about this competitive strategies thing?" And the response from the U.S. delegation was, "Oh, don't take that seriously. That's nothing for you guys to really worry about." Do you remember that?

Marshall: I guess so. I think I remember.

Q3: It was the right thing to say in public, because it drove their paranoia level higher.

Q2: It's also what the guys on the U.S. delegation believed.

Marshall: There are other things along that line. I never really looked into it, but there certainly were several articles and statements by the Russians with a whining "This is unfair" kind of tone, a "You're trying to bankrupt us! That's foul, foul!" kind of thing.[14] And then there was an astounding statement by a German at one of our meetings, where we gave a little presentation on it. It was one of the things on the agenda—and this one officer said, "Well, that could upset the balance. The U.S. shouldn't be trying to get ahead." So there was tied into this whole thing a sense of disrupting this comfortable parity situation.

I don't have anything more on that. It's a kind of interesting episode.

Q4: In a way, do you think the Soviets understood this better than our own people?

Q2: I think they understood it intuitively, because they do that kind of thing.

Marshall: I thought that by the time this happened, early '85 and into '86, there must have been a lot of awareness among these people [the Soviets] that they were really hurting. So I think that any notion that the U.S. was systematically going to do things to drive up their costs or to focus on their weaknesses probably seemed very threatening.

Q4: Not to disagree with you strongly on this, but if perhaps we had done this in 1962, we might not have provoked the same reaction out of the Soviets. In other words, they may not have understood the concept that much; what they understood was that they were, in fact, in a great deal of difficulty, and when they saw something that was so clearly going to be focused on that, it was clearly perceived as a threat.

Marshall: Well, in the late '50s, '60s, they were still feeling pretty optimistic, according to Vladimir Shlapentokh. And this kind of shift to considerable—

Q2: The question was Soviet understanding, not reaction to it. I mean, the understanding, I think, would have been the same in '62 or '85. But the response to it in the '60s…they were much more optimistic about their prospects for success and so they wouldn't have been nearly as whining about all this, they would have gone, "Yeah, that's what we thought, and now we're just going to…"

Q4: It seems to me it raises an interesting question of why our military didn't—generalization—our military had a hard time grasping it, or didn't. Our business people did. But here are the Soviets, whose business acumen is somewhat less but from a military standpoint, they grasped it immediately. There's something about the way they were training or developing their officer corps, for example, and the way we were developing ours which probably—

Marshall: I don't know. It's hard to know. It's something you'd have to talk to John Hines or somebody about.

Q2: For example, they obviously are long-time believers...or not believers, but tried to do this thing, like "reflexive control" has been part of the Soviet military effort. They've had, on the General Staff, offices devoted to deception and perception management and things like that. Clearly, when you have those kinds of organizations that's a manifestation of an inclination to look at your adversary and what your adversary is doing and how he's affected by what you are doing. You set up organizations to try to manage that process. So certainly, it indicates that they're more aware of the importance of doing those kinds of things and trying to do them than we do. We tend to think, "Well, put some loudspeakers on airplanes and fly around and talk to the population, drop some leaflets, by God, that should shape perceptions."

Marshall: Let's go on. I think the next thing we ought to talk about is this nuclear strategy effort.

Q2: Can I say one other thing on competitive strategies?

Marshall: Yes, OK.

Q2: There have been some efforts since the collapse of the Warsaw Pact and the Soviet Union, to characterize the Strategic Defense Initiative as an effort not really to deploy a strategic defense, but really a way to bankrupt the Soviets, which clearly would be a major kind of competitive strategy initiative.

Marshall: Yeah.

Q2: Did you ever get any indication from talking to Weinberger or any of those people that they saw SDI as kind of this final straw to break the Soviet back by being such a huge shift in the strategic relationship?

Marshall: I don't think so. I think a number of them did have this sense that this was something that the Soviets would have trouble matching. But I don't think it had this other aspect.[15] You know, if you were [indistinct] talking to people [indistinct] about that.

OK, nuclear strategy. Here the story is that this was something that Iklé wanted to do and, I think, right after the election started to focus on this.

The idea was the following: In the first years of the first Reagan Administration, there had been this—I forget what it was called—"road map" or "master plan"... Master Plan that had been put in place.

Q2: The NWEAMP? The Nuclear Weapons Employment and Acquisition Master Plan. That one?

Marshall: Yeah. That ran out to '95, I think. And Iklé said, "We haven't done the thinking beyond that, and we ought to start that now." It was also driven by some other things. First, there could be new technologies. There were going to be stealth bombers. The way people had thought that they [indistinct] they didn't take much account of that, and there weren't going to be that many until toward the end of that period, with the Master Plan. Secondly, as a result of the changes made in nuclear targeting as a result of the stuff done during the Carter Administration, signed off toward the end of the last year, really…I guess it was '79, wasn't it, when it was signed off? I forget.

Q2: PD-59 [Presidential Directive 59]?

Marshall: Yeah.

Q2: Yeah. '79 or '80 [25 Jul 1980].

Marshall: The intelligence effort had been reoriented. One of the results of that was that we were finding lots and lots of hardened buried structures. It was a period in the early '80s when we became far more aware of the extent of the underground construction efforts that had been undertaken. So the character of the target system that we were going after was changing. Then, in addition, there was the SDI initiative.

This Master Plan was very much in the mold of strategic nuclear planning, thinking that went back to the '50s and hadn't changed at all, really. Iklé also had always been concerned with and had some questions about the advisability of relying in perpetuity on deterrence, as people had formulated it.[16] In some sense, we were staking everything on having the probability of a large-scale war at zero, really zero, and not really dealing with the issue of, if in some way deterrence fails, what the hell are we really going to do? He was very interested in that and wanted to start this effort.

In the end, the effort was not entirely successful. [Indistinct] there was a lot of work done, but it didn't eventuate in any new, real plan that extended beyond the thing.

I can't remember all of it [the Nuclear Strategy Development Group effort], except that I was put in charge of an analysis steering group, I guess it was, and oversaw various analytic studies that were done in connection with this, although a lot of them were parceled out. One of the things that was done was a whole series of war games that, in particular, looked at the effect of potential SDI programs. If I remember, there were 14 of these things played over about two years. I thought they had some very interesting results. And as people played through them, repeatedly, various situations…

We started the games with people kind of playing Red and Blue as though they were the national command authorities on the two sides. After the first couple of games, I threw that out the window, because, in the first place, I think the kind of people who you get to play these games don't reproduce the kind of behavior you get from really top-level leaders. Secondly, they spend all of their time endlessly on stuff that they're not very good at. So the second kind of game we organized around the notion of, "Alright, you people are essentially the general staff. You are planning what war plans will be, and then you play through the things." And that turned out to be far more effective. They could play these roles and it got them out of the political bullshitting they did otherwise.

And the outcome, I thought, had taught me some things, anyway. First, what the games showed was that contrary to the kind of argumentation that was going on about SDI, the people in the game found the presence of SDI systems stabilizing, not destabilizing. The argument tended to be, "It's going to be destabilizing because one or both sides are going to see that if the other side is going to attack, then I have fewer missiles, and I can't get through, and therefore he's going to do me in, and so on." But, in fact, in the games people had just psychologically the other reaction, and it wasn't that they were stupid or not calculating. Because if both sides had these systems, the first strike, the initiating strike was not going to be as successful. And both sides also felt more secure.

One of the other things, though, of course, that happened in the games was that the players moved toward strategies in which the first thing they tried to do was degrade the other guy's SDI system, not just strike out at his homeland. And, of course, if one side is able to remove the other guy's SDI and keep his own or something like that, you have a very asymmetrical situation and the person without an SDI ought to give up.

So you had a very different kind of situation. And it also convinced me that such things as the "Nitze criteria" are nonsense.[17]

Q3: What is that?

Marshall: The cost-effectiveness kind of thing. Because the insertion of an SDI system into the overall picture changes the whole nature of the game. That's the big thing that happens…

If you continue on as though it's still the same game, you're just not getting it. And so that all of the kind of usual systems analysis— And the RAND guys ran some stuff that was nonsense, took their earlier models and ran them, which just showed me how awful most of the analysis was, that people just don't think about how things change.[18] And I tried to use analogies: OK, here's chess. And you know how to play it. But supposing the board is enlarged a little and two new pieces are put in that have different moves than any of the existing pieces. That's a new game. And the opening might be different. So playing your usual openings might not be a smart thing to do in this new game. And this was the significance of SDI systems, if they could ever get put in place, that is. That's how you ought to think about them.

In those kind of circumstances, it's one of the virtues of war games that they allow you to get at it, get some preliminary notion of how the whole nature of the game has changed and how you ought to be thinking about things like measures and what's important. If you go on with the standard systems analysis, guys who just take the old models and run them with some new things in them, you're probably going to get the wrong answers.

Q2: Was it the NSDG games that also really highlighted this notion of "preferential defense," and the kind of leverage you got off of preferential defense, or was that another series?

Marshall: No. People knew about preferential defense. That came out of the standard calculations.

Q?: [Indistinct] way back in the '60s.

Marshall: Yeah, right.

Q2: On the RSAS [RAND Strategy Assessment System], were you trying to use the RSAS to get at some of these things?

Marshall: The original notion of the RSAS came out of the Defense Science Board Summer Study of '78, where there was a look at the whole issue of the nuclear balance. And, of course, I've had the kind of heretical view that most of the analysis of the nuclear situation is pretty poor. So out of that came a recommendation that we use a much more war-gaming approach to looking at the nuclear balance, and also embed it in a wider war.[19] I think we talked about this last time. Strategic nuclear conflict tends to be treated as though it's a detached weight-lifting contest; here is the football game going on over here and then over to one side there is a weight-lifting contest that's not connected with this ongoing game. And it has a lot of other difficulties, the standard analysis. It just arbitrarily treated nuclear-powered ballistic missile submarines as having no vulnerabilities, which is not true, particularly if there were any kind of extended conventional phase, when the worst that might happen is the attrition of the submarines, particularly on one side. Might make a difference, huh? Not in any of the known calculations of the strategic balance.

Q3: You weren't always [indistinct].

Marshall: Well, wait a minute… Alright, so we… Look, that was the notion for that, though. Then the guys at RAND kind of screwed it up as far as I was concerned, and you could never drag them back to what its original purpose was. They ended up developing it as a way of analyzing regional wars.[20]

Q2: Well, even there it was screwed up.

Marshall: Yeah, we understand that. So we did not use it for… In principle, it could have been if it had ever fulfilled the earlier hopes, yes. Look, they were not good at almost anything. They didn't have the right idea of the regional conventional fight, and they never got on with what was supposedly the fundamental purpose of doing it— to really get at large-scale nuclear warfare in the context of a wider war.

Q3: I just remember this one briefing that we went to. The briefer said, "Well, regional warfare is so complicated, we can't treat everything in equal detail. We're just going to not look at the NATO northern region or the Far East, because it's totally irrelevant for the nuclear war thing." And Andy starts going… "Where the hell do you think those bastions are going to be?"

Q5: I can remember reading advertisements from RAND. At a certain point, this [RSAS] was the flagship of the RAND Corporation.

Marshall: Yeah, right. OK, this nuclear strategy thing went on until about the middle of '86, I think, so it was going on about a year and a half. Then in the middle of '86, Wohlstetter and Iklé sold the idea of this commission [Commission on Integrated Long-Term Strategy]. Originally, it was jointly sponsored by the NSC and by the Defense Department. And as the thing ended, the NSC kind of withdrew. But it was started in the middle of '86.

Q2: Was Bill Cockell on the NSC?

Marshall: Yeah, I think so. So there was the commission. [Indistinct] I forget everybody who was on it, but there were four working groups. Retired General Paul Gorman ran one on low-intensity warfare. Henry Rowen and Fred Hoffman ran one on power projection on the periphery of the Soviet Union. Charlie Herzfeld ran one on technology. And Charlie Wolf and I ran one on the future security environment, in which we tried to provide to the commission a view of what the next 20 years might look like.[21] So we formed a group of people to help us look at that. At RAND, in particular, they did long-term economic projections, which have turned out to hold up very, very well.

They made a big breakthrough. They were the first people to apply purchasing power parity to China. Otherwise, you had these people doing all these things and then this Chinese number prepared on a totally different basis [indistinct].

And at least for me, the study of the future security environment was very formative in terms of my thinking. Really, I haven't changed my views on the basic sort of things going on in the world that much since that time. I came out of that experience feeling that three big things are happening, from a defense point of view. One is this geopolitical transition, which we weren't able to articulate well, because Iklé didn't want us to point out how weak *we* thought the Soviet Union was and was likely to be. Another is the proliferation of weaponry and so on. And another is the military-technical revolution.

But the thing that is, I think, interesting to talk about in the context of this commission was that both Charlie and I were, in fact, very frustrated by… In fact, even though people seemed to agree, and it was included in the more general summary document, I came away with the feeling, and I think Charlie also, that the commission itself, because of the preoccupations of Albert Wohlstetter and to some extent Fred, but particularly Albert, was much nearer term in its focus. While the thing *was* supposed to be about the next 20 years, it was in large part about undoing wrong thinking of the past and talking about what you would do, the forces you needed now. Far more remained very focused on the Russians, didn't really deal with other things, and remained very, very focused on the Russians.

Let me tell you a story. One of the meetings of the full commission was held in the Old Executive Office Building. I remember during one of the breaks going down the hall to the men's room with General [Retired] Jack Vessey and telling him my sense about the commission and the fact that they just weren't getting it in terms of what I thought Charlie Wolf and I were trying to tell them. And he sort of said, "Well, I agree with you. The Russians have had it. It's just a matter of time." This would be the spring of '88 [before Dec 1987], I guess, about that time. I tell you the story mainly because here's Vessey, who says he agrees, but he doesn't say anything [about it to anyone else]. So you can have these situations where a lot of people really know what's happening, but it somehow isn't quite accepted openly to talk about it.

Q4: Why would Vessey have been mute on the subject? Hardly known for being bashful.

Marshall: I don't know. Again, to really say that at that point still was to say something which, as I say, some people might increasingly believe, but maybe people aren't sure enough, whatever it is, but they're not willing to say, "Look, this document ought to be different. We ought to [focus?]..."

And as I mentioned to you, talk about [indistinct]. As we were approaching putting the thing out, at one point Wolf and I made the argument that the base case for the Russian economy ought to be not the Central Intelligence Agency number, but it ought to be a number which puts instead of... The CIA number at the time gave the Soviet GNP as about 55 percent of the U.S. GNP, but we thought it's more like 30 to 35 percent, and that ought to be the base case, and there ought to be this potentially higher one if we're wrong, and then a lower one. But what we ended up doing was, essentially on Iklé's insistence, that the base case be the CIA number, which I thought was totally hokum at that point, and had for some years, and then we had, I think, two other cases that were lower.[22] So that's the way that we handled that.

The other thing where what we felt didn't quite get reflected in the final result was about AIDS, which we felt to be a major sort of thing. It hasn't turned out to be quite as big an economic problem for the United States as I thought it might be, but we felt it was going be a big thing. Iklé was concerned that the report, at least the summary report, really not say much of anything about AIDS because it's the kind of thing that the journalists would grab ahold of and wouldn't pay attention to the real central things.

But at any rate, I thought the future security environment thing was very interesting and certainly formative of my own views.

Q2: How would you rate the other three working group products?

Marshall: I thought Gorman's thing was quite good, the—

Q3: Really? I thought he went really overboard in the application of information technologies to the conduct of the counterinsurgency mission.

Marshall: Umm-hmm.

Q3: I mean, he's a very smart guy and knows a lot, but I always remember being very skeptical about that. In other words, if you had Cray supercomputers and "thinking machine" parallel processors, boy, you could have really done a job in El Salvador. I thought, well...?

Marshall: Yeah. Well, let me say of all the other three, I think he's the best. The technology one—I guess there was a lot of pulling and hauling among the people there.

Q3: People seemed to be using it to push their pet projects.

Marshall: Yeah. And the other one, the power projection report, the other kind of scenarios, was sort of OK, but nothing new.

Well, then that gets into this USSR endgame. Again, over this period I was becoming progressively more convinced that the CIA numbers were wrong; their estimate of the Soviet GNP is too big, and they're underestimating the size of the military expenditures. Dennis Ross has come back to Washington and is at the NSC. I go and have talks with him from time to time. And Dennis is telling me about the discussions he's having with people sort of comparably positioned, I guess, in Russia. I suppose some of them are in the foreign ministry, some of them are in some of the Party committees. He tells me how seemingly frank and open things are. And I keep telling him—and, of course, we're in the midst of various arms control negotiations and other things—one of my themes is, "Look Dennis, you need to think of the people you're talking about as Louis XVI's ministers. They're not going to be there very long. Time is on our side. They're going to come apart. Instead of making long lists of fallback positions in negotiations, we ought to be looking at and putting lists of we want, what are our demands as their position weakens." In fact, we had a meeting here in this room, where Dennis actually came with the idea, "Let's just generate some of this." It didn't turn out to be very successful. People couldn't get into the spirit of things, from my perspective.

Q2: So you didn't have withdrawal from Eastern Europe on that list or anything like that?

Marshall: No.

Q3: Partition of the Soviet Union?

Marshall: No, nothing interesting. We didn't have really detailed scenarios. It probably would have been a good thing to do that. We didn't do that. It was just sort of, "Look, these guys are in deep, deep trouble."

Q3: Just let me ask a point of clarification. Was Dennis, when he was talking with his Soviet counterparts...They were telling him how bad things were in the Soviet Union?

Marshall: No. It was more how open they were. Presumably they talked about some of their own internal problems, but it was more how open, how more approachable, how willing to discuss things, you know, they were.

During this period, the people who were relaying stuff about how things were coming apart and so on were still people like Shlapentokh and Jim Billington and Igor Birman, and occasionally Vladimir Bukovsky would come by and talk. There was a growing set of stories, but mainly corning still from the same people. Also about this time, Charlie Wolf had become convinced... he'd done this stuff on the "cost of empire." I forget the dates, exactly, of when that...

Q3: I think it was a little bit earlier.

Marshall: But he went through a second stage of things. The first set was done in the early '80s, but then there was some other stuff that was done.[23]

Q3: It's the '85–'88 time frame that the Anders Åslund papers started to circulate.

Marshall: Yeah, and Anders Åslund came and we talked with him.[24] He had spent this year [1987–88] at the Wilson Center. So there was a gradual increase of things. And then there was a meeting at Stanford that must have occurred at this time, of some people from the Agency and a lot of the critics about the estimates of the Soviet economy, some of which ended up in this book [*The Impoverished Superpower*].[25]

Also '86 was the time of the meeting of the four of us with President Ronald Reagan. This meeting took place the day [14 Apr 1986] that there was the strike against Muammar Qaddafi, actually. The four of us were Harry Rowen, Charlie Wolf, myself, and the guy from Duke, Vlad Treml. The purpose of the meeting was related to discussions underway about the agenda for the next meeting of the Group of 7. One of the proposals that was being put forward by one of the guys on the NSC—and he was the guy who arranged this meeting with Reagan—was that the topic should be the state of the Soviet economy and where things were going, and then we'd start off the discussion of how are we going to manage this thing.

At the meeting, Rowen talked some about the economy in general; Charlie talked about the "cost of empire" thing; I talked about the military burden and how it ought to be looked at, particularly in view of the reduced estimates that we ought to have of the GNP, the gross underestimates of the resources going into the military establishment; and then Treml talked about the combination of the "second economy," and the corruption, and the mistakes he thought were being made, and the importance of the imported technology from the West, and the distortions in the pricing system, and so on. There were a whole bunch of people at the meeting, other than Reagan himself. Vice President George Bush was there, Secretary of the Treasury James Baker, Secretary of State George Shultz wasn't there but Under Secretary of State for Economic Affairs Allen Wallis was. Some of the other people from the White House I can't remember.

Q2: JCS?

Marshall: No.

Q2: DoD? OSD? CIA?

Marshall: No.

Q3: What did the President say?

Marshall: He didn't really say very much, except he nodded at various points, and said, essentially, "I agree with you." Bush said absolutely nothing. Baker asked a question or two. The meeting was about an hour, I guess. Each of us talked about ten minutes and then there was time for some questions. The whole thing, as I said, was preliminary to a decision about whether it was going to be an agenda item for the G7. In the end, it turned out not to be that. Because the notion of this NSC guy was, "Look, if we've been making these mistakes in these estimates, we ought to see what other people's views are and get into a discussion of, 'What if they're really in all of this trouble?'"

Q3: Can you say who the NSC guy was who made the meeting [indistinct]?

Marshall: Lou Pugliaresi

Q3: He was brought on as sort of one of the senior economic analysts [indistinct].

Marshall: That's right, yeah. He's the one who arranged it, because he became aware of the stuff that I was doing, Harry and Charlie…

Q3: Lou was also tight with Dennis.

Marshall: Yeah. The last thing to say about this USSR endgame is that as this period wore on, more and more my view was, look, we've won. Really, it's like in chess.

You've come out of the middle game. You have this tremendous advantage in material, but you've got to not screw up the endgame. You still got to win. So the whole thing to really focus on is being bright about the endgame.

I'd say the last year of this period was very much time [indistinct] into the first part of the Bush Administration. One of the things I did, I wrote a memo to Secretary of Defense Frank Carlucci. Weinberger left and Carlucci came in. I would never have bothered to send it to Weinberger, but I did send it then to Carlucci, about six months before the end of the Reagan Administration. It was a memo that was aimed at influencing what the last posture statement would say. The memo said to him that there's a very substantial chance that the Soviet Union will not be able to keep up, that they will have to cut back on their military establishment, that the country may founder. We ought to start thinking about positioning ourselves about how we're going to defend the defense budget when this happens.

Q3: You're so sly, Andy. How'd you get to be so sly?

Marshall: I don't know. It's all obvious. My view about it is it's all so obvious that—

Q3: To me, in the mid-1980s, it was not so obvious that the Soviet Union was going to collapse [indistinct] serious difficulties [indistinct].

Marshall: Well, this would be in the middle of '88. Yes, I wrote in this memo…

Q3: I'm not arguing with you.

Marshall: Well, I didn't envisage the rapidity of the collapse. But by that time General Secretary Mikhail Gorbachev is really screwing everything up and things are really coming apart. Anyway, I felt that…

Q2: You saw some major shift going on.

Marshall: Oh yes. At any rate, I said that it was dangerous for the Defense Department… As I say, the memo was targeted on how they ought to be beginning to talk about things and not write a posture statement that…

Q2: Was threat-driven, Soviet-driven defense budget. What was Carlucci's reaction?

Marshall: I never heard a word.

Q2: That was his reaction.

Marshall: That's right.

Q4: And he's a businessman.

Q2: That's not clear from his experience at Sears.

Q3: Let me ask you kind of a contrarian question. I remember at least as late as '84, maybe '85, that one of the things that you and I always used to also talk about was the need to think through, "What if we really had to quadruple American defense spending?"

Marshall: Yeah, right.

Q3: Because, you know, we're thinking… I heard of specific scenarios [indistinct] but the chances of a really radical American foreign policy catastrophe were one in four, as I recall, [consensus?] figure—

Marshall: Yes.

Q3: —and that, therefore, one of the things we have to get people doing is, forget about this, "How do we justify the two-percent increase over what we've got?" kind of stuff, but what if we have to go into a 1950–51-style buildup?

Marshall: Yeah, right.

Q3: Now, clearly the two are logical alternatives: the collapse of the Soviet Union and [indistinct]. But these are two really interesting scenarios that we have to be prepared to think about. Or was one dominant and the other was not, or one was dominant in one period and the other was dominant in a later period?

Marshall: I would put it this way: In the early '80s, on the one hand, you started to get all this information about the Soviets, and they're really probably in real trouble, and they're pessimistic, and this and that, and so on. On the other hand, one thing I did have apprehension about was that the Reagan kind of buildup was too brute force and couldn't be sustained, that you probably would have been wiser—which probably was a mistake on my part—that it would have been better to have tried to have less of a kind of brute-force buildup and try to spend the money better in the competitive strategies sense. Certainly you should have some buildup, but you probably weren't going to be able to sustain this kind of brute force, just-pouring-money-at-it kind of thing. I was concerned about that.

Then I think the other thing was that I always felt that U.S. actual military preparations were never very comprehensive or well-balanced. On the upper end, we were really relying very much on deterrence and that really big wars were never going to happen. And you might get into a situation where your judgment about that would change, in which case you're going to have to undertake really big expenditure efforts. And we ought to be thinking about that, about how we're going to do that.

I think in some ways we were very lucky, very fortunate that we had this deterrent focus. And, of course, we were much richer than the Russians, richer than we generally understood at the time. But we were betting a great deal on there not really being a war. The Russians, on the other hand, although they certainly wasted a lot of money, also had far more comprehensive preparations for actually fighting than we ever did. And it's one of the sources of our underestimating it, the size of their efforts.

If you talk to the Germans now, who during much of this period I was trying to tell that the Russians and the East Germans had various serious preparations and so on, they relate with surprise what they've found as they've gotten into discussions and [gained] insight into what the East Germans were really doing. They're surprised at the level of readiness that was maintained, at the huge ammunition stockpiles compared to them; of course, they [the West Germans] had a systematic policy of only buying a few days of ammunition.

And they're surprised at the preparations for early use of nuclear weapons.

Q2: Actual plans for conducting…

Marshall: Right, for really conducting war.[26] The West Germans didn't have plans for really conducting war.

Q2: And even surprised at a technical level. The West Germans got their hands on some East German equipment, I think in the armor/anti-armor area. They were very surprised how it was much more capable, both in terms of armor protection and in terms of ballistics, than they had thought it was. Now, we for some reason always figured that the next generation of tank that we put out wasn't as good as the next generation that they put out. We always had this "bathtub of ignorance" in the armor/anti-armor arena, we were always lagging them. The Germans never ever really quite subscribed to that. And they always kind of thought, well, the Leopard and Leopard II and all that is keeping pace very well with what they're doing on the other side. And then when the collapse came and they got their hands on it, it wasn't true, it wasn't true. They actually did live fires of tank on tank and Leopard IIs were getting killed and they weren't able to kill the bad guys.

Q3: Just to come back to this USSR endgame stuff. I was trying to figure out... What seems to be really useful and unusual and smart about the way you were thinking about this was that you were really willing to entertain simultaneously outliers on both sides.

Marshall: Yes.

Q3: That the endgame might be very messy and nasty and violent and you had to not necessarily spend a lot of money, but you think very hard about that.

Marshall: Yes.

Q3: But also that it might be extremely advantageous to us, but we could still blow it if we didn't think about it. And my problem was always that I would tend to bet my previous positions, or rather, that I would bet against my fears. I would bet against the worst outcome and I'd say that's what we have to plan against. And not take the upside seriously, so I would never be in a position to exploit opportunities when they came. So that's useful,

Marshall: OK, let's talk about Goldwater-Nichols as it deals with this office. Goldwater Nichols came along in, what, the middle of '86?[27]

Q2: After the Packard Commission.

Marshall: Yeah, right. Well, there are two things. One is, I took the trouble later to try to understand why the Packard Commission had these recommendations in the commission report about net assessment.[28] According to Gorman, who was on the Commission, he and Carlucci were the ones who put that in, because they thought that a broad kind of look at where things were going would be very, very important and very useful. But what they had in mind when they tasked or suggested that the Chairman of the Joint Chiefs of Staff come forward with this net assessment... They did not wish... Since I haven't talked to Carlucci, maybe this is just Gorman's view. His view was that this was something that should happen at the beginning of the cycle of setting the budget's programs. It is something that should be prepared in a few weeks by the very top-level people, that is, by the Chairman in consultation with the theater commanders in chief, and should represent their distilled view of what the current situation is, where things were going. And then, that should be presented to the President and should assist him in deciding on the budget level that would be laid down for the next couple of years.

What happened when that was received over here [the Pentagon], was that the JCS or the Joint Staff, who had been preparing something that they called a "Risk Assessment" at the end of the process, would relabel that, rework that, and that would be the response. So it became a staff document, not the appreciation of the top people.[29]

One of the things that happened at another level was this Net Assessment Coordinating Committee was created by Iklé, to be jointly chaired by Iklé and the Director of the Joint Staff. It was to be, under the initial notion, a place where an exchange of views could take place between, say, this office and the Joint Staff representatives about what each one was planning to do, which areas they were going to be covering, and so on. It met a couple of times and didn't do anything. It got revived later for very different reasons, essentially to staff the net assessment that would be sent over to the Congress, because the Congress stipulated that it wanted net assessments. And also there were some net assessment studies that were tasked to the Department by the Congress, and this was a forum for deciding who within the Building would answer, and to review what was done.

Q2: A lot of it was driven by Tony Cordesman.[30]

Marshall: Yes, right.

Q2: ...who kept putting this requirement for net assessments in the law.

Marshall: So they've now died off—

Q2: Cordesman's back on Senator John McCain's staff.

Marshall: —and don't happen anymore, for the moment anyway.

The other thing to talk about the Goldwater-Nichols business is that I think it led, initially, to kind of delusions of grandeur or competence on the part of the Joint Staff, who thought they...were going to take over this whole area and so on. And it also, I think, was complicated in the initial stages by Admiral William Crowe's [Chairman of the Joint Chiefs of Staff] relations with Weinberger that, I think, were not very good. In fact, I remember sitting in a meeting with Crowe and talking with him about these various net assessment problems. In fact, I prepared a paper, I think, at one point for him about what I thought the state of net assessment was and the kind of things that might be usefully attempted.

But I remember him telling me that he wanted to do an assessment of his own, and he didn't want the results dictated by politicians and so on. And I don't know, on the one hand, he certainly, I think, had some doubts about the competence of the Joint Staff to really do this sort of thing. On the other hand, he allowed this Marine colonel, I forget what his...

Q2: James Sims.

Marshall: Yeah, Sims, to become overly ambitious and so on. Although it was interesting, I remember being out on the West Coast and receiving a call from Crowe. He was going to have to go up on Capitol Hill to give some sort of presentation and wanted to know if we could help. And I called back to the office and had people here put together what became his kind of overall assessment of the situation in the world. That was done overnight for him.

Q2: It's the only time I've ever seen this office go into a crash mode like you see on the Army Staff. Really, it was amazing. Scott Fisher was the Assistant Director at the time or something.

Marshall: Yeah.

Q2: And they parsed the whole thing out and gave everybody writing assignments. And we went into this crisis mode, everybody writing a section and bringing it in and David Epstein editing it and going back out again and all this other stuff. And we put the thing together in 72 hours or something like that. I think it was a Top Secret-level document. But it's the kind of thing you always do on a Service staff, never in this office. And here we were doing this for the Chairman of the Joint Chiefs of Staff. And he used it.

Marshall: Yeah, he did. Right.

Q2: It was really the basis for his testimony.[31]

Q3: I'm still kind of puzzled, because I know that Crowe thought very highly of you.

Marshall: Yes.

Q3: Remember that he brought you and Rosen out [indistinct] that East Asian balance when he was CINCPAC [Commander in Chief, Pacific Command]?

Marshall: Yes, right. Well, I don't know. I was puzzled. I would call him occasionally. I think he thought well of us, and yet I think that one of the bad aspects of the Goldwater-Nichols Act was that it tended to move the Joint Staff to seal itself off, to have these kind of ambitions to take over things, which has kind of gone away, to some extent.[32] But there was this initial, as I say, period of delusions of competence. Which is unfortunate that you have to say that in a way, but... right.

OK. Maybe the final thing to talk about is the Summer Studies. The motive for that, in a way, came out of some discussions between Jim Roche and I, and Dennis, maybe in a minor way, I can't remember, but mainly Jim and I. There was this sense that there weren't enough young people getting into this business, and there wasn't enough communication between various groups—the academics, the people here in this building, the people on the congressional staffs—and that we would try to help that through these Summer Studies, and we might get something useful out of it. So we organized, and the first one was held out in Los Angeles at RDA [Eds: a defense contractor], which kind of organized it.[33] This first Summer Study probably was in '85. I don't know that that's the right date, it could have been '84.

The focus of the first one was rather different than the later ones, as I remember it. There were, I forget, a couple of groups. One of the groups, that I think Dennis chaired, was to come up with ideas of issues, topics that deserved study, analysis, and so on. I don't know. What's your memory of that?

Q3: Dennis talked about the Middle East.

Marshall: Yeah. But there was an effort to kind of catalog a whole bunch of issues.

Q3: No, I remember. His group's task was to develop a research agenda.

Marshall: Yes, right.

Q3: And I must admit—I did the other working group—I can't for the life of me remember what that working group did, but I do remember Dennis talking only about the Middle East in his brief.

Marshall: Well, I think both of them were focused on generating agendas, in a way.

Q3: Yeah. I remember my group sort of did this tour of the horizon focusing on problems of warring regions, but beyond that I can't [indistinct].

Q2: If the original ideas was to get young people involved, who did you have involved that's now a senior-level strategic thinker?

Q3: I don't know. It was heavier on the academics, because I remember…

Marshall: There were a lot of academics, and there were some people from the RAND Graduate Institute, and then there were a few people from the Congress.

Q2: Any staffers of note?

Q3: Aaron Friedberg was there, Steve Rosen was there, Steve David was there—these are all academics, Princeton and Johns Hopkins. I think there were more, a lot of people from the RAND Graduate School because Charlie Wolf was heavily involved. And then there were a bunch of Albert's protégés from Pan Heuristics.

Marshall: Yeah.

Q3: And there were some guys from the Army. Captain John Rose. A nuclear artillery guy.

Q2: Yeah, he's a brigadier general now over in SHAPE [Supreme Headquarters Allied Powers Europe], in charge of their strategic planning.

Marshall: Yeah.

Q2: Yeah, he was [indistinct]. But I can't remember any staff guys.

Marshall: I think probably the staff people, maybe…

Q3: It was in California, it was hard for them to get there.

Marshall: Yeah, but I think we had one or two.

Q3: Oh, the guy who wound up running anti-ballistic missile stuff … He was on the Senate staff and then he came over [indistinct].

Marshall: Oh, Brian Dailey?

Q3: No, I'm sorry I can't remember his name.

Marshall: At any rate…

Q3: Yeah, there was one [indistinct].

Marshall: Yeah. And then the next one was up in Newport. We decided that's a lot better. And we had also probably the biggest set of people from the Hill.

Q3: Right. Jackie Clegg [indistinct]. That was when we did long-range, nonnuclear strikes, war planning against the Soviet Union involving long-range, nonnuclear strike capabilities and how they could be used to degrade Soviet nuclear capability. That relates to your [Marshall] point about how the nonnuclear phase affects nuclear capabilities.

Q2: Do you have Henry Sokolski at that next one?

Marshall: I think so, yeah.

Q2: Sokolski, and then there's this other guy who was a Senator Robert Dole staffer.

Marshall: Yes. I forget his name.

Q2: He wasn't very helpful. He just wanted to go run around Newport.

Marshall: And Janne Nolan was there?

Q2: Not Janne Nolan. The one who went to Harvard to work with Samuel Huntington. Kim Nolan. Kimberly Nolan.

Marshall: Yeah, OK. Then, we didn't have one in '87, because of the meeting in San Diego connected with the CILTS thing. And we've had one every year since then.

Q2: Well, if the objective is to get young people engaged in this so they're going to be set on the right path, so to speak…

Marshall: That's kind of drifted away from that. We're not doing that so much, I'd say. It has become more, "OK Here's a topic. Can we get some really good work done on it?" as against the emphasis on getting people to mingle. Now, we still would like to have gotten more people from the Hill, but since we're going to the Naval War College, this first week or so in August fits the best with their schedule and so on, and that tends to preclude many people from the Hill, because they haven't adjourned by that time.

Q2: But you know, given what happened in November ['94 elections], on your next one, you might really try much harder to get some young people that haven't been up there before and don't know what's going on. And maybe you ought to think about, unlike the last couple of these things, really getting some Hill staff up there that you can start shaping their minds…

Q3: [Indistinct] I don't know exactly how you would do it, but I was talking to Bill Kristol. Right now, one of the things he's doing is he's placing people in House staff positions. He says that these guys, in 90 percent of the cases, have just no foreign policy, no defense policy background, nothing. So it's a chance, an opportunity.

Q4: Something a little tinier than Newport, maybe it could be in the Washington area [indistinct], if you really wanted to try give them a sense [indistinct], hold it in Baltimore [indistinct].

Q3: Right. But right now, I mean, there's just… A lot of them are just trying to figure out, where's my desk? How do I make the phones work? And they're almost exclusively concentrating on the "Contract with America."

Did you want to talk about the bilats [bilateral talks] at all? Any of those things worth raising? Any of the…

Marshall: I don't think so, especially. About the only thing that's new during this period is we start the arrangements with the Japanese in the last year [1988]. That had come about because of Richard Armitage [Assistant Secretary of Defense for International Security Affairs], when he more or less replaced Iklé in the last year of the period. He asked me to go and talk with Nishihiro, who was the second guy in the Japanese Defense Agency, to talk to him about the possibility of some discussion with them. And then Nishihiro wanted to create something like a net assessment activity in the Japanese establishment. So we started to do that that fall, the fall of '88. I think I went there initially in August, then we went back in October, November, something like that.

Q2: We had that one little brief thing with the Brits, where they wanted to start a net assessment activity.

Marshall: Well, we've had a couple of those, three or four of those, with the Brits. But there's nothing particularly notable about those.

Outline for 25 January 1995 Talk

Reagan second term

1. Underground command centers/costing of, long history of, mid '40s—> ~1–2% of GNP/part of cost of military power/military burden
2. Earner/strategic planning activity—maybe mid-'50s/papers in office for Guthe to look at
   — Major theme/part of long-term competition focus of Marshall's intellectual life

1985–1988[34]

Marshall: There are two aspects of the Reagan second term I thought it would be important to go back and pick up. Maybe the first thing to do would be to do that, just to add to the earlier discussion.

As I thought about it, there were two things that we didn't really talk about. One was that in the middle of the Reagan period, in part because Bill Earner was here in the office, we made another effort in this strategic planning notion that James Roche and I had first worked on in the middle '70s. A number of papers were written and some specific areas, particularly tactical air, were discussed. Dick DeLauer [Under Secretary of Defense for Research and Engineering], who was around at the time—so this may merge into the last part of the first Reagan term—was very receptive and we spent a lot of time talking with him.

And there are a whole bunch of papers here. There are at least three bound volumes of various kind of papers that were written. It also was the background, in part, for sponsoring research on cost-imposing strategies,[35] some continuing work in that, and things that I think I had Jim Martin do on fleshing out this notion of long-term competition and strategies for that.[36] So there was a whole set of things that went on that need to be covered.

The second thing that I had forgotten to mention was that one of the things that came out of the shift in the intelligence community's focus as a result of PD-59 [Presidential Directive 59] was this discovery, which I think was mentioned last time, of the various protection constructions for the Soviet elite and so on, but in particular, the underground command centers that were discovered.[37] Also discovered was their very long history, and the Soviet efforts at concealment and, in fact, active deception that went along with them.[38] And one of the things that we had done, only partially successfully, was a project that was undertaken jointly for us by the people in intelligence and at the Defense Nuclear Agency to try to cost this major effort [Soviet underground facilities]. As I remember, it came out to be that from the mid'40s through the mid-'80s or so, the thing had run at one or two percent of gross national product, just for that—probably. These are kind of shaky numbers, but…

Q2: Is that the real GNP or the "pseudo" GNP?

Marshall: No, I think the real GNP.

Q2: So this is one or two percent of the overall defense burden, which you thought was between 25 and 33 percent of GNP.

Marshall: Yeah, right. Or 35 percent of GNP.

Q2: So it was one to two percent out of that.

Marshall: Yeah, right. So it was an enormous effort.

Q3: Was that incorporated into the formal, published estimates of Soviet defense burdens?

Marshall: No. You see, the Central Intelligence Agency people, I think, never did these things or didn't cost them and there was no… I don't think it's the kind of thing they ever really reflected very well in their estimates.

Q3: So they didn't accept it, even when…

Marshall: I don't know. At that point I didn't care.

Q8: They probably didn't care either.

Endnotes

1 Eds: This chapter includes two sections. The first section covers the period from 1985 to 1988 and is an interview conducted on 16 December 1994. The second section also covers the period from 1985 to 1988, and is excerpted from an interview conducted on 25 January 1995. The first outline was prepared for the first interview while the second outline is an excerpt of the relevant sections from the outline prepared for the second interview.

2 Eds: Interview conducted on 16 December 1994.

3 Andrew W. Marshall and Jeffrey S. McKitrick, testimony in "Stenographic Transcript of Hearings before the Committee on Armed Services, United States Senate, Hearings to Receive Testimony on NATO-Warsaw Pact Military Balance (U)," 4 Feb 1988, SECRET.

4 Andrew W. Marshall in House Armed Services Committee, *Competitive Strategies,* 101st Cong., 1st sess. (Washington: GPO, 1989), 2–3, 5–8.

5 A.W. Marshall, *Long-Term Competition with the Soviets: A Framework for Strategic Analysis (U),* R-862-PR (Santa Monica: RAND Corp., Apr 1972), SECRET. See also A.W. Marshall, *Long Term Competition with the Soviets: A Framework for Strategic Analysis,* IN-21542-PR (Santa Monica: RAND Corp., 15 Apr 1971) (CONFIDENTIAL; Declassified 7 May 1980). Although published in 1972, R-862-PR "was actually written in the period from the summer of 1969 through the end of 1970."

6 Memo, Caspar W. Weinberger to Secretaries of the Military Departments, Chairman of the Joint Chiefs of Staff, Under Secretaries of Defense, Assistant Secretaries of Defense, General Counsel, Assistants to the Secretary of Defense, Directors of Defense Agencies, Commanders-in-Chief of Unified and Specified Commands, subj: FY 1983-87 Defense Guidance (U), 10 Jun 1981, with attachment, "Competing with the Soviet Union in Peacetime (U)," SECRET, 13. See also Memo, Caspar W. Weinberger to Secretaries of the Military Departments, Chairman of the Joint Chiefs of Staff, Under Secretaries of Defense, Assistant Secretaries of Defense, General Counsel, Assistants to the Secretary of Defense, Directors of Defense Agencies, Commanders-in-Chief of Unified and Specified Commands, subj: FY 1984-88 Defense Guidance (U), 22 Mar 1982, with attachments, "D. U.S. Strengths and Opportunities (U)," 5; "F. Competing with the Soviet Union in Peacetime (U)," SECRET, 24–26; and Richard Halloran, "Pentagon Draws Up First Strategy For Fighting a Long Nuclear War," *New York Times,* 30 May 1982, 1.

7 Memo, Paul Thayer to Secretaries of the Military Departments; Chairman, Joint Chiefs of Staff, subj: U.S. Weapons Development Policy (U), 10 May 1983, CONFIDENTIAL; Memo, John Lehman to Deputy Secretary of Defense, subj: U.S. Weapons Development Policy (U) - Info Memorandum, 18 Jul 1983, SECRET; Memo, Verne Orr to Deputy Secretary of Defense, subj: U.S. Weapons Development Policy (U) - Information Memorandum, 20 Jul 1983, SECRET.

8 Prepared statement of Graham Allison, in *Competitive Strategies,* 15; Caspar W. Weinberger, *Department of Defense Annual Report to the Congress for Fiscal Year 1987* (Washington: GPO, 1986), 85–88. For a history of the competitive strategies effort within the Department of Defense, see David J. Andre, *New Competitive Strategies Tools and Methodologies, Vol. I: Review of the Department of Defense Competitive Strategies Initiative, 1986-1990,* SAIC-90/1506 (McLean: Science Applications International Corp., 30 Nov 1990).

9 *National Security Strategy of the United States* (Washington: GPO, Jan 1987), 20; *National Security Strategy of the United States* (Washington: GPO, Jan 1988), 39; *National Security Strategy of the United States* (Washington: GPO, 90), 241.

10 On intelligence support for competitive strategies, see *New Competitive Strategies Tools and Methodologies, Vol. I: Review of the Department of Defense Competitive Strategies Initiative, 1986-1990,* 73–76, 131.

11 James G. Roche, "Competitive Strategies and Arms Control" (briefing slides), presentation at the American Defense Preparedness Association Conference on Competitive Strategies—Impact of New Technologies, 17 Feb 1988.

12 The third speaker at the meeting in California, Alan B. Goldstayn (Director, Plans and Programs, DCS/Science and Technology, HQ Air Force Systems Command), gave a talk titled "New High Leverage Technology Policy" (information from symposium brochure).

13 Assistant Chief of Staff, Studies & Analysis/USAF, subj: The Cost to the USSR to Defense Against US Low Altitude Penetrators and Cruise Missiles, Time - 1992 (U), 16 Jun 1982, SECRET when with attachment, "Memo, John M. Porter to Under Secretary of Defense for Research and Engineering, [no subject], 1 Jun 1982, SECRET, with briefing slides, AOC/OSD Electronic Warfare Technical Study Committee, "The Cost to U.S.S.R. to Defense Against U.S. Low Altitude Penetrators and Cruise Missiles, Time - 1992 (U)," 17 May 1982, SECRET.

14 See, for example, Col. V. Alekseyev, "The Pentagon's 'Strategies of Rivalry'," *Krasnaya zvezda,* 4 May 1988, 3, in Foreign Broadcast Information Service, *Daily Report: Soviet Union,* FB-JS-SOV-88-091, 11 May 1988, 1–2; Lt. Gen. V. Pavlov, "And Now 'Competitive Strategies…'," *Krasnaya zvezda,* 30 Nov 1988, 5, in FBIS-SOV-88-232, 2 Dec 1988, 1–2; Lt. Gen. Viktor Starodubov, "From Differing Standpoints," *Pravda,* 27 Jan 1989, 7, in FBIS-SOV-89-020, 1 Feb 1989, 4–5; Maj. Gen. Yu. Lebedev and A. Podberezkin, "What Lies Behind the 'Strategies of Competition' Concept?" *Krasnaya zvezda,* 14 Mar 1989, 3, in Joint Publications Research Service, *Soviet Union: Military Affairs,* JPRS-UMA-89-013, 26 May 1989, 62–63; and Col. N. Ruzayev, "U.S. Military-Political 'Competitive' Strategy," *Zarubezhnoye voyennoye obozreniye,* no. 8 (Aug 1989): 3–8, in Joint Publications Research Service, *Soviet Union: Foreign Military Review,* JPRS-UFM-90-001, 18 Jan 1990, 1–4. According to Dennis Kloske, when Secretary of Defense Frank Carlucci "visited the Soviet Union in August of [1988], [Chief of the General Staff Sergei Akhromeyev and [Minister of Defense Dmitry] Yazov pointed out to him, [on] top of all the modernization carried out in NATO the introduction of competitive strategies was, quote, 'unfair'." *Competitive Strategies,* 61.

15 Robert McFarlane claims in his memoirs that during his tenure as President Reagan's Assistant for National Security Affairs he viewed the SDI "as a bargaining chip—the President never would have conceded that, he really meant to build this shield—but that was its primary value at the time, a means of imposing a fearsome financial burden on the Soviets and leveraging their behavior at the bargaining table. And this is precisely the effect that SDI had." Robert McFarlane and Zofia Smardz, *Special Trust* (New York: Cadell & Davies, 1994), 234.

16 Fred Charles Iklé, "Can Nuclear Deterrence Last Out the Century?" *Foreign Affairs* 51, no. 2 (Jan 1973): 267–285.

17 "The criteria by which we will judge the feasibility of [defensive] technologies will be demanding. The technologies must produce defensive systems that are survivable; if not, the defenses would themselves be tempting targets for a first strike. They would decrease rather than enhance stability." "New defensive technologies must also be cost-effective at the margin—that is, they must be cheap enough to add additional defensive capability so that the other side has no incentive to add additional offensive capability to overcome the defense. If this criterion is not met, the defensive systems could encourage a proliferation of countermeasures and additional offensive weapons to overcome deployed defenses instead of a redirection of effort from offense to defense." Paul H. Nitze, Address before the World Affairs Council in Philadelphia, 20 Feb 1985, in *Department of State Bulletin* 85, no. 2097 (Apr 1985): 28. These criteria later were incorporated in Presidential guidance on the SDI program. See National Security Decision Directive 172, subj: Presenting the Strategic Defense Initiative, 30 May 1985 (SECRET; Sanitized 27 Nov 1992) NSDD 172, National Security Decision Directives, Box 2, Records of the National Security Council, Record Group 273, National Archives.

18 Glenn A. Kent and Randall J. DeValk, *Strategic Defenses and the Transition to Assured Survival,* R-3369-AF (Santa Monica: RAND Corp., Oct 1986); Dean Wilkening and Kenneth Watman, *Strategic Defense and First-Strike Stability,* R-3412-FF/RC (Santa Monica: RAND Corp., Nov 1986); Dean Wilkening, Kenneth Watman, Michael Kennedy, and Richard Darilek, *Strategic Defenses and Crisis Stability,* N-2511-AF (Santa Monica: RAND Corp., Apr 1989); Glenn A. Kent and David E. Thaler, *First-Strike Stability and Strategic Defenses: Part II of a Methodology for Evaluating Strategic Forces,* R-3918-AF (Santa Monica: RAND Corp., Oct 1990).

19 A.W. Marshall, "A Program to Improve Analytic Methods related to Strategic Forces," *Policy Sciences* 15, no. 1 (Nov 1982): 47–50.

20 For a discussion of how the RSAS has been used, see Bruce Bennett, Daniel Fox, Carl Jones, Lawrence S. Painter, John Schrader, Robert Weissler, and Bruce Wilson, *RSAS 4.6 Summary,* N-3534-NA (Santa Monica: RAND Corp., 1992), 13–17.

21 *Discriminate Deterrence,* Report of the Commission on Integrated Long-Term Strategy (Fred C. Iklé and Albert Wohlstetter, Co-Chairmen) (Washington: GPO, Jan 1988); *Commitment to Freedom: Security Assistance as a U.S. Policy Instrument in the Third World,* Paper by the Regional Conflict Working Group (Paul F. Gorman, Chairman) (Washington: Department of Defense, May 1988); *Supporting U.S. Strategy for Third World Conflict,* Report by the Regional Conflict Working Group (Washington: Department of Defense, Jun 1988); *The Future of Containment: America's Options for Defending Its Interest on the Soviet Periphery,* Report by the Offense-Defense Working Group (Fred S. Hoffman and Henry S. Rowen, Chairmen) (Washington: Department of Defense, Oct 1988) ; *Extended-Range Smart Conventional Weapon Systems,* Paper by the Standoff Weapons Panel, Offense-Defense Working Group (Washington: Department of Defense, Oct 1988); *Technology for National Security,* Report by the Working Group on Technology (Charles Herzfeld, Chairman) (Washington: Department of Defense, Oct 1988); *Recommended Changes in U.S. Military Space Policies and Programs,* Paper by the Working Group on Technology (Washington: Department of Defense, Oct 1988); *Sources of Change in the Future Security Environment,* Paper by the Future Security Environment Working Croup (Andrew W. Marshall and Charles Wolf, Chairmen) (Washington: Department of Defense, Apr 1988); *The Future Security Environment,* Report of the Future Security Environment Working Group (Washington: Department of Defense, Oct 1988).

22 *The Future Security Environment,* 6–7, 92–95.

23 Charles Wolf, Jr., K.C. Yeh, Edmund Brunner, Jr., Aaron Gunvitz, and Marilee Lawrence, *The Costs of the Soviet Empire,* R-3073/1-NA (Santa Monica: RAND Corp., Sep 1983); Charles Wolf, Jr., Keith Crane, K.C. Yeh, Susan Anderson, and Ed Brunner, *The Costs and Benefits of the Soviet Empire, 1981-1983,* R-3419-NA (Santa Monica: RAND Corp., Aug 1986); Charles Wolf, Jr., "The Costs and Benefits of the Soviet Empire," in Henry S. Rowen and Charles Wolf, Jr., eds., *The Future of the Soviet Empire* (New York: St. Martin's Press, 1987), 121–140.

24 Anders Åslund, "How Small Is Soviet National Income?" in Henry S. Rowen and Charles Wolf, Jr., eds., *The Impoverished Superpower: Perestroika and the Soviet Military Burden* (San Francisco: Institute for Contemporary Studies, 1990), 13–61; and Anders Åslund, *Gorbachev's Struggle for Economic Reform: The Soviet Reform Process, 1985-1988* (Ithaca: Cornell University Press, 1989).

25 *The Impoverished Superpower: Perestroika and the Soviet Military Burden* ("This book developed from a conference held at the Hoover Institution on March 22–24, 1988, on the role of the defense sector in the Soviet economy," v.).

26 Ministry of Defense, Federal Republic of Germany, *Military Planning of the Warsaw Pact in Central Europe: A Study,* Feb 1992, translated by Mark Kramer and reprinted in Cold War International History Project *Bulletin,* no. *2* (Fall 1992): 1, 13–19; Harald Nielsen, "The East German Armed Forces in Warsaw Pact Nuclear Operations," prepared for Sandia National Laboratories, Jul 1993; Harald Nielsen, "Employment and Handling of Nuclear Weapons in the Warsaw Pact," prepared for Sandia National Laboratories, Jun 1994.

27 Goldwater-Nichols Department of Defense Reorganization Act of 1986, Public Law 99-433, 1 Oct 1986, Title II, Part A, Section 201.

28 President's Blue Ribbon Commission on Defense Management, *A Quest for Excellence,* Final Report to the President (Washington: GPO, Jun 1986), xix, 12–13, 15, 18–20; *Appendix,* 5, 7. On the implementation of the Commission's recommendation regarding net assessment, see Office of the White House Press Secretary, Fact Sheet: Summary of a Directive Implementing the Recommendations of the Blue Ribbon Commission on Defense Management, 2 Apr 1986, 1–2, NSDD 219, National Security Decision Directives, Box 2, Records of the National Security Council, Record Group 273, National Archives; National Security Decision Directive 219, subj: Implementation of the Recommendations of the President's Commission

on Defense Management (U), Memo, Fred C. Iklé to Director, Joint Staff, OJCS, subj: Implementation of Recommendations of the President's Commission on Defense Management, 15 May 1987, with attachment, Memo, William H. Taft TV to Under Secretary of Defense for Policy and Director, Joint Staff, subj: Implementation of Recommendations of the President's Commission on Defense Management, 4 May 1987; Memo, Vice Adm. P.F. Carter, Jr., USN to [?], subj: Implementation of Recommendations of the President's Commission on Defense Management, 29 May 1987, DJSM-1075-87.

29 See prepared statement of Adm. William Crowe, in Senate Armed Services Committee, *Department of Defense Authorization for Appropriations for Fiscal Year 1989,* Part 1, 100th Cong., 2nd sess. (Washington: GPO, 1988), 40–59 (18 Feb 1988). This statement includes both an unclassified version of the "first in-depth net assessment" by the Joint Staff (43–54) and a "risk assessment" (57).

30 Memo, Anthony H. Cordesman to Sen. John McCain, subj: Developing an Adequate Net Assessment for Use by the Congress, 24 Feb 1988; Letter, Sen. John McCain to Frank Carlucci, 16 Feb 1988.

31 In early 1987, a few months after the passage of the Goldwater-Nichols Department of Defense Reorganization Act, which made the Chairman of the Joint Chiefs of Staff responsible for performing net assessments, Admiral Crowe, in testifying before a number of congressional committees, submitted an unclassified prepared statement that included his net assessment of the East-West balance in six categories: "Strategic Forces," "Maritime," "NATO/Warsaw Pact," "East Asia," "Power Projection," and "Special Operations Forces." He characterized this analysis as "my own personal assessment" rather than that of the Joint Staff. This statement was submitted first to the Senate Budget Committee on 13 Jan 1987. See Senate Budget Committee, *Concurrent Resolution on the Budget for Fiscal Year 1988,* Vol. II, 100th Cong., 1st sess. (Washington: GPO, 1987), 145–151. The statement also can be found in House Budget Committee, *Review of Budget Proposals for Fiscal Year 1988,* Part 1, 100th Cong., 1st sess. (Washington: GPO, 1987), 151–157 (21 Jan 1987); Senate Armed Services Committee, *National Security Strategy,* 100th Cong., 1st sess. (Washington: GPO, 1987), 402–405 (21 Jan 87); House Appropriations Committee, *Department of Defense Appropriations for 1988,* Part 1, 100th Cong., 1st sess. (Washington: GPO, 1987), 61–64 (4 Feb 1987); House Armed Services Committee, *Defense Department Authorization And Oversight,* 100th Cong., 1st sess. (Washington: GPO, 1987), 46–51 (24 Mar 1987); and Senate Appropriations Committee, *Department of Defense Appropriations for Fiscal Year 1988,* Part 4, 100th Cong., 1st sess. (Washington: GPO, 1988), 353–357 (19 May 1987).

32 Memo, Lt. Gen. Robert W. RisCassi to Assistant Secretary of Defense (International Security Affairs), subj: Release of JCS/Joint Staff Planning Data, 12 Jul 1988, DJSM-1758-88.

33 R&D Associates, *Summer Study on Strategic Policy (U),* RDA-TR-132903-001 (Marina Del Rey: R&D Associates, Jan 1986), SECRET.

34 Excerpted from an interview conducted 25 January 1995. The remainder of the interview covered the George H.W. Bush and William Clinton Administrations and can be found in Chapter 7.

35 *Enhancement of the Efficiency of U.S. Military Programs Through the Application of "Cost- Imposing" Strategies (U),* DSA Report 485 (Arlington: Decision-Science Applications, 24 Jun 1983), SECRET (executive summary only); *Cruise missile Cost-Imposing Effectiveness: A Case Study of the Impact of the U.S. Air Launched Cruise Missile (ALCM) on Soviet Air Defense Costs (U)* (Arlington: Decision-Science Applications, date?), SECRET; G.E. Pugh and R.R. Hubbard, *Cost Imposing Effectiveness of an Attack RPV: A Case Study of a Remotely Piloted Airframe for Ground Combat Support,* Partial Draft (includes summary of findings and conclusions only), DSA Report 521 (Arlington: Decision-Science Applications, Jan 1985).

36 J.J. Martin et al., *The U.S.-Soviet Long-Term Competition, Vol. I: Concepts; Vol. II: Planning and Analysis; Vol. III: Appendices* (San Diego: Science Applications International Corp., 5 Jun 1990).

37 Desmond Ball and Robert C. Toth, "Revising the SIOP: Taking War-Fighting to Dangerous Extremes," *International Security* 14, no. 4 (Spring 1990): 73–75; Department of Defense, *Soviet Military Power: An Assessment of the Threat, 1988* (Washington: GPO, 1988), 59–62.

38 Kenneth E. deGraffenreid, "The Art of Deception," *Global Affairs* 4, no. 4 (Fall 1989): 168.

7 The Bush and Clinton Administrations: 1989–1993[1]

Introduction

The election of George H.W. Bush was viewed by many as signaling an extension of the national security policy of Ronald Reagan, under whom Bush served as vice president. But Bush was immediately confronted with Chinese repression of civil unrest, exemplified by the use of military force against civilians at Tiananmen Square in June 1989. In addition to condemning the response of the Chinese leadership, the United States and the European Union imposed an embargo on arms sales to China, and the subsequent growth of political and economic ties with China slowed.

The United States faced a new problem when Saddam Hussein's Iraq invaded Kuwait in August 1990. Bush quickly assembled an international coalition of forces to expel the Iraqi forces in Operation Desert Storm in 1991, but the first Gulf War led to the years-long deployment of U.S. forces to the region.

At the same time, Bush had to deal with the ongoing collapse of the Warsaw Pact. Free elections were held in Hungary in March 1990, and in Czechoslovakia in June. East and West Germany reunified in October 1990, and Lech Walesa was elected president of Poland in December. The dissolution of the Soviet Union continued, with the final transfer of power from Mikhail Gorbachev to Boris Yeltsin in December 1991. Yeltsin busied himself with trying to shift to a free-enterprise economy and a democratic government—both efforts were doomed to failure.

The START I treaty with the Soviet Union, reducing strategic nuclear weapons, had been signed in July 1991. It was followed shortly by Bush's Presidential Nuclear Initiative in September, which called for unilaterally withdrawing tactical nuclear weapons from overseas American forces and U.S. Navy ships. Gorbachev, in one of his last official acts, made a similar pledge of tactical nuclear force reductions in the Soviet Armed Forces, a pledge Yeltsin reaffirmed when he came to power.

William Clinton came to the Presidency with little experience in foreign affairs. He inherited a humanitarian mission in Somalia from the Bush Administration that quickly turned sour as American soldiers were dragged through the streets of Mogadishu, the Somali capital. The resulting outcry from the American people led to the resignation of Defense Secretary Les Aspin.

Clinton sought to promote a policy of "enlargement" designed to work through international organizations as much as possible and avoid American unilateralism.[2]

Poland, Hungary, and the Czech Republic joined NATO and Russia was persuaded to withdraw its forces from Estonia and Latvia.

Despite his desire to avoid U.S. unilateralism, in 1996 Clinton deployed two aircraft carriers to the vicinity of Taiwan in response to the People's Republic of China's attempt to intimidate Taiwan by shooting missiles into the waters around the island.

During Clinton's presidency he was forced to spend a great deal of time addressing problems in the Balkans. The break-up of Yugoslavia in the early 1990s created six separate countries and produced regional instability. Serbian atrocities against Bosnian civilians eventually led to NATO bombing of Serb positions, a situation that only escalated over the years until Clinton sent in 20,000 American troops as part of a NATO deployment. A US-led operation in 1999—Operation Allied Force—eventually brought relative peace to the region.

Outline for 25 January 1995 Talk

[Note: Italicized items were not discussed during the interview.]

Bush/Clinton
1. Revolution in military affairs work begins about '90/'91
2. Shift in focus of office
3. Endgame
4. Monitoring USSR Chapter 11, especially for Cheney
5. Wolfowitz success in managing Office of the Under Secretary of Defense for Policy—Bridge Game effort
6. Desert Storm
   - — Summer Study antecedent/New problems in deterrence
   - — French story
   - — Early games/Mark Herman
   - — Panels for Paul Wolfowitz/Harry Rowen
   - — War termination effort/Fred Iklé
     - *C3I modeling/Mark Herman*

Bush/Clinton other themes
1. Gaskin/Korean balance [saga?]
   - — Marshall's view
     - Good on our problems (inspector general)
     - Weak on North/Intelligence problem
   - — Suppression/Carl Ford
     - Political hot potato
     - Consequences
2. Office focus
   - — Northern Flank *([______?]/Air Force officer)*
   - — Power projection
   - — Korea
   - — NATO-Warsaw Pact/wrap-up
   - — Military-technical revolution
   - — Soviet Union Chapter 11
3. *Office problems*
   - — *Gaskin—>other Air Force officer ([______?])/[Ceil?]/fill in/Jo Ann Koerner Xmas '90*

Jeff McKitrick with Quayle
1. Andy Krepinevich replaces to take up NATO-Warsaw Pact Balance (mid-'89)
   - — One assessment
   - — Start new modeling/meeting engagement (Information war aspects)
2. Andy Krepinevich then shifted to military-technical revolution effort in early '90—>Jul '92 assessment

Cheney/Wolfowitz
1. Soviet Union Chapter 11/Scooter Libby/*State guy*
2. Early meeting—Jim Roche, Paul Wolfowitz, *Steve Hadley*
3. New focus of competitive strategies. Gouré/strategic competence (failure)
4. Bridge Club effort
   - — Good
   - — Nuclear failure (Marshall's view/no progress since '50s)

1989–1993

Marshall: Let me begin. The idea is to talk about the Bush and Clinton periods, and talk mainly about the Bush period.

The first thing is to say that the Bush period was very favorable for the office, principally because of Paul Wolfowitz [Under Secretary of Defense for Policy]. In the first place, Wolfowitz was the first person to have that job who got to choose his own deputies. And it really ran as a team. Also, my own past association with him, and everything, just made everything really work really very, very well. Indeed, when he was offered the thing, he called me up to ask whether he should take the job. Well, he talked to Jim Roche and me, and, actually, I suppose, other people. Jim and I had some meetings with him prior to him actually coming on board to tell him what our view of the world was and so on. A lot of it came out of the Future Security Environment stuff that had been finished shortly before that.[3] And then, also, Paul had Roche prepare for him a plan for the organization and function of the whole thing.

Another thing that happened was this issue of what to do about the competitive strategies effort. A group of people was put together to look at that. A proposal was made that Daniel Gouré was to head the thing up. Some good ideas were put forward as to how it ought to be refocused, but it fundamentally failed, I think. I didn't pay enough attention to know exactly why, but it never quite made much progress on some of the things that were proposed to it.

Q2: I think there were a couple of things involved. One was that I think Dan really turned out not really to be the right guy for the thing.

Marshall: Yeah.

Q2: It's not clear anybody could have done it at that time. But Gouré is really a much more policy-oriented kind of guy, than trying to think more long term. The other thing is, at that point in time, it was very, very difficult, because you had the whole transition with the Soviet Union, etc., and so, what did you do about all that, and how did you change the competitive strategy initiative if your main competitor was evolving or changing or maybe, as some said, going away. So Gouré had this sort of competitive strategy portfolio with no competitor. So there were these meetings…

Marshall: It took a long time to get agreement that, in fact, the thing would continue, and kind of a rump thing was kept in existence.

Q2: Right. And there were these sessions that eventually made the recommendation to Secretary of Defense Dick Cheney—

Marshall: Right, right. We had a session with Cheney.

Q2: —to keep the thing going. But there was this off-line, ad hoc group, that I guess you were the nominal co-chair of with Jim Roche, with Michael Duval…

Marshall: Yeah, Duval and Graham Allison.

Q2: And initially Lieutenant General [Air Force] Lee Butler came, but then he ended up sending a brigadier in his place.

Marshall: Yeah.

Q2: Butler was the J-5 [Director for Strategic Plans and Policy, Joint Staff] at the time. And [William] Cockell was supposed to be involved in the thing, but he never could quite make it out from the West Coast.

Marshall: Right.

Q2: But the notion was, "How do you change all this, or should you even keep it going?" So the recommendation went to Cheney to keep it going.

Marshall: And to refocus it.

Q2: Refocus the effort, etc. But then, I think that nobody ever got a new focus that was sufficiently compelling to keep the initiative going the way that it had been when you had this real peer competitor out there like the Soviet Union to focus on. So, it got caught up in the circumstances, never really got a new raison d'être that allowed it to keep going, although we did keep putting the words in the *National Security Strategy of the United States* report,[4] that the National Security Council and the President put together, and then put it in the Defense Department annual report, and those kinds of things. But there really wasn't much there. We couldn't really figure out what to do with it. And then I think, eventually, it devolved in this reconstitution thing. Being able to reconstitute our forces faster than potential emerging competitors, which I think really wasn't a very useful way to think about it. The mobilization guys in the Department, seemed to me, got their hands on it and it just kind of withered away into nothingness.

Marshall: Right. During the Bush years, we had an abortive effort to relook at the Northern Flank. We continued to do some looking at power projection. We did a new assessment of Korea that I'll come back to, the Robert Gaskin thing. Andy Krepinevich replaced Jeff McKitrick, with the initial idea that he was going to continue the Warsaw Pact thing, but within a year, there was a wrap-up effort that was undertaken to understand what had happened to the balance as the result of the change of status of the Soviet forces in Eastern Europe. We also did work related to the beginning of the more evident breaking up or coming down of the Soviet Union, the Soviet Union "Chapter 11." And then there was the start-up of this effort on the MTR [military-technical revolution], which began probably in mid-'89, I would think.

The first thing to talk a little bit about is this Soviet Union Chapter 11. The main pitch I made to Paul Wolfowitz, even before he took over, was, "Look, the Soviet Union is in real trouble." I used to use chess analogues of …

Q3: Middle game [indistinct].

Marshall: Yeah. "We've emerged from the middle game with a won game. Now the big thing is not to screw it up, but really be astute. That's the frame of mind people ought to be in. And it's not only that we have a [indistinct] won game, we have a tremendous advantage." This view was based on my feeling that most of the people in the U.S. government didn't understand that and, secondly, that the nature of the changes likely to be taking place, although they all happened much more rapidly than I ever expected at that time, was such that you shouldn't count on President Mikhail Gorbachev being there or any of the people that you were talking with or dealing with being there. And I probably kept going back to things that I was saying in the middle '80s to Dennis Ross, that you ought to think of yourself as talking to Louis XVI's ministers, and that relationships you might develop might not last too long, because they might not be there too much longer. I don't know that any of that had too much effect, but I think Paul was convinced and took it seriously.

Of the things that were done during the Bush period, in terms of assessments, I'd say only two, maybe the beginning of the third thing... But one was the Korean case that we looked at. And the second was the stuff that ended up with the first MTR assessment in July of '92. So let me talk about the Korean thing...I guess the other thing we were looking at was how proliferation would change the character of the power projection problem, and we did do some things, gaming and so on concerning that. But the two big things, I guess, were the Korean assessment and the MTR thing.

The Korean thing really came about because Gaskin— It was appropriate to pick it up because, again, we had this pattern of after people either not watching closely enough or finally they were looking closely all the time, but it finally became clear to them that the North Koreans had gone through a whole new period of investment and change, and that significant improvements had taken place in their force posture.

Q2: Did this come out of a National Intelligence Estimate or something?

Marshall: No, it was some work by some guys in the Defense Intelligence Agency, fundamentally. And we came to know about that, and Gaskin, who had served, I guess, a couple of times in Korea, was very interested, and so volunteered to pick up that and do an assessment.

The assessment, I thought, always had far too much this character of an inspector general's report, rather than a real assessment, because it was heavy on the problems on our side. I kept pushing Gaskin about, "What are their problems?" and to get it more even-balanced, but was never fully successful. He always had the excuse, "Well, we don't know enough about them and some of their things." That's true in a way, but nonetheless, it had this fairly aggressive inspector general-like aspect to much of it, which I got toned down as much as I could. Maybe I could have persevered and got it toned down still further. But I thought that he did turn up some really interesting stuff.

Now, the assessment was extremely unwelcomed.

Q8: You're thinking of Carl Ford [indistinct].

Marshall: Carl Ford [Principal Deputy Assistant Secretary of Defense for International Security Affairs], well...

Q8: Who had worked on this project while he was in earlier jobs. He was a stakeholder.

Marshall: There's that, but also he was a stakeholder for a different reason in that in the early part of the Bush Administration, there was some pressure on Capitol Hill to withdraw more troops from Korea, and a big defense had been put up by the Administration about why that wasn't a good idea. And there also was this overall strategy that was supposed to be developed for Asia, and it was very awkward to have an assessment whose major message was that the South Koreans really weren't trying very hard.[5] On the one hand, there were these new developments that the North had gone through—the repositioning of the forces, the great increase in artillery, and a number of other things—to which the South had not really responded. Instead, the South was off starting to build a submarine fleet and other naval things to deal with its real and future enemy, the Japanese, or whoever else in the region they were focused on. And so, the assessment was very unwelcome.

It was not entirely suppressed, but it had very limited distribution. It went to the Chairman of the Joint Chiefs of Staff, and General Robert RisCassi, who was about to go out and take command of U.S. forces in Korea, got a copy and read it, and talked with both Gaskin and myself, and he took it very seriously. And it had an impact, both in terms of, I think, some things that RisCassi did when he was out there, and also certainly his request for a relook at the whole plan and so on, and some modifications were made in the war plans to beef up the forces and so on.

When it was sent out, I put a memo on top of it which pointed out what I thought was the main import of the thing. In the first place, this thing just dealt with the conventional kind of attack, and I guess the first thing I mentioned was that these people may have, somewhere along the line, nuclear weapons and that's something we need to be thinking about, and so the thing is not the whole story about what the situation is there. Secondly, I wrote that I found the story of the South's limited response to what the North was doing, and indeed the whole orientation of their efforts, to suggest that the South was, in fact, much like our other feckless allies, relying on us to deter the opponent, while they didn't do the things they ought to do to fully play their role and their part. I said that this was not a novel problem, because we'd been through that in Europe and the outcome had been OK, but it was a dangerous game. Therefore, in the current situation, on the one hand, we ought to be talking to the South and trying to motivate them to do what they should do to deal with the threat. On the other hand, we should be focusing a lot on the issue of how we, in fact, deter the North. And that's not so much a matter of bringing in necessarily additional ground forces or so on, but what threats are important, how do we influence their calculations sorts of thing. They don't do it, because in the longer run, the whole game is running against them for economic and other reasons.

My belief is that nothing at this level ever happened. I don't know that there was any effort to really seriously deal with the South and say, "What the hell are you doing spending your money the way you're doing?" I'm not sure that we did things to increase deterrence, although I did have a long discussion with Wolfowitz about that, because there were some things that I'd known about in the past. In the late '70s or so [Jul 1979], I remember General John Wickham, who was then commander out there, had talked to me about, when I visited him—the only time I've ever been in Korea—about various actions that were being taken to impress on the North the capabilities that we might employ and so on, possibilities of deception, a whole bunch of other things that could be used, and that people ought to be being imaginative about that. But I have no idea whether anything ever came of that.

Q3: These discussions were Wickham's…had some things to do with the presence of B-52s over Korea at the time of the tree-cutting incident [Aug 1976]?[6]

Marshall: Yeah.

Q3: And his observations about North Korean reactions to that?

Marshall: Yeah.

Q2: RisCassi says that one of the first things he did when he got over there was he asked to see the war plan. And he took a look at the war plan and he said, "This is great, it doesn't say anything about how we're going to win the war." And so he told his war planners to go back and construct a war plan in which the political-military objective was not just for the restoration of the status quo ante (even that was not made an explicit objective in the existing plans), but was, in fact, reunification of the Korean peninsula.[7] And he said it was interesting, because when they did that and rewrote the war plan, it came out to be a totally different kind of war plan than the kind that he had when he walked in there. He told me this as he left being Commander in Chief, ROK/U.S. Combined Forces Command, but I don't know if you had any discussions with him as to the nature of the changes he made in his war plans over there or how that might have related to this.

Marshall: No. I heard a little bit about that. I did have lunch with him when he was retired, in fact, about a year ago or so, when all of this thing flared up again, got leaked by Gaskin, among others, I suppose.[8] I wanted to talk with him just to ask him how he felt about what the real state of the situation was, because one of the other things, you see, that came out clearly was our limited intelligence about the North, and the whole issue really in the end came…well, anyway, one thing that would be very important would be, "How well do you think these guys are really going to fight?" And—

Q2: Actually, that was a question about both sides, wasn't it?

Marshall: Well, but you see, you had a lot of this evidence that maybe the South wasn't going to do very well. On the other hand, what are you going to assume about the North? Are they going to be very effective? If you're dealing with the equipment of the panzer divisions, you're in a hell of a lot of trouble. And on the other hand, if you're dealing with…[gap for tape change] …skills and so on, that's another thing. So I wanted to talk to RisCassi about that.

One of the things we did after the assessment was done was we met with people in the intelligence community, both from CIA and the National Security Agency, to see what could be done to try to increase our information about the North. And it really all came down to the ground forces and the special operations forces. The air force, it's clear, is terrible. They don't fly very much and so on. It really comes down to, "What's the operational effectiveness of the North's ground forces?" And, in fact, I don't think that that's an impossible task, to say something about the existing bases, which really were limited. I don't think things have gotten any better. I haven't looked at it, but I don't think so. Part of my reason for having lunch with RisCassi was to just get his judgment about what he really thought about that.

Q2: I know one of the other things that RisCassi did while he was over there was he had this massive improvement effort, which really became a model for most of the theater commanders, on a C4I [command, control, communications, computers, and intelligence] architecture, because he was convinced that he was going to have to do all this on the move, on the run, from the get-go. He was going to have to flow the force from CONUS [Eds: the continental United States] that he needed to fight and win the war over there, at the same time that he was fighting and trying not to lose a war over there. So he developed a seamless architecture from CONUS to Korea that allowed him to fight on the move, because he really didn't feel that the in-place forces he had over there were going to be sufficient and adequate to actually win the war and achieve these new wartime objectives of the reunification of Korea.

Marshall: Right. And I think some changes were made in the pace of reinforcement, or they hoped for that. So at any rate, that's one of the—

Q2: It may be one of the few instances, actually, or at least one of the few identifiable instances, where one of your assessments directly affected—

Marshall: Right.

Q2: —a CINC [theater commander in chief] OPLAN [operation plan], force structure, key elements of the force structure, reinforcement plan, etc. I don't know of any other case—

Marshall: No...

Q2: —with the possible exception of the stuff you did on antisubmarine warfare.

Marshall: Right.

Q2: But that wasn't really a theater campaign plan. That was a smaller-scale thing.

Marshall: Right.

Q2: So this may be one of the few cases where there was a direct effect on a CINC OPLAN and force structure.

Marshall: Right.

Before we go to the MTR thing, one of the things that happened in the wrapping up of the NATO-Warsaw Pact effort was that where the initial effort was to just look at what had happened as a result of the withdrawals, or what would be the case when the Soviet forces had withdrawn from Eastern Europe and the reductions had taken place and so on, we also wanted to look at future contingencies. That led to a major investment at BDM in modeling, which also was an effort to... because you wanted to look at a different kind of engagement, in effect, a large meeting engagement in the middle of Poland. And it also was, if you looked into the future, to look at the role of reconnaissance-strike complexlike systems, long-range systems that would begin engaging targets before the forces themselves actually got very close to one another. And, in fact, we knew from the Russian writings that they had talked about this new phase in the engagement of...what did they call it? "Range fires," I guess.

Q2: "Range fires"?

Marshall: Yes.

Q2: Talking about "maneuvering fires"?

Marshall: No. I mean just this whole notion that as the forces closed, they would become engaged through these longer-range fires, and attrition would begin before the forces came to meet. So instead of the forces starting up against one another, now you had to take account of this fact that...

Q2: Long-range fires could engage and attrit before the ground forces even engaged.

Marshall: Yeah, before they engaged. So this was one phase. Then there was the actual meeting engagement, and then the close engagement of the forces.

Q2: In fact, your meeting engagement in the middle of Poland ended up being a Defense Planning Guidance scenario and a kind of a cause célèbre in *The New York Times,* didn't it? Remember when that DPG scenario leaked [Patrick Tyler], as it always does, and this thing about this battle in the middle of Poland was a big issue.[9]

Marshall: Could have been. Right. Also, the attempt was made, I think for the first time, to build in to this model some of the information aspects of this engagement. Because if you're going to have these long-range fires, one of the things that needs to be reflected is the degree to which one side or the other would be better at that, of finding targets, and also of knowing where the other forces really were as the meeting engagement came about.

Q2: Who did that?

Marshall: That was built into the thing, and I think we played several...

Q2: Built into BDM's [model]?

Marshall: Yes, right. And, in fact, we got the SOLIC [special operations and low-intensity conflict] people involved, because one of the ways in which information might be obtained was not through sensor systems, but that [special operations]. So anyway, that was one of the good things that happened.

I started to talk with Krepinevich about the MTR, again, coming out of the middle '80s stuff on the future security environment [for the Commission on Integrated Long-Term Strategy]. So he began to look at that, as I said, probably in mid-'89 or certainly by early '90, and put out the assessment in '92.

From the beginning, I made a decision that we were going to do that assessment in a very, very different way. The usual thing had been that the assessments were done really in-house, just by people in this office and with inputs from the outside of various sorts. But because it seemed to me that if you really thought of this as the beginning of a period of major change, like the early '20s, we ought to conduct this thing in a way that tended to co-opt people, and we would be in the process of generating a vocabulary that we wanted to be adopted. And so, totally differently than we'd done any other assessments, it was done through a series of panels. So people were addressing various topic areas and so on, such as, "What are the technologies which you think are going to be important? How might warfare change? Here are the Russian ideas, what ideas do people have?" So it was conducted, as I say, in a very, very different way.

Q2: What was your plan for the allies?

Marshall: I didn't have any plan for the allies. The critical questions really were, "Do you really feel that we're in one of these periods where there will be major changes?" In other words, are the Russians right when they say that we're in this new period of a military-technical revolution? Then, "Why has it happened?" What are the technologies or other things that will lead to changes? Are we at the beginning, middle, or end of this process? Is the thing mostly over, or are we at the very beginning? So we posed those questions, and a number of others which we didn't answer very well, like, "Who else is going to be involved in this? What should our strategy be?" Various things of that sort. And one of the purposes of going for these kind of panel discussions was to get a sense of how widely shared was some judgment that, in fact, we're in such a period. Because if that were so, then in some sense, the main issue would be, "How do you make people take seriously what it is you ought to do if you really are in such a period?" And my feeling now is that those people tend to sign up to the fact that you're in such a period, but then don't draw very striking conclusions as to what you ought to do. But as to the allies, that matter emerged as we got around to really writing the assessment, and when we came to the issues, it was really one of the four major issues that we surfaced for people to really start thinking about, which hasn't happened, at least on that. So…

Q2: Who does the assessment go to?

Marshall: The assessment initially went to people like Wolfowitz, to the heads of the Services… One of the things, again, that was striking to me was that I expected that we would have a lot of resistance and flak, but we didn't actually encounter that much. I was surprised at the degree to which people tended to agree that we were probably in such a period, but since implicitly you would be threatening platforms that people currently love and other things, I thought we would have a lot more overt resistance—I'm sure that we have lots of foot-dragging and other things.

The other thing that struck me was that the top-level people often were quite supportive. Indeed General Gordon Sullivan [Army Vice Chief of Staff at the time] had somebody call down and asked to see me. My guess is that that probably would have been, say, the summer of '91. And, in fact, his office and the people who support him didn't ever seem to get it through their minds that it was that he wanted to see me, because they kept calling down here about what it was that I wanted to talk to him about. And all I knew was that he wanted to see me. They wanted to know, was there a read-ahead? Was there this, was there that so they could prepare him. So I went there, and he by that time knew that we were working on the MTR assessment, I guess because the Army guys had been involved in a number of these panels that we were running. What he had to say to me was that he was very supportive, and, in fact, asked, "What could the Army do? How could he help me?" and so on. And then a little later—probably in very late '91 or early '92—General Merrill McPeak [Air Force Chief of Staff] called up, made an appointment, and came to see me in my office. He's the only Chief of Staff who's ever done that. He said he wanted to be the point of contact in the Air Force for whatever it was I was doing, and that he was very supportive and so on. So I'd been struck by the support from top-level people. Or somebody like Rear Admiral William Owens [later promoted to four star and appointed Vice Chairman of the JCS]—he was Cheney's military assistant—was always very supportive all along and helpful. So at the very top level you had an amazing and favorable kind of response.

Q2: Then again, recall Sullivan's comments about, "We're downsizing too fast."

Marshall: Yes.

Q2: "We got all these Serb hordes. We're kicking people out of the Army left and right. You don't know who the threat is. I can't really ask the Army to get engaged in an RMA right now because it will break them. They just can't handle it. That's one more straw they can't handle. So I'm going to try to get through the transition period and the next guy will have to worry about RMA."

Marshall: Right.

Q2: So he took it on board intellectually, but he wasn't prepared to do anything bureaucratically to change the Army in that direction.

Marshall: Yes, right.

Q2: And in fact, you recall last year, he had gotten wind, I guess as a result of the Army roundtable that we had down at the Training and Doctrine Command, that there was some disappointment on your—

Marshall: Yes [indistinct].

Q2: —or at least he viewed that as meaning you were disappointed that he hadn't done much, so he wanted to talk to you and reassure you that he still thought this was very important and everything, just too hard to do right now, not on my watch, you know, the next guy is going to have to do it.

Marshall: Yeah, right, right. Well, I know, but I think Sullivan has a lot of good arguments. I think it is very difficult to do. And so…

Q2: And, in fact, I think I told you in his most recent meeting with James Blackwell—I think it's been like two months ago now—but he laid out in front of Jim his new slide, I mean, he's got a set of slides about his vision for the Army.

Marshall: Yes.

Q2: And it now has RMA on there, after Force XXI.

Marshall: Yes, right.

Q2: Force XXI is his thing, modernize the force, etc., etc. But he says now there is an RMA out there and it's coming after…

Marshall: Right. Well, I'm a patient man…

Q2: The Army will be around for a long time.

Marshall: Yes. And…

Q3: And he… it was on [indistinct] initiative, from what I understand—you know more about this than I do—that things like the new 525-5 [indistinct].[10]

Q2: Well, it was, and General Freddie Franks [Commander of U.S. Army Training and Doctrine Command] also was a key player on that. He personally was a big believer in information warfare. He thought, as did Sullivan, that that was really a key component of the Army [indistinct], and so they were really trying to get the 525 thing. In the new Field Manual 100-5, as well.

Q3: Less than that.

Q2: And the new Field Manual 100-6, which deals with information warfare. I think that's the only Service that has a manual out on information warfare doctrine. Is it great? No. But, it's a first step, and so… There clearly were a whole bunch of other things that the Army could have done. And it's not just the Army…

Marshall: Yeah, right.

Q2: All the Services could have done a lot more. So it's nice for Sullivan to say, you know, "Hey [indistinct]…"

Marshall: Well, look, compared to what I thought the problems would be…

Q2: That's right. They were going to firebomb your house, blow up your car, [indistinct]. "You're threatening our organizational essence and all of our toys."

Marshall: Right, right.

Q2: On the other hand, don't forget what they did on competitive strategies, which is, in some way, what they're doing now on RMA. "We are the RMA. We are competitive strategies."

Marshall: Yes, right, right.

Q2: F-22—RMA!

Marshall: Yes, right.

Q2: I really think the more interesting thing is not what's happening at the top levels, but what's happening sort of at the mid and younger levels…

Marshall: Yes, right.

Q2: We really are developing, within the Army, anyway, and, in some ways, even as much within the Marine Corps, some within the Navy, less so in the Air Force, these O-4s, -5s, and -6s that are thinking about this, talking about this, and putting, you know, in place things—like Gary Anderson in the Marine Corps, Chuck Miller in the Air Force—you know, that are putting in place these kinds of things, where they're trying to develop these things intellectually, whether with games, or workshops, or whatever.

Marshall: Right. As I say, I'm a patient person. I don't expect these things to happen very quickly. I'm pleased with results thus far, on the whole.

Q2: Your assessment—you said it went to Paul Wolfowitz and the Service chiefs. Did it go…

Marshall: It went to Cheney. In fact, I had a meeting with Cheney. It was a long time getting it scheduled, and by the time I met with him—it probably was in August or September [1992]—it really was too late for him to really do anything. He listened, found it interesting, but at that point I think he wasn't going to try to do anything.

Q2: What about General Colin Powell [Chairman of the Joint Chiefs of Staff]? Did it go to Powell?

Marshall: Yes, it went to Powell.

Q2: And what was his reaction?

Marshall: I can't remember exactly, but it was along the lines of, yes, he thought that was interesting, but… A number of things were sent to Powell. He says some of the right things, but I don't think… It was not the kind of thing he was really interested in.

Q2: Did he ever meet with you on it?

Marshall: No.

Q2: So these were just written responses back to you?

Marshall: Yes, or a phone call, I think.

OK. Let me go back and pick up some other things that were going on. One of the things that we clearly were continuing on was this monitoring of the Soviet Union as it continued to go down the tubes. Cheney was really very interested in that. For example, he was very interested in the stuff that Vladimir Shlapentokh was putting out, or other things that we had people put out, people like Vlad Kontorovich and others. He was also, I thought, very good in the sense that he was not only interested in the thing, but he had the right instincts, as against the people in the White House, who had the notion that the Russians had completely changed and there were never going to be any problems of—

Q2: I remember Jonathan Howe [Deputy Assistant to the President for National Security Affairs, 1991–92] making that argument.

Marshall: —"Gorbachev was wonderful" and so on. Cheney was skeptical, in fact, more skeptical than the people in the Services; for instance, the Joint Staff, who kind of eliminated all Russian problems—a little premature, from his perspective.

He also was very interested in getting people from the outside in, and not confining himself to, "What does the CIA say?" etc. And so there were these meetings that used to take place on Saturday mornings, of people from the outside. Initially, they were really run by, I guess, Scooter Libby [Principal Deputy Under Secretary of Defense for Strategy and Resources] [indistinct]. But I think before they got started, I picked the people, based on the guys who'd been right instead of those who'd been wrong for 20 years. There was some pressure, "Why don't we have a more representative sample?" And I said, "The hell with that. Why don't we pick the people who have been right and listen to them for a while?" And I used to attend them, but gradually other people took over the actual arrangements and so on.

One of the flaws in the thing was that after the first one, they invited a couple of the people from CIA, who then tended to talk too much, take up too much of the time, rather than listening to these other people. My own view was that it was a mistake to invite them. I mean they could be there like the other people to listen, but they tried to play their role as experts.

Another general point I would make, which I've already alluded to, is that I thought that under Wolfowitz the [Office of the Under Secretary of Defense for] Policy operation in this building ran better than I have ever seen it run before or since, really. Also, I think the effort that he undertook that was called the "Bridge Game," which was the effort in the last year of the Administration to really address the question of, "How should we change our policy, strategy over the longer term? Taking a longer-term look, what should our policy, strategy really be?" It was the best general effort that I've seen of that sort. The one major failure area, I thought, had to do with the nuclear area, where people, I think, continually have trouble getting beyond the thinking of the 1950s.

Q3: Who were the participants in the "Bridge Game"?

Marshall: There were a whole bunch of them. It involved a whole bunch of people, although it was... The people who came to the meetings depended some on... Some part of it proceeded on a regional basis, and then the regional people would come in and so on. But there was, I guess, a core group of people. I forget who all that was, but there was a group of about maybe eight or ten people.

Q3: You were a member?

Marshall: Yes.

Q2: Dennis Ross?

Q3: Jim Roche?

Marshall: No, Jim wasn't involved. It was people in the government. So it was, like, Scooter Libby and the heads of all the units and some of the other...

Q2: Someone like Harry Rowen [Assistant Secretary of Defense for International Security Affairs], Steve Hadley [Assistant Secretary of Defense for International Security Policy]?

Marshall: Well, Rowen had gone by that time, I think.

Q2: Was Dennis involved? Ross? Bring anybody over from State?

Marshall: No, it was entirely a...DoD [group], people from [the Office of the Under Secretary of Defense for] Policy. Now, it came a cropper when it got leaked to the newspapers and people backed off...

Q8: What was the leak?

Q3: This was the DPG leak I assume? Draft?

Marshall: No... well, it was a draft... No, it wasn't a leak of a document... Well, there was a thing about some of the scenarios, but it was also more this kind of...[gap for tape change]

Marshall: Then I guess we should talk about Desert Storm.

Q2: I don't know, we went all the way through the '60s and '70s and didn't talk about Vietnam, so we could...

Marshall: Right. We didn't have very much to do with Vietnam.

Q8: A distraction. Some said the Vietnamese War and the contras' war were distractions from major strategy.

Marshall: Yeah, right. Well, I have two or three things, not about DESERT STORM in general, but just about the involvement of this office in some respects. One thing, of course, that was interesting was that we were in the midst of the Summer Study up at Newport when Iraq's invasion of Kuwait happened [August 1990].

Q2: Did you have a panel on the Middle East and on Iraq going on at the time?

Marshall: Well we had a panel on new problems in deterrence and we had a panel on...

Q3: Dealing with regional nuclear actors, and we were planning air strikes against Iraqi nuclear facilities at the time, as I recall. [Indistinct]

Marshall: Yes, right. So two of our panels were very relevant. The reason we'd had the panel on deterrence was—I don't think much of deterrence theory, anyway—first, that we had had the good luck that the deterrees were the Bolsheviks, with their focus on rationality and so on, so the theory had a little bit more congruence with reality than it might otherwise have had, and that in the future, however, a Russian government might well be a far more traditional, normal political elite, and there's no reason to believe that they would be especially characterized by putting a high value on rationality, calculations, and so on.

Secondly, some of the people we would be most interested in deterring were people from other cultures who had nuclear weapons and so on, and we needed to think about that and what special problems came with that. And, of course, a lot of that was focused on particularly the people in the Middle East coming from Arabic cultures. So there was that aspect.

Then there was the thing that Steve Rosen ran, which was looking at, "What do you really do when these people are about to use them [nuclear weapons], have just used them, and so on?"

The other thing that happened fairly shortly after that was that I took a trip to Paris in late September '90. I forget whether I told this story earlier, but Henri Conze [then a former French Ministry of Defense official; the French armaments director at the time of the talk] invited me out to his apartment in Versailles and had a number of other people for dinner, men from the French military industry. Of course, we talked about the situation in the Middle East and Iraq. And, of course, these people had had a lot to do with the Iraqis. They told me of their appraisal of the Iraqis, which was that they thought very poorly of the army, the military, but had great respect for the technical people, the engineers, and the people that they met in the plants and the laboratories and so on. So their view was, first, that given this latter characteristic, we shouldn't underestimate them, and that they were very capable of doing surprising things, of putting together systems, technologies, and so on, and so we needed to be wary about that. On the other hand, they thought the army would perform poorly, especially after attack or if cut off or encircled.

Conze came over to the United States about a couple weeks later, and I had a chance to talk to him again, and asked for more details about why they had reached their appraisal about the army. He told me that it had come about, or at least the reason why the people in industry had this view, was that, of course, the Iraqis were using a lot of French weaponry in the war with Iran, and the French industry people had wanted to know a lot more about the performance of their weaponry. The Iraqis weren't telling them as much as they wanted, so they opened up their own effort to find out exactly how their weapons had been used and how they had performed. It was the by-product of this effort of theirs that they came to this appraisal of the general competence, performance, morale, and so on of the Iraqi military establishment.

I passed on both of these things to the people in the Joint Staff J-2, the intelligence people. I don't think it had much impact on anybody's judgment. You see, I had urged, "Look, the French seem to know a lot. Why don't you have some people go talk to them?" I asked later [in France] whether they had—not all of them—but asked whether there had been any inquiries or any discussion with these people on the part of Americans, and they said no.

Q2: Do you know what the assessments were in the Pentagon of the Iraqi military capabilities?

Marshall: They tended not to rate them as tremendous, but I think they were seen as comparable.

Q2: Just let me rephrase that. Cheney and Powell—Do you know their assessments of the Iraqi military capabilities?

Marshall: I don't. I don't know.

Q2: I don't either, but I do know Brent Scowcroft's [Assistant to the President or National Security Affairs]—

Marshall: Yes.

Q2: —Which was the same as mine, which was that we were going to beat these guys very quickly—

Marshall: Yes, right.

Q2: —with very few casualties.

Marshall: Yes. Now, I think people must have felt some that, for whatever reason, whether it was based on a low view of the Iraqis, or whether it was intrinsically that they weren't competent, or whether we just had overwhelming capabilities. I do know in talking with Wolfowitz that he said to me at one point, just a few days before the thing got started, I mean our attack, that he hoped that things were going to turn out as well as people were assuring him they thought they would. And yet, even then, of course, it turned out better.

Q2: Yeah, even Scowcroft's projections were not as good as it really turned out.

Marshall: Right. Another thing was that early on we undertook [some quick games] and turned them over to the people on the Joint Staff. We had put in place a contract with Mark Herman [at Booz Allen Hamilton] for quick-turnaround games. The guy who was working on the Middle East here, the Army…

Q8: Fred Downey?

Marshall: No.

Q2: The colonel? Ned Cabaniss?

Marshall: Cabaniss, yeah. He quickly organized the things and so we had some games done fairly early. I think their orientation was very much, you know, "What happens if they (the Iraqis] continue on? How do things go?" And those were turned over to people.

But the most direct involvement that we really had was that both Rowen and Wolfowitz were so busy they asked me to run some panels for them, where we had people from the outside brought in, experts, to discuss various things. One panel that had several meetings—this would be in probably the October, November, December period—were regional and country experts. And the issue was how well the U.S. was doing in influencing the views of various publics and elites. After all, the U.S. was trying to explain why it was there, what the issues were, and the view of the Iraqis as the bad guys. And at the same time, the Iraqis were trying to reach the same kind of publics and elites. The issue was, "How was that going? Is our view doing reasonably well? What are the problems?" and so on. So we had, I think, probably three meetings of that.

They were kind of interesting, but the main thing I came away with was a sense that there are experts and then there are experts, and some of them were obviously a lot more interesting, a lot more insightful than others.

Q2: Were you actually trying to develop a strategy for shaping perceptions?

Marshall: No, it was mainly… Well, other people were…we were just gathering kind of input and appraisals and so on for them.

We also had some meetings and put together some papers as an input to trying to look at alternative outcomes of the war, and what the longer-term consequences in the region would be. What would be the distribution of power? What would the region look like depending on…? And three potential outcomes were posited.

Q2: Were any of them the actual outcome?

Marshall: No.

Q2: You needed more—four possible outcomes.

Marshall: Right. Tried to supply that.

The last thing, which maybe would have been potentially the most useful, was once the ground war began, and it looked very quickly as though we were obviously going to win, Fred Iklé suggested that we ought to really think about the war termination process. So we had two meetings in which Iklé was one of the participants, but there were some other people. And one aspect of it was, looking at the whole literature that exists on war termination and the problems you can get into or the things to be avoided, we wanted to try to put together some tailored set of suggestions that came from the literature on things that ought to be avoided and so on in whatever termination process we got into.

Then there was also some set of suggestions about things that might be part of an attractive termination strategy for us. And those were sent up to Rowen and Wolfowitz. I don't know, in all of the scramble at the time, and it all happened so fast, but my guess is it didn't have much effect. But it was kind of interesting. It was a useful thing to do.

Q2: Did in any of your possible outcomes or your war termination strategies did you call for the destruction of the Republican Guard as a condition for war termination?

Marshall: We assumed that that was more or less going to happen, then it was one of the things that didn't happen.

Another aspect of it was that I remember people were proposing that we remain and occupy the southern part of Iraq, because you would have a hold over a major part of the oil fields, and that you would keep this as leverage for the future. Now, in some sense, as things have worked out, that may be not too important, because we have been able to maintain the embargo on shipping.

Q2: What about weapons of mass destruction? Does your war termination strategy call for destroying all their capabilities?

Marshall: No, because we'd sort of assumed that it was already destroyed. We didn't know that there were the undiscovered parts of the thing. No, you see, most of it was procedural and dealt with what not to get sucked into, and things of that sort.

Q3: No talks while fighting, that kind of stuff.

Marshall: Right. Now this talk has addressed mainly activities during the Bush Administration. I don't really have a lot to say about the Clinton period.

Q2: Before you leave the Bush thing... Are you done on Bush?

Marshall: Yeah.

Q2: One of the things at the start of the Bush Administration that you were involved with, of course, was the NSR-12 [National Security Review 12] process.

Marshall: Oh yes, that's true.[11]

Q2: That's not to say that the results of the NSR-12 process were anything we really want to highlight as being great thinking on the future of the world or anything.

Marshall: Right.

Q2: Nonetheless, it was interesting that you were invited over to participate in that as a member—

Marshall: Yeah, right.

Q2: —of the assistant secretary-level group that Paul Wolfowitz chaired while he was bouncing back and forth between Indonesia and Washington waiting for confirmation as the Under Secretary of Defense for Policy—

Marshall: Right.

Q2: —that you participated in that effort. And I was going to ask you, were there other, setting aside the Kissinger era, were there other NSC-level efforts comparable that you directly participated in in terms of attempts to do strategy formulation at the NSC level?

Marshall: I would say in the Carter period. I mean the thing [Presidential Review Memorandum 10, "Comprehensive Net Assessment"] that Sam Huntington ran and so on, which we were very much involved in.[12] But I guess those are the [indistinct].

Q2: Nothing under Reagan?

Marshall: No. Nothing under Reagan.

Q2: And nothing under Clinton.

Marshall: And this time I don't think there was anything. In fact, one comment I would make about the Clinton period is my sense that it's probably the only administration that I know of that came in that didn't somehow address these issues in some systematic way, whatever you thought about how well or poorly they [the other administrations] did it. I think this is the only one that I know of that didn't really seem to address these issues. Now, maybe this was done, say, as far as this building was concerned, and in much more informal ways through Secretary of Defense Les Aspin, but…

Q2: Or maybe it wasn't done.

Marshall: Yeah.

Q3: In DoD it wasn't done.

Marshall: One other thing to say about the Bush thing is that it was in that administration that because of what was happening with respect to the Soviet Union, the whole focus of attention of this office really changed. It started out where we were still looking at some aspects of the situation relative to the Russians, and then fairly quickly moved on to a very different focus on this power projection problem, on Asia, and on the MTR and RMA effort.

Q3: I just have one small question about your effort to study the Soviet Union in Chapter 11. Did you sponsor John Hines's work with the Soviet general officers?

Marshall: Yeah.

Q3: And it was as part of that effort that he got some insights into the Soviet nuclear command procedures.

Marshall: Yeah. In fact, we finally have the draft…

Q2: 160 pages.

Marshall: Yeah, it needs a little editing, but I read the thing and it's really pretty good.

Q3: Yeah, you just mentioned a couple of things out of it. It seems to be really interesting.

Marshall: Yeah, right. Of course, it's gone on for quite a long time. It looked to me like there was this period where there would be an opportunity to go and talk with these people, and it wouldn't last forever. So I started John Hines on that as a way of trying to understand, "What did they really do? What were their assessments? What can we find about the actual methods, techniques that they really used in the General Staff for making assessments?"

Q2: Focused on the nuclear issue.

Marshall: Well, this is on the nuclear issue. But there were earlier things that were broader: How did they look at the correlation of forces and so on.

Q2: And have you got stuff on that? I mean, did you get stuff—

Marshall: Yeah, got some of that.

Q2: —that confirmed that we were on the right track, or we weren't on the right track, or…?

Marshall: We were sort of, I think, on the right track. Although, again, it's like one of these things when you start looking into it you say, "Ah yes, right. It's a big government with people with different views" and so on. But it turned out to be an interesting effort. I can't quite remember when, in fact, we started that, but it was certainly four or five years ago.

We also had John Hines speculating based on whether we really get on the next… Well, that same summer, he was running that other group. That was the summer of '90, when we had five groups for the Summer Study at Newport, and John was doing this 20-year look-ahead on Russia: After it [the Soviet Union] went down, when did it recover? What kind of a state would it be? And he then continued to do some things in that area.

That's pretty much all I have. Are there other things that ought to be talked about?

Q2: Did you want to talk about the RMA stuff you've done so far in the Clinton Administration?

Marshall: The thing to say about that is that following the MTR assessment in '92, there wasn't time for anything to happen. When the new group came in, fairly early Graham Allison [Assistant Secretary of Defense for Policy and Plans] was interested in having Aspin become aware of this and use and support it. Clark Murdock [deputy to the Under Secretary of Defense for Policy—Planning] saw that as, I guess, a thing that Aspin ought to grab ahold of, and was important in trying to formulate actions he could take to institutionalize some effort: And it's really through that that this RMA Senior Steering Group that John Deutch [Deputy Secretary of Defense at the time of the talk] heads came about. I think a memorandum setting that all up was signed by Aspin in September or October of '93.

Q2: Any sense for how Powell reacted to that? Or Allison or Murdock or Aspin or William Perry [Deputy Secretary, then Secretary of Defense] talking to Powell or his people about it?

Marshall: No. Now, Admiral David Jeremiah, who was the Vice [Chairman of the Joint Chiefs of Staff], was very interested and we talked some with him. Of course, he felt, I think, he was kind of a "lone warrior" looking ahead and looking more imaginatively at things and the changes that could be made. So he was always very supportive, but he had his own narrow, little effort going, the "2025" study.[13]

Q2: Well, and then, of course, Deutch has been a big supporter of this from the beginning.

Marshall: Yes, right. And early on, when Perry was the Deputy Secretary, I went to see him just to tell him what I was doing. I told him about the RMA, the Asia stuff, the power projection stuff. And he was very satisfied. Those were all the right questions, he thought, to really look at. Of course, he had known about the whole thing, because in probably November or October [Aug 1992]—I don't know whether it was before or after the election—I had held a meeting of all of the then-living DDR&Es [Directors of Defense Research and Engineering], and they all came.[14] The thing I wanted to talk with them about was [that] in the assessment that we put out in July of that year, one of the four issues had to do with fixing the acquisition process, not in general, but in a particular phase to support the experimentation and so on. At that meeting, it was a one-day meeting.

Prior to the meeting, we sent the DDR&Es materials about what our views were. I started off the meeting with a general presentation on what we were about and the four issues. Then the two things I really wanted to talk with them about and get their reaction to was, one, did they also believe we were in a period where big changes would take place, and then, secondly, this matter of the acquisition system and what ideas did they have, what might you do. And Perry was at that and spoke about several things that interested him.

Q3: It was around that time that he published that Brookings Institution study [*A New Concept of Cooperative Security*][15] along with Ash Carter and…

Marshall: Yes, right.

Q3: Who else?

Marshall: John Steinbruner.

Q3: They got some of the elements of the RMA.

Marshall: Yeah. Of course, he [Perry] has tended to be a bit of the view that the revolution has largely happened, in a way.

Q2: Is that Deutch's view? I don't think that's Deutch's view.

Marshall: No.

Q3: One other small thing. It was in the Clinton Administration that this business on innovation…

Marshall: Yeah, right

Q3: And I gather [that's] still in progress.

Marshall: Yes, right. They're all connected with the same thing.

Q2: Two bureaucratic notes.

Marshall: Yes?

Q2: Since you're non-career SES [Senior Executive Service]…

Marshall: Yes.

Q2: ...and every change of political administration you tender your resignation?

Marshall: No, I've never...well, I forget... I may have actually only once been actually asked to tender...

Q2: When was that?

Marshall: I think under Bush.

Q2: That Cheney asked for it?

Marshall: I think the White House [asked for it] or something.

Q2: I thought at the start of this administration all non-career SESs and Schedule Cs were asked to tender their resignations.

Marshall: I can't remember.

Q2: Well, I do remember that at the start of this administration, there was a movement within the DoD Transition Team to replace you.

Marshall: Yes, right.

Q2: And that's the only time I know of that there really was an active effort—

Marshall: Right.

Q2: —to replace you by an incoming administration. And I don't think this thing, when Bush came in, that that was an active effort to replace you.

Marshall: No.

Q3: They asked everybody.

Marshall: That's right.

Q2: But this [under Clinton] was an active effort to replace you.

Marshall: Yes.

Q2: And they had to go to Larry Smith [Counselor to the Secretary of Defense and the Deputy Secretary of Defense] and he went to Aspin to bring it to his attention—

Marshall: Yes.

Q2: —and Aspin said that he was not going to be the only Secretary of Defense to have fired Andy Marshall, or he was not going to be known as the Secretary of Defense that fired Andy Marshall.

Marshall: Yeah.

Q2: And so he wanted the word passed to you that you were not to leave and wanted very much for you to stay.

Marshall: Yes, right.

Q2: The other bureaucratic note is that they subordinated this office to an Assistant Secretary of Defense [the Assistant Secretary of Defense for Policy and Plans].

Marshall: Yes.

Q2: The only time in the history of the office, since it was created in 1973, that it had been subordinated to an Assistant Secretary of Defense. Now, we know that you have been all over the place, sometimes reporting directly to the Secretary,[16] sometimes reporting to the Under Secretary,[17] but never before to an assistant secretary, Graham Allison—

Marshall: Right.

Q2: —which Allison, of course, wanted—

Marshall: Yes.

Q2: —which, if I were him, I would want it, too. But, it says something in some ways about how this OSD [Office of the Secretary of Defense] has viewed this office. As I say, those are bureaucratic notes, because, as usual, you've outmaneuvered all those guys. Allison's gone, you're working for Deutch, God's in his heaven and all's right with the world, you know.

Marshall: Right.

Q2: But, the bureaucratic tendencies initially were not going the right way on this one.

Marshall: Yes.

Q2: They probably were the most negative, I would say, of your entire experience here.

Marshall: Yeah.

Q2: Although, now I think it's a fairly good working situation.

Marshall: Right.

Q3: I seem to recall you mentioning that there was some initial skepticism or questions at the beginning of Harold Brown's tenure as Secretary of Defense and then very quickly it's the stuff that you were doing for him…

Marshall: No, I don't think so. Early, I was told I should stay, and never… No. Well, what did happen was, as we talked about earlier, Brown did have Eugene Fubini do this effort looking at how OSD ought to be organized. Brown also didn't want so many people reporting to himself. And they also created this Under Secretary for Policy and stuck Net Assessment over there?[18]

Right. Well, maybe we should adjourn.

Q2: One of the key questions people have always and always will ask is, what influence has this office had? I mean, how can you tell that's ever made a difference?

Marshall: Yes, right.

Q2: And since you work in such strange and indirect ways, that's difficult to do.

Marshall: Hard, hard, hard to say.

Q3: But there are people around now. You could go and interview them.

Q2: Well, that's what I'm suggesting, is that there are people, like maybe RisCassi, that we could go and just ask them, what influence did net assessment have on the kinds of things you did and how did it happen? Now, maybe he doesn't recollect, but, we should at least go ask or somebody should go ask those people to see what they say. Because it's one thing for us to say, "Well, this happened and that happened, so there might have been some linkage." It's another thing for the guy that actually did the changes to say, you know, "Yeah, by God, that was really a great thing." And you've still got…you know, you've got all these guys still around.

Q3: That's right.

Q2: One of the things that you always made me do in the NATO [Warsaw] Pact Balance was to project into the future.

Marshall: Yes.

Q2: Have you tried to do any of that?

Marshall: About the office? About the future?

Q2: Or, maybe, the strategic environment?

Marshall: Oh, yes. Yes, right.

Q2: I mean, I think the projection about the office is fairly straightforward: when you leave, it dies.

Marshall: You generally know what I think about the future, but we could do something on that.

Q8: I agree on the issue of the influence of the office and the need to have outsiders make observations.

Marshall: Yes.

Q8: And I'm not sure the tape is still rolling, but just from today there are several things you haven't said that I know about or heard about.

Marshall: Yes, right.

Q8: You know, in Korea, the South Koreans buying submarines. That became a big issue in direct talks with the South Koreans, and they had to give explanations, and then we made them stop doing it. But it became… It was put on the agenda, "Why aren't you improving your forces where you should be, according to the Gaskin suggestions?"

Also, this is such a small list of things. Maybe you're not… your self-effacing modesty is giving [indistinct].

Marshall: well, I don't really gather those things…

Q2: A lot of times you don't get…

Marshall: I don't care a lot. I don't get very much feedback. My view has tended to be, you put things out and people make whatever use they do of them. Often you don't know, and it doesn't make much difference.

Q8: Cheney announced that he would freeze the cuts in South Korea. And there was an internal deliberation. And the Gaskin piece was… I mean, I saw this kind of thing happening, I was part of that. And ISA [International Security Affairs office in OSD] had actually written out the units to be pulled out of South Korea.

Larry Smith told me that your RMA stuff had been used by Aspin in a meeting with Clinton, and how happy they were with... Remember I told you I complained, "'Why hasn't Secretary Aspin had a long meeting with Mr. Marshall yet?" And Larry Smith told me, "What do you mean? We've used his stuff. Aspin took it over to tell Clinton about it in some big kind of NSC-type meeting." So I said, "Well, Mr. Marshall hasn't been told that." Smith says, "Oh, you know [indistinct]" and walked on down the corridor.

Q2: That's an interesting point, though, because there are almost, maybe two levels, or two categories that... Because if you went and asked Harold Brown, or former Secretary of Defense James Schlesinger, "What impact did Marshall have?" They wouldn't have those kinds of anecdotes. They would be at a much more macro, strategic level, although they might have one or two things, like this ASW thing or something like that. But then it might be interesting, also, to get to that kind of stuff, where you can say, look it's not just that, but even in day-to-day policy deliberations—we ought to hammer the South Koreans about submarines. Because I think for those that have always doubted the utility of the office, "What the hell does that office do, anyway?" it would be useful to have some of those old, "And we hammered the South Koreans on submarines." But also these larger things. Harold Brown says, "Well, we shaped the entire U.S.-Soviet competition, and by going after the {...}" etc.

Q8: The discovery that the Russians were writing these bizarre things in '82 and so forth about the revolution in military affairs; that didn't have to happen. I mean, if you hadn't had Notra Trulock and whoever else was on this in the mid-80s...[19]

Q2: John Hines.[20]

Q8: ...Hines... That's not considered to be a CIA Soviet military matter, to say, "Oh, by the way, look at this." And yet we don't even know what the consequences are yet of that having been discovered, Andy Krepinevich, MTR, RMA. You would probably say it's early, only ten years since Marshal Nikolai Ogarkov's first books. So just in terms of the process, of how strategic planning takes place or doesn't take place [indistinct].

Marshall: But in things of this sort, it's all so complex, and there are so many causes that it's hard to trace.

Q2: Well, wait a second. One thing that we do know is that neither the CIA, nor the DIA, nor any of the Service intelligence agencies, even to this date, have been jumping up and down about Soviet and Russian military writings about the MTR, RMA, and how this is really going to shape the future of warfare and we'd better get serious about it. None of those intelligence agencies for the last 13—first book was in '82, Ogarkov—none of them for the last 13 years have seized on this thing to say, "By God, we ought to get off..." And what's the purpose of an intelligence agency to begin with?

So this thing about your being kind of this window on the policy world, and between the policy and the intel worlds, where you try to grab stuff out of intel that you think is really important for policymakers, I mean, that's where this kind of stuff comes in.

So you can say, "Well, jeez, we don't really know cause and effect." We do know that there hasn't been a single intelligence agency out there that's been jumping up and down about this. Been a couple people like Lieutenant General Bill Odom [Retired],[21] you know, when he was over at Army intel [Assistant Chief of Staff for Intelligence] and then Director of NSA, or something like that, but not institutions, not organizations, [indistinct]. So I think you can claim credit for some things, not everything.

Marshall: Right. OK.

Q3: Well, let me say one thing I mentioned to Kurt. I feel that we now have a first cut at an oral history.

Marshall: Yeah.

Q3: And it took a while, but it's now all down on paper. So we can develop it more and so forth, but now whatever happens, we've got some legacy. So I'm…

Marshall: You're happy.

Q3: I'm pleased. And there's more to be done certainly. Well, look, I mean [indistinct].

Q8: It's a start. Where are your remarks on China and Asia, though, and the future security environment stuff you began in '87?

Marshall: Well, a couple of years before that, when we really started working on that.

Q8: A big theme is missing here.

Marshall: Yes, I agree. Right. That's presumably an issue of the future.

Q8: It's also an area where Wolfowitz really took off. You finally got an Under Secretary for Policy who had some personal background in Asia. From Robert Komer [Under Secretary of Defense for Policy in the Carter Administration] on, you have these NATO guys.

Marshall: Yes, right.

Q8: That's another reason to have Wolfowitz interviewed. There's a big process story here. For the "organizational process" man himself to not be interested in organizational processes is quite an irony.

Marshall: No, I am, I am. But I see myself in the business of trying to understand things, and have not been tracing every little thing that happens as…

Q8: It's not little. War termination meetings and the memos—who knows what happened? All we know, it was a very different approach than…

Marshall: Again, you'd have to talk with Paul Wolfowitz and to Harry Rowen to know whether anything came of that. But there was such a rush and the whole thing came to such a quick conclusion that there wasn't time, I think, to really absorb that stuff.

Q8: On the Iraqi estimates, there's a book that has a whole bunch of quotes in it about people who thought we'd probably lose against Iraq, or take at least 20,000 American casualties before we eventually won. It's Jeffrey Record. And his book, for various reasons, has a whole chapter…He assembles all these quotations…[22]

Q2: Used to be at the *Washington Times.* Used to have this "Wall of Shame" or "Hall of Shame" thing, that would have one featured, one a week, or something like that, you know, Ed Luttwak, [(Steven) Canby?]. One of the few outsiders that I know of that got it right was John Mearsheimer. In fact, I think he's the only outsider, I mean outside government. Most people inside the government… I remember Luttwak coming over and going, "Oh God! Don't go to war! They're 40,000 American lives! It'll be terrible! They'll just kill us!"

Q8: Mr. Marshall's understating, in other words. On the French estimates of the Iraqis being weak, that's quite important as it turns out. On the—What's the other one that you mentioned? It was something about this meeting taking place and—You're very much understating these things. In fact, the conventional wisdom at the time was the other way. So I'm struck by the difficulties Kurt faces, given a presenter who is almost trying to deceive you by understating the influence of his office and this kind of, "Oh, gee" quality, "It didn't really matter." I made about seven or eight notes here, so I'll just type them up and give them to you. But if you could look at the conventional wisdom at the time, in newspapers and so forth, you'll see it's quite the opposite from what he is sending forward.

Endnotes

1 Eds: This chapter is based on an interview conducted 25 January 1995.

2 Eds: Anthony Lake, Assistant to the President for National Security Affairs, "From Containment to Enlargement," (prepared remarks, Johns Hopkins University School of Advanced International Studies, Washington, D.C., Sept. 21, 1993); William J. Clinton, "Remarks to the 48th Session of the United Nations General Assembly in New York City," Sept. 27, 1993, *Weekly Compilation of Presidential Documents,* vol. 29, iss. 39 (Oct. 4, 1993), 1903. William J. Clinton, *A National Security Strategy of Engagement and Enlargement* (The White House, February 1996).

3 The *Future Security Environment,* Report of the Future Security Environment Working Group, submitted to the Commission on Integrated Long-Term Strategy (Washington: Department of Defense, Oct 1988).

4 *National Security Strategy of the United States* (Washington: GPO, Mar 1990), 24.

5 Department of Defense, *A Strategic Framework for the Asian Pacific Rim: Looking Toward the 21st Century* (Washington: Department of Defense, 1990), reprinted in Senate Armed Services Committee, *The President's Report on the U.S. Military Presence in East Asia,* 101st Cong., 2nd sess. (Washington: GPO, 1990), 28–55.

6 On the tree-cutting incident and the U.S. response, see Richard G. Head, Frisco W. Short, and Robert C. McFarlane, *Crisis Resolution: Presidential Decision Making in the Mayaguez and Korean Confrontations* (Boulder: Westview Press, 1978), 149–215.

7 In early 1994, an article in the Korean press reported that the extant operation plan for the defense of South Korea called for allied forces to remove the regime in Pyongyang if it attacked the South. South Korean Defense Minister Yi Pyong-tae was quoted as telling a parliamentary defense committee that, "We have a strategic concept that in case [the] North Koreans launch a preemptive attack we will use our retaliatory operations [to create] an opportunity [for] national reunification." *Korea Herald,* 25 Mar 1994, 3, in Foreign Broadcast Information Service, *Daily Report: East Asia,* FBIS-EAS-94-058, 25 Mar 1994, 20–21.

8 Michael R. Gordon, "Pentagon Studies Plans to Bolster U.S.-Korea Forces," *New York Times,* 2 Dec 1993, A1, A11; Barton Gellman, "Trepidation at Root Of U.S. Korea Policy," *Washington Post,* 12 Dec 1993, A1, A49.

9 Patrick E. Tyler, "7 Hypothetical Conflicts Foreseen by Pentagon" *New York Times,* 17 Feb 1992, A8; Barton Gellman, "Pentagon War Scenario Spotlights Russia," *Washington Post,* 20 Feb 1992, Al, A21.

10 *Force XXI Operations: A Concept for the Evolution of Full-Dimensional Operations for the Strategic Army of the Early Twenty-First Century,* TRADOC Pamphlet 525-5 (Fort Monroe, VA: U.S. Army Training and Doctrine Command, 1994) outlined the Army's vision of military operations in the early 21st Century and provided the conceptual basis for the Army's future evolution.

11 See Richard Halloran, "Calling for a Redesign Of National Security," *New York Times,* 2 Apr 1989, E4; David Ignatius, "Life After 'Containment'—Muddling Through," *Washington Post,* 9 Apr 1989, B2; and George C. Wilson, "Military Budget to Precede Strategic Review," *Washington Post,* 10 Apr 1989, A17.

12 Presidential Review Memorandum/NSC-10, Jimmy Carter to Vice President; Secretary of State; Secretary of Defense; Director, Office of Management and Budget; Director, Arms Control and Disarmament Agency; Chairman, Joint Chiefs of Staff; Director of Central Intelligence; U.S. Representative to the United Nations, subj: Comprehensive Net Assessment and Military Force Posture Review, 18 Feb 1977 (SECRET; Declassified 6 Jan 1992) [PRM/NSC-10, Presidential Review1 Memorandum (PRM), Box 1, Records of the National Security Council, Record Group 273, National Archives]. The NSC study for PRM-10, "Comprehensive Net Assessment," Jul 1977 has not been declassified (telecon with James Yancey, Jimmy Carter Library). Unclassified discussions of the study can be found in Memo, Zbigniew Brzezinski to Secretary of State; Secretary of Defense; Director, Office of Management and Budget;

Director, Arms Control and Disarmament Agency; Director of Central Intelligence, subj: SCC Meeting on PRM-10, July 7, 1977, 25 Jul 1977 (SECRET; Declassified 29 Apr 1994) [Declassified Documents Collection, 1995/0598]; Zbigniew Brzezinski, *Power and Principle: Memoirs of the National Security Adviser, 1977-1981* (New York: Farrar, Straus, and Giroux, 1983), 177; Samuel P. Huntington, *American Military Strategy,* Policy Paper in International Affairs No. 28 (Berkeley: Institute of International Studies, University of California, Berkeley, 1986), 28–29; Hedrick Smith, "Carter Study Takes More Hopeful View of Strategy of U.S.," *New* York *Times*, 8 Jul 1977, A.1. A sanitized version of the DoD study for PRM-10 has been released. See *PRM/NSC-10 Military Strategy and Force Posture Review Final Report,* Jun 1977 (TOP SECRET; Sanitized 13 Aug 1993) Jimmy Carter Library.

13 See Institute for National Strategic Studies, National Defense University, *Project 2025* (Washington: INSS, 6 Nov 1991).

14 Meeting of Former DDR&Es on Competition in the Military-Technical Revolution, SAIC, Tysons Corner, VA, 31 Aug 1992.

15 Ashton B. Carter, William J. Perry, and John D. Steinbruner, *A New Concept of Cooperative Security* (Washington: Brookings Institution, 1992).

16 Department of Defense Directive 5105.39, Director of Net Assessment, 6 Dec 1971.

17 Department of Defense Directive 5105.39, Director of Net Assessment, 27 Sep 1985.

18 Alice C. Cole, Alfred Goldberg, Samuel A. Tucker, and Rudolph A. Winnacker, eds., *The Department of Defense: Documents on Establishment and Organization, 1944-1978* (Washington: GPO, 1978), 259–261;

19 Notra Trulock, III, "Emerging Technologies and Future War: A Soviet View," Appendix B in *The Future Security Environment,* 97–163; Notra Trulock, III, Kerry L. Hines, and Anne D. Herr, *Soviet Military Thought in Transition: Implications for the Long-Term Competition,* PSR Report No. 1831 (Arlington: Pacific-Sierra Research Corp., May 1988).

20 John C. Hines, Phillip A. Petersen, and Notra Trulock, Ill, "Soviet Military Theory from 1945–2000: Implications for NATO," *The Washington Quarterly,* 9, no. 4 (Fall 1986): 117–137.

21 William E. Odom, "Soviet Force Posture: Dilemmas and Directions," *Problems of Communism,* 34, no. 4 (Jul-Aug 1985): 1–14.

22 Jeffrey Record, *Hollow Victory: A Contrary View of the Gulf War* (Washington: Brassey's (US), 1993), 57–60.

Index

A

Allen, Lew 5, 145, 178, 230, 259
Allison, Graham 74, 82, 196, 254, 279, 286, 303, 305
Ardrey, Robert 10, 20, 25, 43, 74, 97, 132, 145, 184, 185
Armitage, Richard 275
Arnold, Henry "Hap" 86
Aspin, Les 283, 302

B

Badgett, Lee 128
Baker, James 268
Battilega, John 241
Behavioral Theory of the Firm 43, 73, 96
Berenson, Paul 250
Bergson, Abe 79
Bethe, Hans 21, 36
Billington, James 197, 215, 218, 220, 245, 267
Birman, Igor 190, 197, 215, 217, 267
Blue Ribbon Defense Panel 105, 116, 164, 182, 238, 245, 249, 281
Bower, Joseph 119, 120, 183
Brezhnev, Leonid 85, 187, 213
Brodie, Bernard 42, 50, 51, 66, 67, 71, 81, 89, 91, 92, 94, 95, 98
Bross, John Adams 104, 109, 110, 113, 117, 154, 180, 182
Brown, Harold 108, 112, 142, 149, 189, 190, 211, 212, 225, 306, 308
Brzezinski, Zbigniew 151, 189, 202, 212, 311, 312
Bukovsky, Vladimir 197, 217, 267
Bureaucratic Phenomenon, The 6, 42, 59, 73, 93
Burke, Jerry 226
Bush, George H.W. 184, 185, 212, 237, 240, 250, 251, 255, 268, 269, 282, 283, 285, 286, 287, 288, 301, 302, 305

C

Cabaniss, Ned 226, 251, 300
Central Intelligence Agency (CIA) 3, 4, 14, 18, 19, 37, 59, 68, 69, 78, 97, 109, 110, 112, 114–117, 121, 123, 138, 142, 148, 153, 154, 159, 168, 169, 171, 172, 183, 211, 216, 220, 225, 226, 235, 266, 290, 308
Cheney, Dick 285–287, 286, 287, 294–296, 299, 305, 307
Christensen, C. Roland 38, 105, 119, 183
Clements, Williams 191
Cockell, William 264, 287

Collbohm, Frank 85, 86, 95, 96, 182
Commission on Integrated Long-Term Strategy (CILTS) 249, 252, 275, 321
Conze, Henri 298, 299
Cordesman, Anthony 135, 149, 150, 239, 240, 272, 282
Cotter, Donald 133, 134, 148, 185
Crecine, Pat 56, 74, 158
Crowe, William 239, 272, 273, 282
Crozier, Michel 2, 6, 42, 56, 59, 73, 75, 93
Cyert, Richard 43, 56, 73, 96

D

Dalkey, Norman 3, 13
Defense Advanced Research Projects Agency (DARPA) 132, 134, 158, 321
Defense Intelligence Agency (DIA) 7, 16, 122, 131, 146, 194, 211, 288, 308
Defense Nuclear Agency (DNA) 134, 147, 148, 277
DeLauer, Dick 145, 277
Delicate Balance of Terror, The 82
Deterrence and Strategy of Total War, The 49, 90, 96
Deutch, John 233, 234, 303, 304, 306
DeWeerd, Harvey 49, 83, 89, 98
Digby, James 67, 83, 88, 89, 91, 91–99, 104, 108, 180–183
Discriminate Deterrence 234, 281
Dostoyevsky, Fyodor 29
Dubos, Rene 20, 25, 44
Duckett, Carl 106, 126
Duncan, Charles 193
Durnan, Jaymie 256

E

Earner, Bill 277
Einstein, Albert 29
Eisenhower, Dwight 22, 46, 57, 76, 77, 82
Ellsworth, Robert 150, 183, 197
Enthoven, Alain 31, 76, 111, 112, 182
Epstein, David 212, 226, 239, 243, 253, 273
Ermarth, Fritz 138, 241
Essence of Decision: Explaining the Cuban Missile Crisis 74, 97, 120, 183

F

Feshbach, Murray 5, 16, 23, 140, 190, 197, 198, 218
Fisher, Scott 273
Fletcher, Joe 2, 8
Flood, Merrill 13, 51, 56, 70
Ford, Carl 285, 288
Ford, Gerald 184
Fortier, Don 215, 228–230

Fox, Robin 3, 11, 14, 20, 43, 44, 184
Friedberg, Aaron 274
Friedman, Milton 220
Fubini, Eugene 306

G

Gaither Committee 49, 68, 73, 82, 90, 98, 104, 110, 180
Game Theory 2, 6, 13, 24, 30, 32, 51, 52, 55, 56, 60, 65, 92
Gaskin, Robert 285, 287–290, 307
Gates, Robert 229
Gershwin, Lawrence 156, 162, 165, 215, 225
Giessler, Fred 129
Glomar Explorer 127, 184
Goldhamer, Herbert 3, 11, 13, 15, 21, 35, 36, 43, 44, 48, 49, 55, 56, 58, 62, 65, 71, 88, 90, 92, 92–96, 105, 111, 132, 157, 184–186
Goldwater-Nichols Department of Defense Reorganization Act 238, 239, 245, 247–249, 271–273, 281, 282
Gorbachev, Mikhail 211, 243, 247, 269, 281, 283, 287, 296
Gorman, Paul 136, 137, 143, 165, 185, 198, 265, 266, 271, 281
Gouré, Daniel 285, 286
Graham, Daniel 7, 131
Grange, Judy 241

H

Haas, Peter 134, 148, 242
Haig, Alexander 105, 106, 113, 117, 126, 152, 161, 164, 172, 224
Harvard Business School 8, 12, 38, 40, 74, 76, 104, 119
Hayden, Michael 255
Hayek, Friedrich 219
Helmer, Olaf 3, 13, 44, 94
Helms, Richard 105, 109, 117, 118, 125, 153, 161, 165, 177, 182, 183
Herman, Mark 17, 21, 26, 27, 35, 36, 56, 63, 65, 81, 92, 94, 285, 299
Herzfeld, Charles 105, 116, 128, 176, 234, 265, 281
Heymann, Hans 42, 55, 60, 90, 93, 94, 180, 182
Hines, John 3, 10, 26, 85, 226, 241, 251, 252, 261, 302, 303, 308, 312
Hirshleifer, Jack 42, 56, 70, 88, 89, 93
Hitch, Charles 42, 51, 67, 70, 76, 79, 83, 88, 89, 92–94, 110, 182, 206
Hoeffding, Oleg 74, 96, 181
Hoffman, Fred 74, 105, 116, 233, 265
Hoffmann, Marty 133
How Much Is Enough? 112
Hunerwadel, Jo 118, 166
Huntington, Sam 202, 212, 228, 245, 275, 302, 312
Hyland, William 57, 77, 97, 193

I

Iklé, Fred 152, 215, 220, 222, 224–226, 230–232, 235, 236, 245, 249, 254, 259, 261, 262, 264–266, 272, 275, 280, 281, 285, 300
Imperial Animal, The 11, 43, 132, 184
Impoverished Superpower, The 212, 243, 267, 281

J

Jederlinic, John 216
Jeremiah, David 303
Johnson, Lyndon 47, 244
Jones, David 202, 223

K

Kahn, Herman 17, 21, 26, 35, 36, 56, 63, 81, 92, 94
Kahneman, Daniel 2, 9, 11, 20, 25, 43
Karber, Phil 137, 175, 222–224
Kaufmann, William 66, 69, 75, 78, 95, 105, 116, 117, 161
Kaysen, Carl 18
Kellogg, William 50, 89, 90
Kennan, George 8
Kennedy, John 46
Khrushchev, Nikita 46, 54
Kissinger, Henry 10, 38, 55, 57, 58, 71, 77–79, 85, 103–106, 108, 112, 113, 115–117, 121–124, 126, 133, 140, 146, 150, 152–158, 161, 162, 163, 164, 167, 170–172, 174, 176–178, 182–184, 193, 301
Klein, Burton 74, 96, 97, 181
Kloske, Dennis 254, 280
Knight, Frank 33, 218, 220
Komer, Robert 78, 98, 211, 235, 309
Kontorovich, Vlad 296
Kosygin, Alexei 85
Kraus, George 208, 249
Krepinevich, Andrew 179, 285, 287, 292, 308
Kriegspiel 49

L

Laird, Melvin 105, 116, 117, 119, 121, 143, 164, 175, 179
Leites, Nathan 36, 37, 49, 56, 61, 90, 96, 122, 132, 141, 147, 168, 185
Levine, Robert 132, 186
Lewis, Millard "Poppy" 83, 108
Libby, Scooter 285, 296, 297
Loftus, Joseph 4, 20, 25, 42, 58, 96, 109, 219
Long, Bob 233
Long-Term Competition with the Soviets: A Framework for Strategic Analysis 42, 76, 97, 279
Lorenz, Konrad 31, 44, 88

Lukasik, Stephen 132, 134
Luttwak, Edward 205, 309
Lynn, Larry 105, 116

M

M-60 (tank) 131, 135, 136
Manthorpe, William 130, 221
March, James 73
March of Literature, The 29
Marshall, Don 134, 135
Marshall, S.L.A. 150
Martin, Jim 277
May, Ernest 74, 120
McCone, John 109, 125, 154
McFarlane, Robert "Bud" 215, 228, 229, 231, 245, 280, 311
McGiffert, David 151, 190
McGuire, Michael 2, 9, 43
McKean, Roland 42, 88, 93, 182
McKitrick, Jeff 3, 4, 144, 226, 256, 279, 285, 287
McNamara, Robert 46, 73, 112
McPeak, Merrill 293
Mearsheimer, John 309
Men in Groups 25, 132, 184
Merrill, Phil 238
Mettler, Ruben 164, 182
Military-Technical Revolution (MTR) 287, 288, 292, 293, 302, 308
Mood, Alex 51, 67, 70, 92, 94
Murdock, Clark 303

N

National Reconnaissance Office (NRO) 178, 249, 250, 321
National Security Agency (NSA) 114, 118, 153, 255, 290
Negus, Gordon 241
Neustadt, Richard 56, 74
Next Ten Years, The 42, 49, 51, 52, 67, 89, 92, 94
Nichols, Russ 48, 89, 94, 96, 180, 238, 239, 245, 247, 248, 249, 271–273, 281, 282
Nitze, Paul 3, 110, 181, 201, 263, 280
Nixon, Richard 38, 75, 101–103, 105, 108, 113, 114, 117, 119, 121, 122, 125, 126, 140, 142, 146, 152, 155, 161, 164, 167, 175, 177, 179, 184, 185, 211, 218
Norstad, Lauris 50
North Atlantic Treaty Organization (NATO) 45, 58, 61, 73, 102, 110, 128, 146, 147, 188, 191, 192, 203, 213, 230, 231, 264, 280, 284, 309
Nuclear Strategy Development Group (NSDG) 3, 236, 249, 263, 322
Nutter, Warren 79, 98, 218, 219, 243

O

Odom, William 202, 228, 241, 308, 312
Ogarkov, Marshal Nikolai 85, 252, 253, 308
Operation Desert Storm 184, 185, 283, 285, 297, 298
Osgood, Robert 224
Owens, William 294

P

Parker, Pat 105, 106, 120, 124, 125, 137
Pearl Harbor 82, 98, 239
Pease, Charlie 215, 225, 236
Perle, Richard 215, 231, 237
Perry, William 193, 203, 257, 303, 304, 312
Pickett, George "Chip" 38, 105, 118–122, 128, 129, 131, 135, 136, 142, 163, 166, 168, 215, 221, 256
Pillsbury, Michael 229, 239
Pipes, Richard 215, 217
Pirie, Robert "Robin" 105, 118, 120, 121, 128, 129, 166, 170
Poindexter, John 228
Polk, James 140, 143, 166, 167
Ponomareff, Dmitry 251
Post, Jerry 123, 157
Powell, Colin 208, 256, 257, 296, 299, 303
Problems of Estimating Military Power 75, 97, 173
Proctor, Edward 106, 122, 125, 130, 159, 160
Project 186 124, 137
Project Eager 137, 241
Project Solarium 69
Project Sovoy 42, 49, 56, 72, 109, 219
Pugh, George 215, 221, 222, 244, 282
Pugliaresi, Lou 268

R

Reagan, Ronald 213, 243, 267, 283
Reality and Belief in Military Affairs 11, 43, 184
Reddaway, Peter 215, 217
Revolution in Military Affairs (RMA) 252, 253, 255, 294, 295, 302–304, 307, 308
Rickover, Hyman 120
RisCassi, Robert 282, 289–291, 306
Road to Serfdom, The 219
Roche, Jim 2, 9, 14, 39, 40, 43, 107, 129, 140, 141, 149, 192, 193, 199, 200, 209, 215, 220, 221, 223, 226, 228, 256, 273, 277, 279, 285, 286, 297
Rosen, Steve 215, 226, 228, 229, 232, 241, 245, 273, 274, 298
Ross, Dennis 94, 215, 226, 230, 234, 248, 266, 288, 297

Rowen, Henry "Harry" 44, 57, 75, 89, 91, 94, 97, 104, 110, 156, 162, 173, 174, 180–182, 201, 207, 212, 215, 225, 232, 233, 243, 265, 268, 281, 285, 297, 300, 301, 309
Rumsfeld, Donald 17, 19, 40, 104, 106, 108, 135, 139–142, 146, 148–151, 179, 191, 192, 194, 198, 200, 223

S

Santayana, George 29
Schama, Simon 25, 26, 44
Schelling, Thomas 30, 65, 66, 69, 94, 110, 181
Schlesinger, James 2, 3, 10, 11, 14, 15, 18, 19, 25, 44, 56, 57, 74, 75, 78, 87, 97, 104–106, 108, 110, 112, 116, 117, 122–134, 137–140, 142, 145–150, 161, 163, 164, 172–174, 176, 178, 179, 182–186, 194, 195, 197, 198, 200, 207, 209, 308
Schwartz, Peter 3, 12, 44
Scott, Bruce 3, 12
Scowcroft, Brent 184, 191, 299
Selin, Ivan 119
Sharfman, Peter 129
Shlapentokh, Vladimir 190, 197, 215, 217, 260, 267, 296
Shubik, Martin 65, 95
Shultz, George 117, 159, 161, 175, 224, 268
Simon, Herbert 73
Sims, James 272
Slocombe, Walter 202, 212
Sloss, Leon 185, 202, 212
Smith, K. Wayne 112, 182
Smith, Larry 305, 307
Sorley, Robert 129, 130
Soviet Military Power 242, 282
SS-9 Study 105, 115, 155, 156, 157, 175, 182
SS-19 (missile) 106, 126, 172, 183
Stalin, Josef 45, 46, 54
St. Martin, Ron 221
Strategic Air War (SAW) 51, 56, 70, 90, 181
Strategic Defense Initiative (SDI) 13, 213, 236, 237, 261–263, 280
"Strategy as a Profession for Future Generations" 17, 44, 94
Sullivan, Gordon 293–295

T

T-62 (tank) 131, 135, 136
T-64 (tank) 252
Taft, Will 230, 231, 282
Teller, Edward 21, 27
Territorial Imperative, The 74, 132, 145
Thayer, Paul 222, 223, 244, 254, 279
Tiger, Lionel 3, 9, 11, 15, 20, 21, 25, 36, 43, 50, 69, 91, 95, 132, 161, 184

Toynbee, Arnold 24, 29
Treml, Vladimir 215, 268
Trulock, Notra 245, 308, 312
Tversky, Amos 2, 9, 11, 20, 25, 43
Tweedy, Bronson 105, 117, 118, 161, 163, 177, 182

U

U-2 shootdown 44, 76, 97, 185
University of Chicago 33, 42, 48, 55, 58, 93, 94, 116, 218

V

Vance, Cyrus 151, 187, 202
Vandenberg, Hoyt 50, 147
Vessey, John 233, 248, 265
Vietnam War 18, 47, 78, 98, 101, 102, 123, 179, 188, 208, 209, 211, 297, 298
Vogt, John 215, 232, 233
Voice of America (VOA) 5
von Neumann, John 18, 30, 50, 64, 69, 91, 95

W

Wack, Pierre 3, 12, 44
Waddington, C.H. 20, 25, 44
Wallis, Allen 48, 58, 79, 268
"Warbo" (Warning and Bombing study) 50
Warnke, Paul 202
Watkins, James 221, 243
Watts, Barry vii, 2, 9, 43, 144, 226, 256
Weinberger, Caspar 196, 197, 204, 208, 215, 216, 221, 223–226, 230–232, 234, 236–240, 244, 245, 249, 254, 256, 259, 261, 269, 272, 279
Weiss, Sy 185, 215, 232, 233
Welander, Robert 118, 192, 211
Whitehead, Alfred North 29
Wickham, John 240, 289, 290
Williams, John 49–51, 67, 69, 91, 92, 94, 95, 97
Wimbush, Enders 3, 16, 44, 197, 212, 218, 243
Wohlstetter, Albert 66, 82, 90, 98, 181, 207, 228, 233, 249, 265, 281
Wohlstetter, Roberta 82, 89, 91, 94, 96, 98, 180
Wolf, Charles 111, 216, 235, 265–268, 274
Wolfowitz, Paul 215, 224, 226, 232, 233, 285–287, 289, 293, 295, 297, 299, 300, 301, 309

Y

Yom Kippur War 122, 149

Z

ZSU-23/4 (anti-aircraft gun) 136
Zumwalt, Elmo 130, 134, 207, 208

Acronyms

ABM	anti-ballistic missile
ADCAP	advanced capability
AE	atomic energy
AEC	Atomic Energy Commission
AFNorth	Allied Forces Northern Europe (a NATO Command including Norway, Denmark, Germany north of the Elbe, plus the Baltic Sea)
AIDS	auto-immune deficiency syndrome
ARPA	Advanced Research Projects Agency (forerunner to DARPA)
ASD	Assistant Secretary of Defense
ASW	anti-submarine warfare
C3I	Command, Control, Communications and intelligence
C4I	command, control, communications, computers, and intelligence
CENTCOM	U.S. Central Command
CIA	Central Intelligence Agency
CILTS	Commission on Integrated Long-Term Strategy
CINCPAC	Commander in Chief, Pacific Command
CNO	Chief of Naval Operations
CONUS	continental United States
DARPA	Defense Advanced Research Projects Agency
DCI	Director of Central Intelligence
DCSOPS	Deputy Chief of Staff for Operations and Plans
DDR&E	Director of Defense Research and Engineering
DIA	Defense Intelligence Agency
DPG	Defense Planning Guidance
GATT	General Agreement on Tariffs and Trade
GNP	gross national product
IC	Intelligence Community
ICBM	inter-continental ballistic missile
IRBM	ballistic missile
ISA	International Security Affairs (Office of the Secretary of Defense)
ITAG	Interagency Technology Assessment Group
JCS	Joint Chiefs of Staff
MBFR	mutual and balanced force reductions
MIRV	multiple independent reentry vehicle
MIT	Massachusetts Institute of Technology
MRBM	medium range ballistic missile
MTR	military-technical revolution
MVD	Ministry of Internal Affairs
NATO	North Atlantic Treaty Organization
NIC	National Intelligence Council
NIE	National Intelligence Estimates
NIO	National Intelligence Officers
NIPE	National Intelligence Programs Evaluation
NRO	National Reconnaissance Office

NSA	National Security Agency
NSC	National Security Council
NSCIC	National Security Council Intelligence Committee
NSCID	National Security Intelligence Directive
NSDD	National Security Decision Directive
NSDG	Nuclear Strategy Development Group
NSDM	National Security Decision Memorandum
NSSM	National Security Study Memorandum
NWEAMP	Nuclear Weapons Employment and Acquisition Master Plan
O-5, O-6…	Administrative classifications used primarily to standardize compensation across military services. "O" indicates a commissioned officer.
OA	Office of Analysis
OEB	Old Executive Office Building
OMB	Office of Management and Budget (called Bureau of the Budget before 1 July 1970)
OPLAN	operational plan
OSD	Office of the Secretary of Defense
PA&E	Program Analysis & Evaluation (Office of the Secretary of Defense)
PD	Presidential Directive
PDB	President's daily brief
PFIAB	President's Foreign Intelligence Advisory Board
Pk	probability of kill
PRC	People's Republic of China
PRM	Presidential Review Memorandum
RDA	R&D Associates
RDJTF	Rapid Deploymnet Joint Task Force
RMA	Revolution in Military Affairs
ROK	Republic of Korea
RSAS	RAND Strategy Assessment System
S&T	scientific and technical
SALT	Strategic Arms Limitations Talks/Treaty
SAR	special access required
SASC	Senate Armed Services Committee
SCDC	Strategic Concepts Development Center
SDC	System Development Corporation
SDI	Strategic Defense Initiative
SES	Senior Executive Service
SHAPE	Supreme Headquarters Allied Powers Europe
SLBM	submarine-launched ballistic missile
SOLIC	special operations and low-intensity conflict
TRADOC	Training and Doctrine Command (U.S. Army)
TS	Top Secret
USA	United States Army
USAF	United States Air Force
USSR	United Soviet Socialist Republics

About the Author

Andrew W. Marshall founded the discipline and the office of net assessment. As Director, Net Assessment, in the Office of the Secretary of Defense from 1973 to 2015, Marshall advised eight successive presidents and thirteen secretaries of defense. He developed new analytic perspectives and techniques for analyzing long-term strategic competition and for understanding potential shifts in the future security environment. Under Marshall's direction, the Office of Net Assessment accurately diagnosed the weakness of the Soviet Union years before its collapse, highlighted the potential implications of precision weaponry, pandemic disease, and environmental change, and anticipated China's emergence as a strategic rival to the United States. Before joining the government, Marshall worked for the RAND Corporation, where he formulated his initial thoughts on strategic competition while focusing on nuclear strategy and the Soviet Union.

About the Editors

Jeffrey S. McKitrick is a Research Staff Member at the Institute for Defense Analyses. He served as a Military Assistant to the Director, Net Assessment, from 1985 to 1989, as Military Advisor to the Vice President from 1989 to 1992, and as Special Assistant to the Vice President from 1992 to 1993.

Robert G. Angevine is a Research Staff Member at the Institute for Defense Analyses. He has worked as a contractor for the Office of Net Assessment for more than 25 years, and has written extensively on military history and strategy.

Made in the USA
Las Vegas, NV
19 September 2026

f18d17ed-769d-4a0a-8c72-76b777ee2294R02